THERMODYNAMICS AND STATISTICAL MECHANICS

THERMODYNAMICS AND STATISTICAL MECHANICS

LECTURES ON THEORETICAL PHYSICS, VOL. V

BY ARNOLD SOMMERFELD
UNIVERSITY OF MUNICH

EDITED BY
F. BOPP
UNIVERSITY OF MUNICH

J. MEIXNER
ENGINEERING UNIVERSITY OF AACHEN

TRANSLATED BY
J. KESTIN
BROWN UNIVERSITY

ACADEMIC PRESS
New York   San Francisco   London

A Subsidiary of Harcourt Brace Jovanovich, Publishers

ALL RIGHTS RESERVED BY ACADEMIC PRESS, INC.
NO PART OF THIS PUBLICATION MAY BE REPRODUCED OR
TRANSMITTED IN ANY FORM OR BY ANY MEANS, ELECTRONIC
OR MECHANICAL, INCLUDING PHOTOCOPY, RECORDING, OR ANY
INFORMATION STORAGE AND RETRIEVAL SYSTEM, WITHOUT
PERMISSION IN WRITING FROM THE PUBLISHER.

ACADEMIC PRESS, INC.
111 Fifth Avenue, New York, New York 10003

United Kingdom Edition published by
ACADEMIC PRESS, INC. (LONDON) LTD
24/28 Oval Road, London NW1

LIBRARY OF CONGRESS CATALOG CARD NUMBER: 50-8749

PRINTED IN THE UNITED STATES OF AMERICA

Author's Preface

Thermodynamics is a perfect example of a science which is developed from axioms. In contrast to classical mechanics, thermodynamics has withstood the quantum revolution without having its foundations shaken. In the course of the century of its existence it sprouted only several vigorous offshoots: Nernst's Third Law, Section 12, the theory of dilute solutions, Section 15, the application of the Second Law to electricity and magnetism, Sections 18 and 19. We consider that the thermodynamics of irreversible processes, Section 21, constitutes a promising extension of the classical thermodynamics of equilibria; it is based on Onsager's reciprocal relations and attempts to describe real processes which occur at finite velocities. Even Kirchhoff, as related by Planck in his autobiographical paper (Naturwissenschaften 19), restricted the concept of entropy to reversible processes; the firm belief in the general validity of this concept, which Planck stated as early as his doctoral thesis, led him in 1900 to his law of radiation and to quantum theory.

In any case, we do not propose to adhere so rigorously to the axiomatic mode of presentation as to endeavor to develop the science from the smallest possible number of axioms. This was achieved by Carathéodory in his proof of the Second Law which we shall, it is true, duly outline, but to which we shall not accord any preference over the Carnot-Clausius proof. The latter contains so much that is instructive and thoughtful that we consider it indispensable in an introductory course of lectures. The fact that it makes use of concepts derived from engineering is, in our opinion, an advantage rather than a matter for reproach. After all, thermodynamics did originate from the needs of steam engine builders.

Epistemologically there is a certain rivalry between the consideration of *cycles* and the method of *thermodynamic potentials*. The former are preferred in engineering because of their greater appeal to intuitive imagination. However, we shall almost exclusively make use of the latter method. It is much shorter and less arbitrary because it need not rely on artificially invented cycles. Moreover, we shall accord equal prominence to the four Gibbsian potentials, Section 7, although in the applications, the Gibbs function (also known as free enthalpy, or, simply, as the thermodynamic potential) is by far the most important one.

The experimental material which we include is very incomplete. In the case of real gases we restrict ourselves to the consideration of the van der Waals equation; in spite of its very simple form and in spite of the fact that it makes use of only two empirical constants, it reproduces the main outline of the behavior of liquids and their vapors in a very satisfactory manner. In the field of ferromagnetic phenomena Weiss' theory plays a similar part and succeeds, with its single constant of the internal field, to render similar services. A critical re-examination of these two theories must be left to more specialized treatises.

In my original University lectures I used to devote more time to *statistical mechanics* as compared with *classical thermodynamics* because I was personally drawn more to the former on account of its relation to quantum theory. In the present formulation quantum theory must, in principle, be left out and can only occasionally be drawn in as a supplement to *Boltzmann's statistics*. For this reason, the chapters dealing with thermodynamics, Chapters I and II, occupy an increased proportion of space, as compared with Chapters III, IV, and V. *Fermi's statistics* comes into the fore only on the occasion of a short account on metal electrons.

Chapter III contains a preliminary introduction to statistical mechanics, as far, that is, as is possible with elementary methods. The examples quoted in this connection (van der Waals constants, Langevin's theory of paramagnetic phenomena) serve to fill in some of the gaps left open in the sections on thermodynamics. Brownian motion, which is the most important example of statistical fluctuations, is treated together with the theory of the torsional balance. The problems arising in connection with the mean free path are only mentioned and not presented fully because they belong to the most difficult problems in statistical mechanics.

Chapter IV constitutes the summit of our consideration of statistical mechanics. I am of the opinion that Boltzmann's combinatorial method, when it is restricted to stationary processes, surpasses in fruitfulness and boldness its rival, the dynamic method based on Boltzmann's collision equation. In fact, in the first sections of this chapter we shall describe the combinatorial method in the original form given to it by Boltzmann, in which the molecules of a gas are endowed with a physically real existence. We shall free it from the resulting blemishes in Sections 32 to 35 when we shall introduce discrete energy levels of quantum-mechanical origin. However, in this way we are not yet led to quantum statistics proper. Since, in the realm of quantum mechanics, molecules are indistinguishable from each other, the original method due to Boltzmann (distribution of particles over the states) becomes illusory. Moreover, from the point of view of quantum mechanics the states are given first; the various numerical combinations which govern the way in which the mutually indistinguishable particles are distributed over the states constitute the substance of the new statistics. We reach these points in Sections 36 and 37. Suitable examples are given in Section 38, light quantum gas, and in Section 39, metal electrons.

It is perhaps necessary to apologize for the fact that we have not placed this proper quantum-statistical treatment of states at the very beginning, starting instead with the undoubtedly obsolete method of Boltzmann's statistics of particles. The reason for it is purely didactic. The original method due to Boltzmann achieves so much and is so lucid that it still seems to provide the indispensable foundation for the understanding of the new statistics of states.

Chapter V has been kept very short in comparison with Chapter IV. The assumption of molecular models required here is of a much more specialized nature, the resulting calculations are so much more tedious, than those in the combinatorial method. It is true that in the hands of Hilbert they have led to a consistent theory of such irreversible processes as friction, the conduction of heat, etc. which Maxwell and Boltzmann repeatedly tried to achieve without success. In addition to this, the method due to Chapman and Enskog has been developed numerically to a point where comparison with observation becomes possible. However, such applications by far exceed the scope of a general course of lectures; they illustrate the great difficulties attendant on the exact mathematical development of the problems of the mean free path, which were only cursorily mentioned in Chapter III. Our presentation must necessarily restrict itself to an elucidation of the central problem which Boltzmann posed in his work with statistical methods: to clarify the contradiction between reversible mechanics and the Second Law of thermodynamics.

Arnold Sommerfeld

Editors' Preface

Fate prevented Sommerfeld from completing his treatise on theoretical physics. He died following an accident while he was working on Volume V, the volume on thermodynamics and statistical mechanics. The Editors were entrusted with the task of completing and publishing this volume on the express wish of the Author.

The sections on thermodynamics had then been virtually completed. Unfortunately, the Author could not read Section 21 which had been outlined by one of the Editors. Section 8 existed in two formulations and was completely recast.

The sections on kinetic theory and statistical mechanics existed up to and including Section 35; Section 37 had also been nearly completed. It was, however, clear from the many discussions with the Author that he was not completely satisfied with this portion of the book. We have tried to take this into account by including Section 36 on Gibbs' method, but we realize that the Author might have adopted a different course. The subdivisions and the contents of Sections 38 to 40 had been discussed with the Author, but they could not be written down in time.

Except for the remarks in the Author's Preface there were no clues as to Chapter V. The Author had not yet made up his mind about the contents of this chapter and mentioned casually that it could be included by the editor of future editions. The account of the electron theory of metals is based on the well-known article written by Sommerfeld and Bethe for the *Handbuch der Physik*.

Some of the problems have been taken from the Author's collection. Additional problems have been included following his wishes, which he expressed at one time. Some of them had been brought to his attention, but he could not express his views about them.

Professor G. V. Schubert helped the Author both actively and with advice while Chapters I to III were being written. Professor E. Kappler critically examined the section on Brownian motion. The Author discussed with Professor F. Sauter the contents of Section 26 by correspondence. We are indebted to Messrs. Herbert and Baldus for most of the figures in Chapters I to III. Dr. Mann assisted with the work of proof reading, as on many previous occasions, and made valuable suggestions and useful criticisms. It is possible that, unknown to the Editors, additional acknowledgements should be made. We wish to thank the Publishers for their willing cooperation.

November 1952.

F. Bopp   **J. Meixner**

Translator's Preface

The present book constitutes as nearly a literal translation of Sommerfeld's Fifth Volume as I could make, without impairing its fluency. Changes, if any, were slight and unimportant.

I would not have been in a position to undertake the translation of this volume, particularly of the chapters on statistical mechanics, were it not for the generous help and assistance which I received from Dr. G. F. Newell of Brown University. He has carefully read and criticized the manuscript and suggested many changes and improvements. I am also indebted to Professor F. Bopp, one of the editors of the German edition, who kindly consented to read the galley proof and to clarify many difficulties. Mr. J. R. Moszynski of Brown University read the page proof and prepared the Index. The Publishers spared no effort to meet my wishes.

The responsibility for any errors, mistakes, and omissions which still remain is, of course, my own.

Providence, R. I. October, 1955.

J. K.

Contents

Author's Preface v

Editor's Preface ix

Translator's Preface xi

CHAPTER I. THERMODYNAMICS. GENERAL CONSIDERATIONS 1

   1. Temperature as a Property of a System 1
   2. Work and Heat 4
   3. The Perfect Gas 8
      A. Boyle's Law (The Law of Boyle and Mariotte) 8
      B. Charles' Law (The Law of Gay-Lussac) 9
      C. Avogadro's Law and the Universal Gas Constant 10
   4. The First Law. Energy and Enthalpy as Properties 13
      A. Equivalence of Heat and Work 13
      B. The Enthalpy as a Property 16
      C. Digression on the Ratio of Specific Heats c_p and c_v 18
   5. The Reversible and the Irreversible Adiabatic Process 19
      A. The Reversible Adiabatic Process 20
      B. The Irreversible Adiabatic Process 22
      C. The Joule-Kelvin Porous Plug Experiment 23
      D. A Conclusion of Great Consequence 25
   6. The Second Law 26
      A. The Carnot Cycle and Its Efficiency 27
      B. The First Part of the Second Law 29
      C. The Second Part of the Second Law 34
      D. Simplest Numerical Examples 37
      E. Remarks on the Literature of the Second Law 39
      F. On the Relative Rank of Energy and Entropy 40
   7. The Thermodynamic Potentials and the Reciprocity Relations 42
   8. Thermodynamic Equilibria 47
      A. Unconstrained Thermodynamic Equilibrium and Maximum of Entropy 47
      B. An Isothermal and Isobaric System in Unconstrained Thermodynamic Equilibrium 48
      C. Additional Degrees of Freedom in Retarded Equilibrium 49
      D. Extremum Properties of the Thermodynamic Potentials 50
      E. The Theorem on Maximum Work 52
   9. The van der Waals Equation 55
      A. Course of Isotherms 56
      B. Entropy and the Caloric Behavior of the van der Waals Gas 57

CONTENTS

10. Remarks on the Liquefaction of Gases According to van der Waals . . . 60
    A. The Integral and the Differential Joule-Thomson Effect 60
    B. The Inversion Curve and Its Practical Utilization 61
    C. The Boundary of the Region of Co-existing Liquid-Vapor Phases in the p, v Plane 63
11. The Kelvin Temperature Scale 68
12. Nernst's Third Law of Thermodynamics 71

CHAPTER II. THE APPLICATION OF THERMODYNAMICS TO SPECIAL SYSTEMS . . . 77

13. Gaseous Mixtures. Gibbs' Paradox. The Law Due to Guldberg and Waage 77
    A. Reversible Separation of Gases 78
    B. The Increase in Entropy During Diffusion and Gibbs' Paradox 80
    C. The Law of Mass Action Due to Guldberg and Waage 81
14. Chemical Potentials and Chemical Constants 87
    A. The Chemical Potentials μ_i 87
    B. Relation Between the μ_i's and the g_i's for Ideal Mixtures 90
    C. The Chemical Constant of a Perfect Gas 91
15. Dilute Solutions 92
    A. General and Historical Remarks 92
    B. Van 't Hoff's Equation of State for Dilute Solutions 93
16. The Different Phases of Water. Remarks on the Theory of the Steam Engine 96
    A. The Vapor-Pressure Curve and Clapeyron's Equation 97
    B. Phase Equilibrium Between Ice and Water 100
    C. The Specific Heat of Saturated Steam 101
17. General Remarks on the Theory of Phase Equilibria 103
    A. The Triple Point of Water 104
    B. Gibbs' Phase Rule 106
    C. Raoult's Laws for Dilute Solutions 108
    D. Henry's Law of Absorption (1803) 111
18. The Electromotive Force of Galvanic Cells 113
    A. Electrochemical Potentials 113
    B. The Daniell Cell, 1836 115
    C. Contraction of Individual Reactions into a Simplified Overall Reaction 116
    D. The Gibbs-Helmholtz Fundamental Equation 118
    E. Numerical Example 119
    F. Remarks on the Integration of the Fundamental Equation 120
19. Ferro- and Paramagnetism 121
    A. Work of Magnetization and Magnetic Equation of State 121
    B. Langevin's Equation for Paramagnetic Substances 123
    C. The Theory of Ferromagnetic Phenomena Due to Weiss 125
    D. The Specific Heats c_H and c_M 129
    E. The Magneto-Caloric Effect 134
20. Black Body Radiation 135
    A. Kirchhoff's Law 136
    B. The Stefan-Boltzmann Law 139
    C. Wien's Law 140
    D. Planck's Law of Radiation 145

CONTENTS

21. Irreversible Processes. Thermodynamics of Near-Equilibrium Processes . 152
    A. Conduction of Heat and Local Entropy Generation 152
    B. The Conduction of Heat in an Anisotropic Body and Onsager's Reciprocal Relations 155
    C. Thermoelectric Phenomena 157
    D. Internal Transformations 163
    E. General Relations 165
    F. Limitations of the Thermodynamic Theory of Irreversible Processes . 168

CHAPTER III. THE ELEMENTARY KINETIC THEORY OF GASES 169

22. The Equation of State of a Perfect Gas 169
23. The Maxwellian Velocity Distribution 174
    A. The Maxwellian Distribution for a Monatomic Gas. Proof of 1860 . . 174
    B. Numerical Values and Experimental Results 177
    C. General Remarks on the Energy Distribution. The Boltzmann Factor 179
24. Brownian Motion 181
25. Statistical Considerations on Paramagnetic Substances 187
    A. The Classical Langevin Function 188
    B. Modification of Langevin's Function with the Aid of Quantum Mechanics 190
26. The Statistical Significance of the Constants in van der Waals' Equation 192
    A. The Volume of a Molecule and the Constant b 192
    B. The van der Waals Cohesion Forces and the Constant a 194
27. The Problem of the Mean Free Path 197
    A. Calculation of the Mean Free Path in One Special Case 198
    B. Viscosity 200
    C. Thermal Conductivity 203
    D. Some General Remarks on the Problems Associated with the Concept of the Mean Free Path 205

CHAPTER IV. GENERAL STATISTICAL MECHANICS: COMBINATORIAL METHOD . . . 207

28. Liouville's Theorem, Γ-space and μ-space 207
    A. The Multidimensional Γ-space (Phase Space) 208
    B. Liouville's Theorem 209
    C. Equality of Probability for the Perfect Gas 210
29. Boltzmann's Principle 213
    A. Permutability as a Measure of the Probability of a State . . . 214
    B. The Maximum of Probability as a Measure of Entropy 217
    C. The Combining of Elementary Cells 219
30. Comparison with Thermodynamics 221
    A. Constant Volume Process 221
    B. General Process Performed by a Gas in the Absence of External Forces 221
    C. A Gas in a Field of Forces; the Boltzmann Factor 223
    D. The Maxwell-Boltzmann Velocity Distribution Law 224
    E. Gaseous Mixtures 226

31. Specific Heat and Energy of Rigid Molecules 227
   A. The Monatomic Gas 227
   B. Gas Composed of Diatomic Molecules 230
   C. The Polyatomic Gas and Kelvin's Clouds 233
32. The Specific Heat of Vibrating Molecules and of Solid Bodies 234
   A. The Diatomic Molecule 234
   B. Polyatomic Gases 236
   C. The Solid Body and the Dulong-Petit Rule 236
33. The Quantization of Vibrational Energy 237
   A. The Linear Oscillator 237
   B. The Solid Body 240
   C. Generalization to Arbitrary Quantum States 240
34. The Quantization of Rotational Energy 242
35. Supplement to the Theory of Radiation and to that of Solid Bodies . . . 245
   A. Method of Natural Vibrations 246
   B. Debye's Theory of the Specific Heat of a Solid 247
36. Partition Function in the Γ-space 248
   A. The Gibbs Condition 248
   B. Connection with Boltzmann's Method 250
   C. Correction for Quantum Effects 253
   D. Analysis of Gibbs' Hypothesis 256
37. Fundamentals of Quantum Statistics 257
   A. Quantum Statistics of Identical Particles 257
   B. The Method Due to Darwin and Fowler 259
   C. Bose-Einstein and Fermi-Dirac Statistics 261
   D. The Saddle-Point Method 262
38. Degenerate Gases 266
   A. Bose-Einstein and Fermi-Dirac Distribution 266
   B. Degree of Gas Degeneration 270
   C. Highly Degenerate Bose-Einstein Gas 272
39. Electron Gas in Metals 276
   A. Introductory Remark to Drude's Method 276
   B. The Completely Degenerate Fermi-Dirac Gas 277
   C. Almost Complete Degeneracy 280
   D. Special Problems 282
40. The Mean Square of Fluctuations 286

CHAPTER V. OUTLINE OF AN EXACT KINETIC THEORY OF GASES 293

41. The Maxwell-Boltzmann Collision Equation 293
   A. Description of a State in the Kinetic Theory of Gases 293
   B. The Variation of f with Time 296
   C. The Laws of Elastic Collision 297
   D. Boltzmann's Collision Integral 299
   E. Boltzmann's Hypothesis About Molecular Chaos 301

42. The H-theorem and Maxwellian Distribution 302
   A. The H-theorem 302
   B. Maxwellian Distribution 306
   C. Equilibrium Distributions 309
43. Fundamental Equations of Fluid Dynamics 310
   A. Series Expansion for the Distribution Function 310
   B. Maxwell's Transport Equation 312
   C. Conservation of Mass 315
   D. Conservation of Momentum 316
   E. Conservation of Energy 318
   F. Entropy Theorem 320
44. On the Integration of the Collision Equation 323
   A. Integration with the Aid of Moment Equations 323
   B. Transformation of the Equations for Moments 325
   C. Evaluation of Collision Moments 327
   D. Viscosity and Thermal Conductivity 328
45. Conductivity and the Wiedemann-Franz Law 333
   A. The Collision and Transfer Equations for Electrons in Metals 333
   B. Approximate Solution of the Collision Equation 336
   C. Flux of Current and Energy 339
   D. Ohm's Law 341
   E. Thermal Conductivity and Absolute Thermal Electromotive Force . 342
   F. The Wiedemann-Franz Law 343

Problems for Chapter I 347

Problems for Chapter II 350

Problems for Chapter III 351

Problems for Chapter IV 352

Problems for Chapter V 355

Hints for the Solution of Problems 356

Index 395

CHAPTER I

THERMODYNAMICS. GENERAL CONSIDERATIONS

1. Temperature as a property of a system

The science of thermodynamics introduces a new concept, that of *temperature*; it is absent from classical mechanics, as well as from the theory of electricity and magnetism and from atomic physics (with the exception of Joule heat, intensity of spectral lines conceived as interactions between a large number of material particles). Our sense of heat furnishes a qualitative measure, and a quantitative measure, albeit fortuitous to a certain extent, is given by any thermometer. A body which is in thermal equilibrium has the same temperature everywhere. The same is true of two bodies which have remained in thermal contact for a sufficiently long time. *Equality of temperature is a necessary condition of thermodynamic equilibrium.*

Temperature is a *property* or *parameter of state*. It is independent of the previous history of the body and is defined solely by its instantaneous state. It is associated with the behavior of the body at the instant under consideration, or else, it is measured with reference to the instantaneous indication of a thermometer.

The science of thermodynamics, as already stated in the preface, is an *axiomatic science*. In accordance with its spirit we introduce the concept of temperature by stating the following axiom:

There exists a property — temperature. *Equality of temperature is a condition for thermal equilibrium between two systems or between two parts of a single system.*

The preceding statement was purposely formulated in the same way as those which will be used later to state the First and Second Laws of thermodynamics and, following a suggestion by R. H. Fowler,[1] we shall refer to it as to the "Zeroth Law" of thermodynamics.

---

[1] When giving an account of the book on thermodynamics of the great Indian astrophysicist M. N. Saha and his collaborator's, B. N. Srivartava, Allahabad 1931 and 1935.

In order to give a rigorous mathematical definition of the concept of a thermodynamic "property" or "parameter of state" it is necessary to consider its differential. With two independent variables x, y, which must themselves be measurable properties or characteristics of the system (e. g. pressure and volume), we can write it as

$$(1) \qquad dT = X\,dx + Y\,dy; \qquad X = \frac{\partial T}{\partial x}, \qquad Y = \frac{\partial T}{\partial y}.$$

Evidently we then have

$$(2) \qquad \frac{\partial X}{\partial y} = \frac{\partial Y}{\partial x},$$

which is the necessary and sufficient condition for the expression $X\,dx + Y\,dy$ to be a *perfect differential*. It is equivalent to the statement that T is a property.

The same condition can be also written in integral form

$$(3) \qquad \oint dT = 0$$

for any closed path in the x, y-plane. Denoting the two-dimensional vector which is defined by its components X and Y by the symbol **Z** we can apply Stokes' theorem for a two-dimensional field to the expression in eq. (3), obtaining

$$(4) \qquad \oint \mathbf{Z}\,d\mathbf{s} = \int \operatorname{curl} \mathbf{Z}\,dx\,dy.$$

Since curl **Z** vanishes by eq. (2) it is concluded that statement (3) is, in fact, equivalent to the assertion that T is a property.

The condition for a perfect differential with n independent variables is the vanishing of the n-dimensional curl and can be represented by $n(n-1)/2$ equations of the form (2). Statement (1) generalized in this manner is known as "Pfaff's differential." When there are two independent variables it is always possible to transform the expression $X\,dx + Y\,dy$ into a perfect differential, by dividing it by a denominator $N(x, y)$, even if it was not one originally.

With three independent variables x, y, z this is not, generally speaking, possible. The requirement of integrability imposes certain conditions on the components X, Y, Z of a three-dimensional vector **Z** which have been investigated in Problem I.7, Vol. II. It was found then that the vector must be normal to its curl:

$$(4\text{ a}) \qquad \mathbf{Z}\operatorname{curl}\mathbf{Z} = 0.$$

It was further shown on the example of a field of forces and its potential that this requirement did not uniquely determine the "integrating denominator" ("multiplier" as it was then called) and that any function of one was also such a denominator (multiplier).

These preliminary remarks will help to understand the considerations connected with the Second Law in Sec. 6.

We shall regard the new concept of temperature as a fourth dimension in addition to the mechanical quantities of length, mass, and time in the same way as in the science of electrodynamics when we considered the then new concept of *quantity of electricity* or *charge* as a new fourth dimension. Naturally, in problems of electrochemistry we shall have to deal with five fundamental dimensions, i. e. we shall include the charge. We shall denote the dimension of temperature by the abbreviation "deg" rather than by a new symbol.

In Vol. I we have introduced the concept of a "mechanical system" and understood it to mean a collection of material points or bodies which could be described by specifying geometrically definable links or forces. We shall speak of a "thermodynamic system" when, in order to describe its state, it is necessary to specify in addition the temperatures of its components as well as the details of the quantities of heat transferred between them.

A homogeneous fluid affords the simplest example of a thermodynamic system and we might remark here that this definition will include the special cases of gases and vapors. A fluid possesses only one mechanical degree of freedom, its volume, and only one thermal degree of freedom, its temperature. The volume, V, (extensive property) is associated with its canonical conjugate[1] the pressure, p, (an intensive quantity, also known as tension if its sign is reversed). The temperature T is to be regarded as a thermal intensive quantity; the extensive quantity which constitutes its conjugate will be discussed in Sec. 5 D. Generally speaking p is a function of T and V. The relation $p = f(T, V)$ is known as the equation of state, or the characteristic equation.

The three quantities V, p, and T which have just been introduced can be combined in the *coefficient of thermal expansion*, α, and the *coefficient of*

---

[1] The term originates from Hamiltonian mechanics, Vol. I, Sec. 41. The coordinate q (extensive quantity) and momentum (intensive quantity) were there described as canonically conjugate quantities. The term was, further, extended to include the more general pair of quantities Q, P. This note will suffice to explain the corresponding term in the present text. For more detail see Secs. 7 and 14 of the present volume.

tension, β, the two expressions being referred to the instantaneous values of V or p respectively:

$$(5) \qquad \alpha = \frac{1}{V}\left(\frac{\partial V}{\partial T}\right)_p ; \qquad \beta = \frac{1}{p}\left(\frac{\partial p}{\partial T}\right)_V.$$

The suffixes denote that in the process of differentiation with respect to T, p is kept constant in the one case, and V is kept constant in the other. Both coefficients have the dimension 1/deg, and their values for gases will be discussed presently. A further derived quantity, the (isothermal) *compressibility coefficient*, $\varkappa$, is given by the definition

$$(6) \qquad \varkappa = -\frac{1}{V}\left(\frac{\partial V}{\partial p}\right)_T.$$

The coefficients α, β, and $\varkappa$ satisfy a remarkable relation (see Problem I.1).

Processes during which T, p, or V remain constant are usually called an *isothermal*, an *isobaric* and an *isochoric* or *isopiestic* process, respectively.

2. Work and heat

Let a fluid occupy a cylindrical vessel of cross-sectional area A and let the vessel be closed by a piston touching the liquid. The piston is acted upon by the fluid with a force $p\,A$. If the piston is moved by dh, the fluid will perform the work

$$(1) \qquad dW = p\,A\,dh = p\,dV.$$

This expression is valid not only for a positive dV, lifting of the piston, but also when it is lowered, i. e. when dV is negative, not only for a cylindrical vessel, but also for any boundary and for any change of shape of the surface of the fluid, when it is only necessary to perform an algebraic summation of all volume changes and to extend it over the boundary.

Equation (1) defines dW. Does it imply that a *property* W exists? Certainly not, as in such a case dW would have to be a *"perfect differential,"* and according to (1.3) we should obtain

$$(1\text{ a}) \qquad \oint dW = 0$$

when the fluid is subjected to a cycle, i. e. when it is made to reach the initial state after having traversed an arbitrary path. Such cycles can be represented graphically in a plane for the system under consideration, i. e. for one with two degrees of freedom. The system of coordinates will correspond to any of the two properties chosen as independent variables, such as e. g. the pair V, T (one mechanical and one thermal variable), or the pair V, p (two mechanical variables). The latter pair of variables is used in the well-known *indicator diagram* which was introduced by James Watt as early as 150 years ago and which is automatically traced by every reciprocating steam engine, Fig. 1.

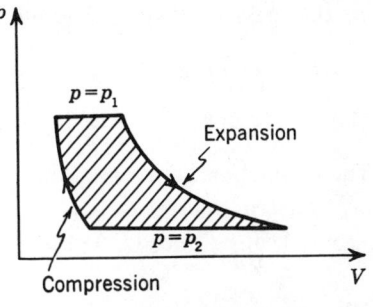

Fig. 1.
Indicator diagram of a steam engine.

The steam cylinder is put into communication with the boiler along the upper horizontal straight line $p = p_1$, whereas along the lower line $p = p_2$ it communicates with the atmosphere or with the low-pressure boiler (condenser). The descending and the ascending branches of the curve correspond to expansion and compression respectively[1]. The abscissa is proportional to the instantaneous distance h between the piston and the dead center and, hence, to the volume of the cylinder which is filled with steam at the moment. The area enclosed by the diagram gives a measure of the quantity

$$\text{(2)} \qquad \oint p\, dV = \oint dW$$

and is, evidently, different from zero. In accordance with (1 a) we must say that a property W which would correspond to dW does not exist.

The steam engine performs the work (2) at the expense of the heat introduced. The reverse transformation of work into heat occurs during every process involving friction. The most impressive and the historically most important experiment illustrating this was performed in Munich by Count Rumford (1798): he caused water to boil when a cannon barrel was being bored.

---

[1] Here it is necessary to overlook the fact that the quantity of steam contained in the cylinder is changed as the valves are opened and that it remains constant only during compression and expansion. In all processes which we shall discuss later the mass of the system will remain constant. The mass can be rendered constant in the steam engine example, actually or in the imagination, by condensing the steam as it leaves the cylinder and by returning it to the boiler. In any case the indicator diagram is a classical example of the representation of a cycle in the p, V- plane.

The instantaneous quantity of heat introduced will be denoted by dQ. As far as its measurement is concerned, it is, as is well known, reduced to that of temperature by adopting the definition: The quantity of heat which raises the temperature of one kilogram of water under atmospheric pressure from 14.5 to 15.5 C is called one (large) *calorie* (denoted kcal or, sometimes, Cal). We now recall the definition of specific heat, which we shall also base on the mass of 1 kg. Denoting the quantity of heat added by dq we put:

(3 a) $\quad dq = c_v \, dT \qquad c_v =$ specific heat at constant volume,

(3 b) $\quad dq = c_p \, dT \qquad c_p =$ specific heat at constant pressure.

The distinction between c_p and c_v is essential in the case of gases. It may be neglected in most cases as far as liquids are concerned. Substituting $dq = 1$ kcal/kg and $dT = 1$ deg into eq. (3.b), we find that for water at 15 C

$$(4) \qquad c_p = 1 \frac{\text{kcal}}{\text{deg} \cdot \text{kg}}.$$

This statement is, evidently, equivalent to our previous definition of one calorie.

It is found that during all processes involving friction the quantity of work used, dW, bears a definite ratio to the quantity of heat generated, dQ, irrespective of the conditions of the experiment. Joule gave a quantitative proof of this statement by numerous, if at first imprecise, experiments. In particular he performed measurements on the heat generated by an electric current (Joule heat). Somewhat earlier Robert Mayer satisfied himself that water becomes heated on being shaken[1]. We write

$$(5) \qquad dW = J \, dQ$$

where J is called the *mechanical equivalent of heat*. Its numerical value is

$$(6) \qquad J = 427 \text{ kg m/kcal}$$

if dQ is measured in kcal and dW is expressed in the engineering units of work – kgm. In this context the word "kilogram" denotes, as is known, the kilogram weight, for which it is preferable to use now the designation

---

[1] In a letter dated September 1841 Mayer states that he performed this experiment on many occasions always obtaining a positive result. See Ostwald "Große Männer", Leipzig, 1905, p. 71. In this connection attention may be drawn to a hypothesis expressed by Albrecht von Haller (1708-1777) in accordance with which animal heat was to have been generated by the friction experienced by the blood in the veins. This hypothesis persisted well into the 19-th century.

"kilopond" = kp [1], reserving the abbreviation kg = kilogram mass for the unit of mass in Giorgi's MKS-system of units. We would thus have

$$1\,\text{kp} = g \times 1\,\text{kg} = 9.81\,\text{MKS}^{-2} = 9.81\,\text{Dyne}.$$

Dyne denotes in this system the unit of force = 10^5 dyne. Following R. W. Pohl (Mechanik p.24) we shall call it the "large dyne". Hence

(7) $$J = 4.19 \times 10^3 \frac{M^2 K S^{-2}}{\text{k cal}} = 4.19 \frac{\text{Erg}}{\text{cal}}.$$

Giorgi's unit of energy Erg = large erg, is equal to 10^7 erg = 1 Joule = = 1 watt sec. The abbreviation cal = small calorie refers to 1 gram of water, in the same way as 1 kcal referred to 1 kilogram of water.

We shall presently revert to the experimental justification of (7). Furthermore, we shall later be in a position to dispense with the use of the special units of heat, kcal or cal, by putting 1 kcal = 427 kgm or 1 cal = 4.19 Erg, in accordance with (6) or (7), respectively, implying that $J = 1$.

At this point it is essential to deduce from eq. (5) that: The quantity dQ is not a perfect differential in the same way as dW was not one. There is no property Q, there is no *characteristic heat content which* would *simply describe* the instantaneous *state of the system*. It is necessary to point out clearly that our previous definition of a calorie constitutes only a rule for the measurement of the *quantity of heat dQ* (or Q when it is a finite quantity) introduced into the system in some way, but *not* for the quantity of heat *contained within the system*. Equations (3 a) and (3 b) show clearly that the manner of introducing heat is important in this connection.

In many text-books the use of the symbols dQ and dW is avoided and, for example, the symbols δQ, δW or $\bar{d}Q$, $\bar{d}W$ are used instead, in order to warn the reader against erroneously regarding them as perfect differentials. We do not think this necessary because we take the view that the existence of a property and of its subordinated perfect differential constitutes a fundamental peculiarity which we shall always stress explicitly, as we have done in Sec. 1.

---

[1] This suggestion emanated from Germany and has not, so far, gained universal recognition (*Transl.*)

3. The perfect gas

A gas is, so to say, the more perfect, the more difficult it is to liquefy it at a normal pressure of 760 mm Hg = 760 torr, i. e. the lower its boiling point. The degree of perfection is illustrated by the following boiling points in deg C at 760 torr:

He	H_2	N_2	O_2	CO_2	H_2O
−269	−259	−210	−218	−78.5	+100

Steam, shown at the end of the list, does not, evidently belong to the class of perfect gases. The perfect gas is a *limiting state* to which a real gas will tend as it is expanded indefinitely. The following laws apply to this ideal, limiting state.

A. Boyle's law (the law of Boyle and Mariotte)

(1) $$pV = \text{const},$$

which is valid provided that the temperature is kept constant. The pressure p is usually measured in atmospheres. One atmosphere denotes either the pressure in the surrounding air when the barometer reads 760 mm = 760 torr (physical atmosphere = 1 atm), or, more recently, one engineering atmosphere has been defined as the pressure exerted by 1 kilopond on 1 cm². It is almost exactly equal to the weight of a column of water 10 meters high acting on an area of 1 cm². We can express this engineering atmosphere as follows:

$$1 \text{ at} = 981 \frac{\text{cm}}{\text{sec}^2} \cdot 1000 \text{ g/cm}^2.$$

The first factor denotes the gravitational acceleration, g, the second denotes the mass of the column of water under consideration, the division by cm² denoting that weight has been replaced by pressure. Thus we have

(2) $$1 \text{ at} = 0.981 \times 10^6 \frac{\text{dyne}}{\text{cm}^2} = 0.981 \text{ bar}.$$

Correspondingly

(2 a) $$1 \text{ atm} = 1.013 \text{ bar}.$$

The unit of 1 bar introduced here denotes

$$1 \text{ bar} = 10^6 \frac{\text{dyne}}{\text{cm}^2}.$$

One thousandth part of this unit constitutes one "millibar," a unit of pressure now often used by meteorologists. In our MKS system, we have

(2 b) $$1 \text{ bar} = 10 \frac{\text{Dyne}}{\text{cm}^2} = 10^5 \frac{\text{Dyne}}{\text{M}^2}.$$

If the density $\rho = \text{mass}/V$ is introduced into eq. (1) instead of V, we have

(3) $$p = \rho \times \text{const.}$$

B. Charles' law (the law of Gay-Lussac)

(4) $$pV = CT.$$

C is a temporary constant which we shall presently express in terms of the universal gas constant, R. We must begin by discussing the temperature scale T defined through eq. (4). According to experience T is the same for all (perfect) gases if C is suitably chosen. To do this we refer to the coefficient of expansion defined in eq. (1.5). Using the same temperature scale as in (4), we find that it can be written as

(4 a) $$\alpha = \frac{1}{V}\left(\frac{\partial V}{\partial T}\right)_p = \frac{1}{V}\frac{C}{p} = \frac{1}{T}.$$

According to this equation, α is independent of the nature of the gas, being only a function of temperature. The same is true of the coefficient of tension

(4 b) $$\beta = \frac{1}{p}\left(\frac{\partial p}{\partial T}\right)_V = \frac{1}{p}\frac{C}{V} = \frac{1}{T}.$$

Equation (4) still leaves the scale unit free. If this is selected so as to obtain the unit on the Celsius (or centigrade[1]) scale, then the melting temperature of ice (at 760 torr) becomes

(5) $$T_0 = 273.15 \text{ deg.}$$

Generally we have

(5 a) $\quad T = T_0 + t, \quad t = $ temperature on Celsius scale,

and the coefficients of thermal expansion and tension become

(6) $$\alpha = \beta \approx 1/273 \text{ deg} = 0.00366 \text{ deg}^{-1}.$$

---

[1] According to the 1948 International Temperature Scale the term "centigrade scale" is now obsolete (*Transl.*).

The temperature scale introduced in our eqs. (4), (5) and (5 a) is the *gas thermometer temperature scale*. Equation (5) shows that its zero-point is shifted by 273 (more precisely by 273.15) deg with respect to the Celsius scale.

The air thermometer (or better still, a hydrogen or helium thermometer) can be arranged to measure at constant pressure or at constant volume, the latter arrangement being more convenient. According to eq. (4) T is then proportional to the pressure, p, of the gas. The pressure difference $p - p_0$ is measured with the aid of a suitable barometric arrangement by noting the position of a column of mercury at a temperature T of the gas as compared with that at a temperature T_0. The definition of a temperature scale with the aid of an air thermometer is sufficient for most practical purposes. The limit of its usefulness is attained at low temperatures when the air ceases to behave like a perfect gas. We shall see later how the temperature scale (Kelvin scale) should then be defined. In the case of a real gas (or in the case of a perfect gas at low temperatures) the Charles-Gay-Lussac equation (4) must be replaced by the already mentioned *general equation of state* of a liquid or gaseous system

$$(6 \text{ a}) \qquad T = F(p, V).$$

C. Avogadro's law and the universal gas constant

Dalton's *law of multiple proportions* which is valid for all chemical compounds is supplemented with Gay-Lussac's *law of integral volume ratios* when gases are involved.

By way of example we take:

1 liter hydrogen + 1 liter chlorine = 2 liters hydrogen chloride, according to the chemical formula

$$H_2 + Cl_2 = 2\,HCl$$

Another example:

2 liters hydrogen + 1 liter oxygen = 2 liters steam, written in chemical form this becomes

$$2\,H_2 + O_2 = 2\,H_2O.$$

The laws of Dalton and Gay-Lussac can be combined into one comprehensive rule due to Avogadro (1811): *Under the same external conditions of pressure and temperature all gases contain equal numbers of molecules in equal volumes* (Avogadro used the term corpuscles instead of the modern term—molecules). This rule remained ignored for a very long time, but since about 1860 it forms the foundation for all determinations of molecular weights. Nernst gave it prominence when he gave his great text-book the title "Theoretische Chemie vom Standpunkt der Avogadroschen Regel und der Thermodynamik").

The atomistic, microscopic point of view is alien to thermodynamics. Consequently, as suggested by Ostwald, it is better to use *mols* (*moles*) rather than molecules. As is well known one mol (see also vol. II, Sec. 7 footnote[1]) represents a mass of as many grams or kilograms (we then differentiate between the gram-mol = mol or gmol, and between the kilogram-mol = kmol) as there are units in the sum of the atomic weights of the constituents of the substance. Thus one gram-mol of O_2 in gaseous form is equal to 32 g, one kilogram-mol of H_2 is approximately equal to 2 kg, one gram-mol of HCl is equal to (1 + 35.5) g in round figures. It is necessary to remember at this point that the fact that hydrogen is diatomic was recognized precisely with the aid of Avogadro's rule; as late as around the year 1850 the chemical formula for water was mostly written HO.

If we introduce the *molar* (or *molal*) volume as a natural unit of volume and if we define it as the volume occupied by exactly one mol of a gas under the given pressure and at the given temperature, then we can put Avogadro's rule in the following simple form: *Under identical external conditions all gases have equal molar volumes.* It is evident that the last statement, just as the preceding ones, is restricted to perfect gases. It may be extended to real gases or even to vapors but only with caution.

We shall calculate the magnitude of this molar volume under a pressure of 1 atm and at a temperature of 0 C. We can utilize the fact that under these conditions the density of H_2 is fairly accurately equal to (see e. g. Vol. IV, eq. (17.14))

$$9.00 \times 10^{-2} \frac{\text{kg}}{\text{m}^3} = 9.00 \times 10^{-2} \frac{\text{g}}{\text{liter}}$$

It follows that 2 g of H_2 occupy a volume of

(7) $$\frac{200}{9.0} = 22.2 \text{ liter.}$$

The accurate value is somewhat higher on account of the atomic weight of H; thus, more precisely

$$V_{mol} = 22.4 \frac{\text{liter}}{\text{mol}} = 22.4 \frac{\text{m}^3}{\text{kmol}} \quad \text{at 760 torr and 0 C.}$$

According to Avogadro this value of the molar volume applies not only to H_2 but is *universally* valid for all perfect gases.

The equation of state also assumes a universal form if the molar volume is used. We can write it as

(8) $$p V_{mol} = R T$$

where R is called the *universal gas constant*. Substituting the value (7) into eq. (8), we can evaluate it as follows:

$$p = 760 \text{ torr} = 1.03323 \, \frac{\text{kp}}{\text{cm}^2} = 9.81 \times 1.03323 \, \frac{\text{Dyne}}{\text{cm}^2}, \text{ see (2) and (2a)}$$

$$T = T_0 = 273.15 \text{ deg} \hspace{4em} \text{see (6)}$$

$$V_{mol} = 22.4 \text{ liter/mol} = 22.4 \text{ m}^3/\text{kmol}.$$

Consequently

(9) $$R = \frac{9.81 \times 1.03323 \times 224}{273.15} \frac{\text{Dyne} \times \text{m}}{\text{deg} \times \text{mol}} = 8.31 \, \frac{\text{Erg}}{\text{deg} \times \text{mol}}.$$

Erg is the unit of work in the MKS system and has already been used in (2.7).

Applying (8) to a volume consisting of n mols we obtain, evidently,

(10) $$p V = n R T$$

so that the value of the gas constant C used in (4) becomes

(10a) $$C = n R.$$

If v denotes the so-called specific volume, i. e. the volume of a unit of mass, we have

(11) $$V_{mol} = \mu v$$

where μ is the molecular or, better, molar "weight" (actually the mass of one mol) of the gas under consideration. Thus, for example, it is 32 kg/kmol for O_2. Substituting (11) into (8), we obtain

(11a) $$p V = \frac{R}{\mu} T$$

4. The First Law. Energy and enthalpy as properties

The so-called "mechanical theory of heat" replaced the theory that regarded heat as a substance, after the latter proved untenable. As the name of the present theory implies, heat is regarded as a manifestation of the random motion of material particles being given by their energy (*vis viva*) or living force. Following these ideas, Helmholtz gave to his book which appeared in 1847 the title: "Über die Erhaltung der Kraft." It is based on the assumption that the whole of the science of physics can be reduced to mechanics and that the interaction between the material particles is due to forces passing through their centers.

The designation "mechanical theory of heat" is evidently too narrow. Solar radiation certainly belongs to the heat balance of the earth and, equally certainly, it is not a mechanical process. For this reason the less vivid designation of "thermodynamics" is preferred nowadays. The ambiguous expression "*vis viva*" has been fortunately replaced by that of "kinetic energy", as suggested by Sir William Thomson. The word *energy* occurs already in Aristotle's writings; it was introduced into the language of science by Rankine (1853); he also used the expression "energetics". Robert Mayer's ("Über die Kräfte der unbelebten Natur," 1842) bold ideas went beyond the framework of classical mechanics and completely corresponded with the modern interpretation of the energy concept, even if he did not yet give them as precise a mathematical formulations as that achieved later by Helmholtz. One of Mayer's outstanding achievements was the emphasis he put on processes involving release (seemingly contradicting the energy principle) which are now so important in the understanding of catalytic phenomena.

We introduce the concept of energy axiomatically and without reference to mechanics and thereby state the *First Law of thermodynamics*: *Every thermodynamic system possesses a characteristic property (parameter of state) — its energy. The energy of the system is increased by the quantity of heat, dQ, absorbed by it and decreased by the external work, dW, performed. In an isolated system, the total amount of energy is preserved.*

A. Equivalence of Heat and Work

Introducing Clausius' symbol U to denote energy we can give the following mathematical formulation to the First Law:

(1) $$dU = dQ - dW.$$

Here dU, unlike dQ or dW, is a *perfect differential*. Hence for any cycle we must have

(1 a) $$\oint dU = 0.$$

The quantity of heat dQ which appears on the right-hand side of (1) need not be measured in calories; we may assume that it has been converted to mechanical units in accordance with eq. (2.6) or (2.7).

We shall now apply eq. (1) to the simplest possible thermodynamic system, namely to a unit of mass of a homogeneous fluid. The corresponding energy is called *specific energy* and is denoted by u, in analogy with the symbol v used to denote the specific volume in eq. (3.11 a), or the specific quantity of heat added, dq in eq. (2.3 a, b). Thus

(2) $$du = dq - p\,dv.$$

First of all we shall use this equation to determine the mechanical equivalent of heat, J, and hence to verify eq. (2.7). In order to do this we consider two processes. The first will take place at *constant v*, changing the state of the system from v, T to $v, T + dT$. The second process will take place at *constant p*, the state changing from v, T to $v + dv, T + dT$ with the same T and dT as in the first process. Taking into account the definitions (2.3 a, b) we have:

(3) $$du_1 = c_v\,dT,$$

(3 a) $$du_2 = c_p\,dT - p\,dv.$$

We now assume that our system is a perfect gas. Then we may apply eq. (3.11 a) to the second process, or

$$p\,dv = \frac{R}{\mu} dT$$

so that eq. (3 a) transforms to

(3 b) $$du_2 = \left(c_p - \frac{R}{\mu}\right) dT.$$

At this stage we supplement the definition of a perfect gas by an additional condition of a caloric nature: *The specific energy u (and, evidently, the total energy U) is a function of temperature T alone*, or, in other words, it is independent of volume or pressure at a given T. Then, according to eq. (3), c_v is also a function of temperature only ($= du/dT = u'(T)$) and eq. (3) can be written at once in integral form

(4) $$u(T) = \int c_v(T)\, dT.$$

Since u is a property, the nature of the path, whether at constant or variable volume, is irrelevant.

An experimental and theoretical justification of our additional caloric requirement in the definition of a perfect gas cannot be given here. We shall revert to this point in Secs. 5 C and 7.

Owing to the assumed equality of T and dT for our two processes, we can now write

(4 a) $$du_1 = du_2 = u'(T)\, dT.$$

From (3) and (3 b) we thus obtain

(5) $$c_v = c_p - \frac{R}{\mu}$$

and also

(5 a) $$\mu(c_p - c_v) = R.$$

The left-hand side contains the difference of the two molar specific heats which has the numerical value of approximately 2 cal/deg × mol for all perfect gases; hence

(5 b) $$(c_p - c_v)_{mol} = 2 \text{ cal/deg} \times \text{mol} = 2\,\frac{\text{kcal}}{\text{kmol} \times \text{deg}}.$$

Substituting this value into eq. (5 a) and using the value of R from eq. (3.9), we obtain

(6) $$1 \text{ cal} = 4.16 \text{ Erg}$$

which agrees with our previous statement to within 1%, the discrepancy being due to the inaccuracy in the value of 2 in (5 b).

We can consider the same example and express the unit of work in terms of calories. In this manner, taking into account eq. (5 b), eq. (5 a) gives

(7) $$R = 2\,\frac{\text{cal}}{\text{deg} \times \text{mol}}$$

and the equation of state (3.8) assumes the rather odd form

$$p\, v_{mol} = 2\, T\, \frac{\text{cal}}{\text{deg} \times \text{mol}} \tag{8}$$

where the pressure is measured in cal/unit of volume.

B. The enthalpy as a property

Along with energy, we introduce a new property which is particularly important in engineering applications; it is given the name of *enthalpy* and is defined as

$$H = U + pV. \tag{9}$$

The term enthalpy means "heat function," and the symbol H (originally the Greek letter η was meant to be used) has been introduced in the American standard text-book on thermodynamics by Lewis and Randall; alternative symbols will be listed in Sec. 7 where we shall also deduce the definition in eq. (9) from a general mathematical concept.

From eq. (9), with $dU = dQ - p\, dV$ we have

$$dH = dQ + V\, dp. \tag{10}$$

At a constant pressure ($dp = 0$) dH is equal to the quantity of heat introduced to the system from an external source, which explains the name of "heat function" (or "total heat") given to it.

The enthalpy per mol (on occasions also that per unit mass) will be denoted by h, so that

$$h = u + p v \tag{9 a}$$

$$dh = dq + v\, dp. \tag{10 a}$$

It follows[1] that the molar specific heat c_p is given by

$$\left(\frac{\partial h}{\partial T}\right)_p = \left.\frac{dq}{dT}\right|_{(p=\text{const})} = c_p. \tag{11}$$

---

[1] In eqs. (11) and (11 a) as well as in succeeding equations we shall avoid writing

$$\left(\frac{dq}{dT}\right)_p \quad \text{or} \quad \left(\frac{dq}{dT}\right)_v$$

because q is not a property.

As a corrollary to eq. (4 a), we can write more fully that

(11 a) $$\left(\frac{\partial u}{\partial T}\right)_v = \frac{dq}{dT}\bigg|_{(v=\text{const})} = c_v.$$

In the case of a perfect gas $p\,v = R\,T$ and hence h is a function of T alone, in the same way as u; consequently, we are justified in dropping the indices p and v on the left-hand sides of eqs. (11) and (11 a) respectively. Subtracting (11 a) from (11), we obtain

$$c_p - c_v = \frac{d(h-u)}{dT},$$

which is identical with (5 a) because $h - u = p\,v = R\,T$.

The concept of enthalpy is particularly important in engineering applications because it is directly connected with the *flux of energy* during a *steady-state* process involving the performance of work. Imagine a steam turbine which receives high-pressure steam at a constant rate per unit time. The steam is expanded, cooled and rejected by the turbine. We shall now consider, quite generally, the energy balance of an arbitrary machine which functions in a steady manner. We shall assume that all thermal quantities and that all quantities of work have been referred to a unit of mass of the gas (in general — the working fluid) supplied to the machine.

Consider a cross-section 1 (area A_1) through the inlet pipe of the machine, and assume that one unit mass has just crossed it. In this way a quantity u_1 of internal energy (the subscript *1* refers to the state of the gas at cross-section 1 of the pipe) has been transported through A_1. The body of gas, which follows and whose pressure is p_1, will have been displaced by a distance v_1/A_1, because the volume of a unit mass of gas at cross-section 1 occupies a volume v_1. The external pressure (e. g. the boiler pressure) has thus performed a quantity of work = force × distance = $(p_1 A_1)(v_1/A_1) = p_1 v_1$. The flux of energy through A_1, neglecting the kinetic energy, is

$$u_1 + p_1 v_1 = h_1.$$

The same reasoning can be applied to cross-section 2 imagined taken through the exhaust pipe. Assume that the machine performs work (per unit mass of gas) at a rate w (useful power) and that, for the sake of generality, it consumes heat at a rate q. (In particular cases q can be equal to 0.)

The energy equation assumes the simple form

(12) $$h_1 + q = w + h_2.$$

This form of the balance equation has the advantage that any specific processes which may be taking place inside the machine do not come into evidence in it. We shall revert to this example in Sec. 5 C when we shall consider a very important physical process.

C. Digression on the ratio of specific heats c_p and c_v

At this stage we are compelled to make a digression which falls outside the field of thermodynamics. The science of thermodynamics can supply *relations between properties only*, such as e. g. eq. (5 a), but not their *absolute values*. In order to obtain the latter it is necessary to adopt microscopic models, as is done in the kinetic theory of gases. According to the law of equipartition of energy of the latter theory (see Sec. 31 B, ahead of eq. (9)), the molar specific heat of gases or vapors is given by

$$(13) \qquad c_v = \tfrac{1}{2} f R.$$

Here f denotes the number of degrees of freedom and is:

$f = 3$ for monatomic molecules; here only linear translations count, rotations being of no importance;

$f = 5$ for diatomic molecules; they can be regarded as possessing the symmetry of a dumb-bell so that it has two rotational degrees of freedom in addition to the three translational ones; rotation about the link of the atoms is unimportant. At the same time the possibility of the two atoms vibrating with respect to each other, which affects the specific heats only at high temperatures, is here disregarded;

$f = 6$ for more general molecular arrangements, i. e. 3 rotational degrees of freedom + 3 translational degrees of freedom, the possibility of internal motions being again disregarded.

Equation (13) shows that c_v is a characteristic *constant* for each gas, i. e. that it is not only independent of v, but also of T. The corresponding value of c_p, also per mol, is obtained from eq. (7) and is

$$(13\text{ a}) \qquad c_p = \left(1 + \tfrac{1}{2} f\right) R.$$

From (13) and (13 a) we find

$$(14) \qquad \frac{c_p}{c_v} = \gamma = 1 + \frac{2}{f}.$$

The numerical value of γ, which is the same for the specific heats referred to a unit mass and a mol, is

f	3	5	6
γ	$1+\frac{2}{3}=1.66$	$1+\frac{2}{5}=1.40$	$1+\frac{2}{6}=1.33$

Examples for $f = 3$: mercury vapor and the noble gases He, Ne, A,....
Examples for $f = 5$: H_2, N_2, O_2,..., air.
Examples for $f = 6$: all polyatomic gases.

The thermodynamic relation (5) is exact and remains unaffected by quantum corrections. On the other hand the values (13) and (13 a) are more or less accurate approximations and must be refined with the aid of quantum theory. In particular, $\gamma = 1.33$ is only a mean value about which the experimental values for polyatomic gases group themselves more or less closely. It is, however, remarkable that the case $f = 4$, $\gamma = 1.50$ which does not correspond to any geometrical model or to any type of molecular symmetry does not occur in nature.

The purpose of the present digression was to throw some light on the strong and weak points of the science of thermodynamics on the one hand, and of the kinetic theory of gases, on the other.

5. The reversible and the irreversible adiabatic process

We shall begin by emphasizing the difference between *reversible and irreversible processes*.

Reversible processes are not, in fact, processes at all, they are sequences of states of equilibrium. The processes which we encounter in real life are always irreversible processes, processes during which disturbed equilibria are being equalized. Instead of using the term "reversible process" we can also speak of infinitely slow, quasi-static processes during which the system's capacity for performing work is fully utilized and no energy is dissipated. In spite of their not being real, reversible processes are most important in thermodynamics because definite equations can be obtained only by considering reversible changes; irreversible changes can only be described with the aid of inequalities when equilibrium thermodynamics is used.

The actual criterion for a process to be reversible states that during its course there are no lasting changes of any sort in the surroundings if the process is allowed to go forward and then back to the original state.

A. The Reversible Adiabatic Process

The term *adiabatic* implies: exclusion of heat transfer to and from the body; in this connection the thermos flask invented by Dewar may be thought of. The opposite case is that of an isothermal process; in order to maintain the temperature it is necessary to allow heat to be transferred; in this connection one may imagine a water bath in which our quantity of gas is immersed.

Consider a unit mass of a perfect gas and substitute

$$dq = 0 \qquad du = c_v \, dT$$

into (4.2), taking into account (4.4). We then have

(1) $$c_v \, dT = - p \, dv.$$

In order to transform this into a relation between v and p we use the equation of state (3.11 a). Instead of (1) we may write

$$\frac{\mu}{R} c_v (p \, dv + v \, dp) + p \, dv = 0$$

$$\left(c_v + \frac{R}{\mu}\right) p \, dv + c_v \, v \, dp = 0$$

so that in view of (4.5)

$$c_p \, p \, dv + c_v \, v \, dp = 0,$$

or, considering (4.14):

(2) $$\frac{dp}{p} + \gamma \frac{dv}{v} = 0$$

We now assume γ to be a constant, see end of Sec. 4, so that we actually exceed the caloric assumption according to which u and hence c_v, c_p and γ depend on T alone. In this case eq. (2) can be integrated directly, so that

$$\log p + \gamma \log v = \text{const.}$$

This is Poisson's equation of a *reversible adiabatic (isentropic)* process. It can be written

(3) $$p \, v^\gamma = \text{const.}$$

Poisson's equation is very important in meteorology. We may also recall the calculation of the velocity of sound in Vol. II, eq. (13.17 a), with the aid of Poisson's equation (described as the equation of a polytrope whose exponent $n = \gamma$). Transforming the equation to T, v or T, p coordinates with the aid of the equation of state (3.11 a), we obtain

(3 a) $\qquad\qquad T v^{\gamma-1} = \text{const}, \quad \text{or} \quad T p^{(1-\gamma)/\gamma} = \text{const},$

respectively. The constants in eqs. (3) and (3 a) can be expressed in terms of the initial state p_0, v_0, T_0, as follows:

$$\text{const} = p_0 v_0^\gamma \quad \text{or} \quad T_0 v_0^{\gamma-1} \quad \text{or} \quad T_0 p_0^{(1-\gamma)/\gamma}$$

According to Boyle's law, isothermals are represented by equilateral hyperbolae in the p, V-plane; on the other hand, according to Poisson's eq. (3), isentropes are steeper downwards (see Fig. 2). In the T, V-plane the isentrope is, evidently, less steep because of the exponent $\gamma - 1$ in eq. (3 a), see Fig. 2 a.

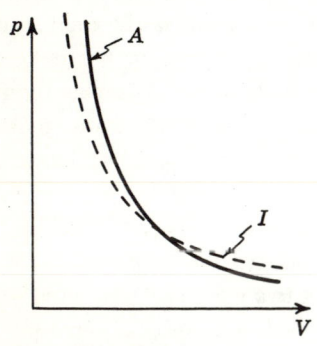
Fig. 2.
Reversible adiabatic (isentrope), A, and isothermal, I, for a perfect gas in the p, V-plane.

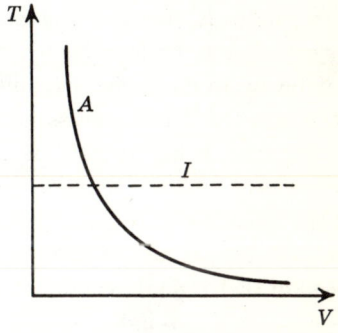
Fig. 2 a.
Reversible adiabatic (isentrope), A, and isothermal, I, for a perfect gas in the T, V-plane.

In order to obtain a clear idea of the *reversible* nature of a process we imagine the gas to be confined in a cylindrical vessel of cross-sectional area A. The vessel is, in turn, enclosed by walls which prevent any exchange of heat, and the gas is contained at the top by a weightless piston. The piston is maintained in equilibrium by a weight $P = p A$ which balances the gas pressure. We imagine P to be subdivided into many small weights δP which will be removed one by one. This causes the piston to rise each time, its pressure p decreasing. Each weight δP is placed outside the vessel at the same level at which it has been removed so that no work is gained or lost in the process. The gas pressure will fall from its initial value p (e. g. 2 kp/cm²) to a final value p_1 (say 1 kp/cm^2), and the volume will increase from an initial value V (e. g. 1 liter) to a final value V_1 (in our example $2^{1/\gamma}$ liter). The center of gravity of each δP has been raised compared with its original level. This work against the forces of gravity stems from the work performed by the gas on the piston. It has not been lost, being found stored in the raised δP's. If we now replace these weights one by one on the piston, the gas will be re-compressed and heated and will revert to its initial state. The *process is reversible* on condition that it has been carried out in *infinitely small steps* and *sufficiently slowly*,[1] i. e. with a sufficiently fine subdivision of P into elements of δP each.

B. The irreversible adiabatic process

If the piston (together with the weight P) is raised *suddenly* the gas will first flow into a vacuum performing no external work. The resulting turbulent motion gradually subsides, the gas coming to rest. What is the final state of the gas? Has it become heated owing to internal friction or has it become cooled owing to its having expanded? None of the two: As far as the final state is concerned the process is not only adiabatic but also isothermal, the approximation being as good as that of the gas under consideration is to a perfect gas.

The preceding experiment was first performed by Gay-Lussac in 1807 (flow experiment) and then repeated by Joule with an increased accuracy. Instead of the original cylindrical arrangement two glass jars were used; they were connected through a narrow tube equipped with a cock. One jar was evacuated and the other was filled with the experimental gas. After the cock had been opened and after equilibrium had set in, it was observed that the final temperature, particularly with air or hydrogen, was substantially the same as that at the beginning.

Anticipating this result we shall first consider the *cycle* which Robert Mayer[2] used for the calculation of the mechanical equivalent of heat and was thus led to the First Law. At the initial state 1, Fig. 3, the gas is under the

---

[1] A reversible process must be carried out infinitely slowly. The reverse is, however, not true, as an infinitely slow process need not be reversible. Example illustrating the latter case: Discharging a condenser through a very large resistance.

[2] The same calculation, based on specific heats, was found in the papers left by Sadi Carnot, (1796–1832) who died at a young age. He was the son of the geometer and general Lazarus Carnot mentioned in Vol. I in connection with eq. (3.28 b). Hence Sadi Carnot can be regarded as having paved the way not only to the Second Law, but also to the first part of the First Law.

atmospheric pressure p_1 and has the volume V_1. It is heated at constant volume V_1 until its pressure is changed to p_2, point 2 in Fig. 3. It now expands to V_3 by being allowed to flow from one vessel to another. Neglecting turbulent deviations, it will reach a state along the isothermal equilateral hyperbola passing through 2. This element of the cycle is shown by a broken line because it is not defined in detail; only the portions of the hyperbola which lie on the other side of points 2 and 3 have been drawn in full. The gas can now be re-compressed at constant pressure p_1 to its initial volume by performing work on it, if V_3 has been so selected as to make the corresponding pressure equal to p_1.

The change in energy per unit mass of gas along the three paths 12, 23 and 31 is

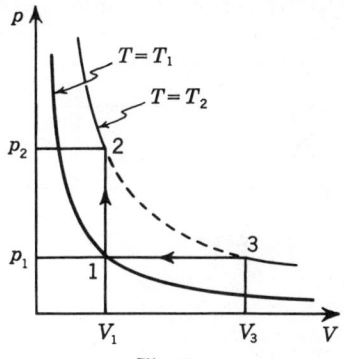

Fig. 3.
Cycle for the determination of the mechanical equivalent of heat.

$$\int_{T_1}^{T_2} c_v \, dT; \quad 0; \quad \int_{T_2}^{T_1} c_p \, dT - p_1 (v_1 - v_3).$$

(4)

According to (4.1 a) the sum of these terms must vanish. Thus we are led to (4.5 a) and to the value of the mechanical equivalent of heat in (4.6) if we, in addition, take into account the equation of state of a perfect gas and allow the temperature difference $T_2 - T_1$ to become vanishingly small.

It is evident that the cycle under consideration must give the same result as our differential method in (4.5 a), because both are based on the same assumption, namely on the premiss that the energy of a perfect gas is a pure function of T.

C. THE JOULE-KELVIN POROUS PLUG EXPERIMENT

In order to refine the experiment in which a gas is allowed to flow into an evacuated vessel, William Thomson devised the porous plug experiment and carried it out experimentally in collaboration with Joule. In the experiment the gas is forced through a plug made of cotton wool, the stream being slow and well regulated and proceeding from a higher pressure ahead of to a lower pressure behind the plug. On passing through the cotton wool plug which was accommodated in a pipe made of beechwood, which is, to all intents and purposes, a heat insulator, the gaseous stream became slowed up. After a steady-state has set in, the temperature in the plug becomes steady, irrespective of how complicated the temperature distribution in its interior is, and the same is true of the temperatures to the left and to the right of the plug.

We shall consider the mass of gas, Fig. 4, contained between an arbitrary section A and the right-hand end of the plug, B, and we shall follow its motion until it reaches the position $A'B'$, when the particles at A have reached the left-hand end of the plug. During the motion the mass of gas is acted upon by a force pA from the left (A = cross-section of the pipe). The opposing force is $p'A$. The path traversed is V/A on the left and $V'A$ on the right, so that the total work performed is

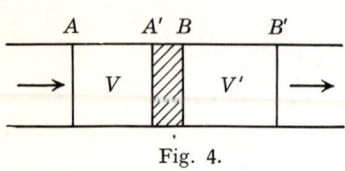

Fig. 4.

The Joule-Kelvin porous plug experiment.

$$\int dW = pV - p'V'. \tag{5}$$

On the other hand there is no transfer of heat either through the gas on the left or on the right, or through the beechwood pipe. Hence

$$\int dQ = 0. \tag{5 a}$$

According to the First Law

$$U - U' = -pV + p'V'. \tag{6}$$

So far our reasoning applies to any gas. Thus we note the following general result of the Joule-Kelvin porous plug experiment:

$$U + pV = U' + p'V' \quad \text{or} \quad H = H'. \tag{7}$$

In words: *The Joule-Kelvin experiment is characterized by the fact that the enthalpy of the gas is preserved as it flows through the porous plug.* We recall here the final remark in Sec. 4 B concerning the flow of energy in the inlet and exhaust pipes of a steam engine. It is evident that the quantity of energy calculated in (7) represents the previously considered energy flux (provided that the unit of energy has been suitably chosen), and our present example can be used as a special illustration of the preceding general theorem.

In particular, for a perfect gas the right-hand side of (6) becomes

$$\frac{M}{\mu} R (T' - T)$$

where M denotes the mass of the gas contained in the volume AB.

In actual fact the Joule-Kelvin experiment showed a very small difference between T and T' in the case of air, whereas for hydrogen the difference was hardly measurable. From this result we conclude that: *In the ideal, limiting case we have*

(8) $$U' = U \quad \text{independently of } V,$$

which is the same result as that from the Gay-Lussac experiment, except that it has now been deduced with a much higher degree of accuracy. It is only now that we have based our additional caloric condition in eq. (4.4) on a sure experimental foundation.

D. A CONCLUSION OF GREAT CONSEQUENCE

We shall now consider the First Law and we shall apply it to a reversible process in a perfect gas, e. g. to a unit mass of gas. In view of the now established relations: $u = u(T)$, $c_v = c_v(T)$, $du = c_v(T) dT$ and the equation of state, we write

(9) $$dq = du + p\,dv = c_v(T)\,dT + \frac{R}{\mu}\frac{T}{v}\,dv.$$

Dividing both sides by T we obtain

(9 a) $$\frac{dq}{T} = c_v(T)\frac{dT}{T} + \frac{R}{\mu}\frac{dv}{v}.$$

We know that dq is *not* a perfect differential, but eq. (9 a) shows that dq/T is integrable. Putting $ds = dq/T$ we obtain by integrating (9 a) that

(10) $$\int_{T_0,v_0}^{T,v} ds = s - s_0 = c_v \log \frac{T}{T_0} + \frac{R}{\mu} \log \frac{v}{v_0}.$$

We have assumed here that $c_v = $ const, which was convenient but not necessary; *s is a property which is independent of the path between the initial and final state* and depends only on the *instantaneous values of the properties T, v* if the initial properties are fixed at an arbitrary state. With Clausius we shall call this new property *entropy*. The term means "transformability."

In order to recognize, at this stage at least, the formal meaning of entropy, we write the energy equation (9) in the form

(11) $$du = T\,ds - p\,dv,$$

since $dq = T\,ds$. We conclude that s is conjugate to T in the same sense as v is to p: *s is the extensive property which corresponds to the intensive property T*, the problem of finding it having been mentioned already in Sec. 1.

It is evident that the definition of entropy (10) can be extended from a unit of mass to one mol and to any mass M (in which case instead of the lower case symbols we use S, V).

The adiabatic processes which were considered in Section A can be also called "isentropic" because $dq = 0$ implies that they are curves of constant entropy. In fact it is easy to convince oneself that the above eq. (3 a) in the T, v-plane is identical with the equation s = const from (10).

6. The Second Law

In order to present the most essential considerations in the science of thermodynamics we shall follow the classical path which was initiated by Sadi Carnot in 1824 and then followed by Rudolf Clausius from 1850, and by William Thomson from 1851 onwards. The title of Carnot's paper "Réflexions sur la puissance motrice du feu et les moyens propres à la développer" gives expression to the historical connection between thermodynamics and the development of the reciprocating steam engine.

Carnot based his considerations on the hydraulic analogy: he thought that the heat substance is capable of performing work on passing from a higher to a lower temperature in the same way as water can perform work when it flows from a higher to a lower level. The weakness of this analogy is evidently derived from the fact that no indestructible heat substance exists. In spite of this, however, Carnot's argument proved to be one of permanent value having become essential in the development of the Second Law which was not discovered until 25 years later.

We shall state the Second Law in an axiomatic way, just as we have done with the First Law in Sec. 4 (and with the "zeroth" in Sec. 1):

All thermodynamic systems possess a property which is called entropy. It is calculated by imagining that the state of the system is changed from an arbitrarily selected reference state to the actual state through a sequence of states of equilibrium and by summing up the quotients of the quantities of heat dQ introduced at each step and the "absolute temperature" T; the latter is to be defined simultaneously in this connection. (First part of Second Law.)

During real (i. e. non-ideal) *processes the entropy of an isolated system increases.* (Second part of Second Law.)

In what follows we shall provide a "proof" of this proposition, but this can only mean that we shall reduce it to simpler, apparently evident, assumptions which, by their nature, cannot be proved in turn. The simplest of these seems to be: *Heat cannot pass spontaneously from a lower to a higher temperature level* (Clausius). In this connection it is necessary clearly to define the meaning of the word "spontaneously," and we shall take it to mean that except for the bodies taking part in the exchange of heat there are no permanent changes of any sort caused by the process. The following postulate, due to Kelvin, is equivalent to that due to Clausius: *It is impossible continuously to produce work by cooling only one body down to a temperature below the coldest part of its surroundings.* If that were not so it would be possible to convert the work into heat, for example through friction, and so to bring it to a higher temperature level. Ostwald expressed this principle in a form in which it is now normally quoted: *It is impossible to design a "perpetual motion engine of the second kind,"* i. e. a machine which would work periodically and which would cause no other changes except the lifting of a weight and the cooling of a heat reservoir.[1] (As is well known the First Law expresses the impossibility of building a *perpetual motion engine of the first kind*.)

A. The Carnot cycle and its efficiency

We shall use an arbitrary, but homogeneous working fluid. The term "homogeneous" denotes that its state is described by indicating only its two mechanical variables, V and p; these in turn determine the thermal variable θ with the aid of some general equation of state. The symbol θ instead of T gives expression to the fact that temperature is, at first, measured with the aid of an arbitrary calibrated thermometer (say a thermocouple, etc.).

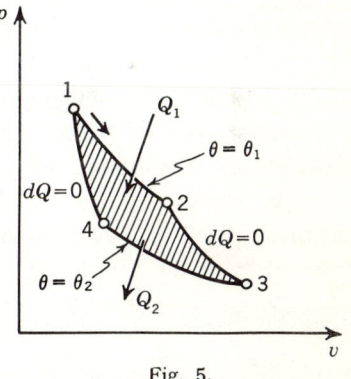

Fig. 5.
The Carnot cycle.

The path of a Carnot cycle (see Fig. 5) consists of two isotherms 1 2 and 3 4 and of two isentropes 2 3 and 4 1. Along 1 2 it is necessary to add a certain quantity of heat Q from the "boiler" (heat reservoir of temperature θ_1) and along 3 4 it is necessary to reject a quantity of heat Q_2 to a cooler (heat reservoir θ_2). The total amount of heat transferred is

$$\oint dQ = Q_1 - Q_2.$$

---

[1] Formulation due to Planck, Thermodynamics, 8-th German ed. Sec. 116, English ed. "Treatise on Thermodynamics," 3rd. ed. p. 89.

The work performed by the working fluid is equal to

$$\oint dW = \oint p\, dV = W$$

in the same way as for the indicator diagram in Sec. 2.

According to the First Law

(1) $$W = Q_1 - Q_2$$

since the internal energy U returns to its initial value at point 1. The efficiency of the cycle is defined as

(2) $$\eta = \frac{\text{work performed}}{\text{heat added}} = \frac{W}{Q_1} = 1 - \frac{Q_2}{Q_1}$$

i. e. in the same way as for a steam engine.

Carnot considers an engine **E** which realizes the process 1 2 3 4 infinitely slowly (without frictional or radiation losses) so that the working fluid is always in thermal equilibrium. (In such a case the isentrope must be qualitatively the same as that described in Sec. 5 A in connection with the special case of a perfect gas.) Such an engine is called *reversible*: It can equally well traverse the sequence of states of equilibrium in the direction 1 4 3 2 in which case it does not operate as a *prime mover* but as a refrigerator ($W < 0, Q_2 > Q_1$; it is now necessary to add the amount of work $|W|$ in order to depress still further the lower temperature level of the cooler).

Carnot shows that the efficiency of such an engine is independent of the properties of the working fluid. In order to do this he considers two engines **E** and **E'** which operate on different working fluids but between the same heat reservoirs θ_1 and θ_2 developing equal power W. The quantities of heat processed by **E'** are denoted by Q_1' and Q_2' respectively. Let us assume that

(3) $$\eta' > \eta.$$

In this case let us arrange **E** and **E'** in such a way that **E** operates as a refrigerator, i. e. in the direction 1 4 3 2, being driven by **E'**. From (2) and (3) we have

$$\frac{|W|}{Q_1'} > \frac{|W|}{Q_1}, \quad \text{i. e.} \quad Q_1 > Q_1'.$$

The hotter reservoir receives more heat from **E** than it loses to **E'**. Owing to the simultaneous operation of **E** and **E'**, this difference $\Delta Q = Q_1 - Q_1'$ is taken from the lower level θ_2. The total effect is to transfer the quantity of heat ΔQ from the lower level θ_1 without the performance of work and without making any permanent changes in E, E', or in the surroundings. According to the preceding postulate this is impossible. Thus assumption (3) is untenable.

The assumption $\eta > \eta'$ is equally untenable: It suffices to interchange the roles of E and E' in oder to arrive once more at a contradiction with our postulate. Consequently we must have

(4) $$\eta = \eta'.$$

All reversible engines which exchange heat only at two temperatures θ_1 and θ_2 have equal efficiencies. In view of (2), eq. (4) can be replaced by

(5) $$\frac{Q_1}{Q_2} = f(\theta_1, \theta_2),$$

where f denotes a universal function which is independent of the working fluid and of the design of the heat engine.

B. The first part of the Second Law

In order to split the function of two variables $f(\theta_1, \theta_2)$ into two functions of one variable each it is necessary to span two reversible Carnot cycles between the two temperature levels θ_1, θ_2 and a heat reservoir of an arbitrary but constant intermediate temperature θ_0 so that the heat reservoir θ_0 acting as a cooler for one cycle absorbs the same quantity of heat, Q_0, as it is forced to reject when serving as a heater for the other. In this manner the reservoir θ_0 will not enter the heat balance equation and the simple cycle (θ_1, θ_2) is seen to operate with the same quantities of heat as the compound cycle $(\theta_1, \theta_0) +$ $+ (\theta_0, \theta_2)$. In addition to eq. (5) we can write the equations

(6) $$\frac{Q_1}{Q_0} = f(\theta_1, \theta_0); \quad \frac{Q_0}{Q_2} = f(\theta_0, \theta_2)$$

in which the same quantities of heat Q_1 and Q_2 appear. On multiplying, we have

(6 a) $$\frac{Q_1}{Q_2} = f(\theta_1, \theta_0) \times f(\theta_0, \theta_2).$$

Comparing with (5), we find

(6 b) $$f(\theta_1, \theta_2) = f(\theta_1, \theta_0) \times f(\theta_0, \theta_2).$$

Inserting $\theta_1 = \theta_2$[1] as a special case, so that according to (5) we also have $f(\theta_1, \theta_1) = 1$, we have

$$f(\theta_0, \theta_2) = 1/f(\theta_2, \theta_0).$$

Consequently, eq. (6 b) can also be written as

(6 c) $$f(\theta_1, \theta_2) = \frac{f(\theta_1, \theta_0)}{f(\theta_2, \theta_0)}.$$

Since θ_0 cancels out, eqs. (5) and (6 c) lead to

(7) $$\frac{Q_1}{Q_2} = \frac{\phi(\theta_1)}{\phi(\theta_2)}.$$

With the intrinsically arbitrary temperature scale θ we can now associate an *absolute scale* in such a way that to each mark on θ there corresponds the mark

(7 a) $$T = \phi(\theta)$$

on the latter scale. We shall see in Sec. 10 how this can be done in practice. At the moment we shall only remark that this *absolute temperature* T coincides with the temperature measured on a gas thermometer over a range in which the thermometric substance behaves like a perfect gas if a suitable value is chosen for the still arbitrary constant factor in $\phi(\theta)$. The proof of this proposition will be advanced in Problem 1.

Equations (7) and (7 a) can be combined into the Carnot ratio:

(8) $$Q_1 : Q_2 = T_1 : T_2.$$

From this we deduce the formula for efficiency, *viz.*

(8 a) $$\eta = \frac{T_1 - T_2}{T_1},$$

and applying it to an infinitely narrow Carnot diagram (finite temperature difference, but infinitely small quantities of heat added and rejected, dQ_1 and dQ_2), we obtain

(8 b) $$\frac{dQ_1}{T_1} = \frac{dQ_2}{T_2}.$$

---

[1] It will be noticed that we now abandon the stipulation that θ_0 is intermediate between θ_1 and θ_2, but this point has no bearing on the result.

We shall consider an arbitrary, but still reversible cycle. We shall represent it with the aid of the continuous contour in the p, V-diagram of Fig. 6, and select two arbitrary points A and B on it. We now replace the process by infinitely narrow Carnot cycles. The fact that the continuous contour is now replaced by a sequence of small saw-teeth, as shown in Fig. 6 at A and B, makes no difference for the integration. If we consider that the rejected heat dQ_2 is negative, which is entirely consistent, we obtain at once from (8 a) that

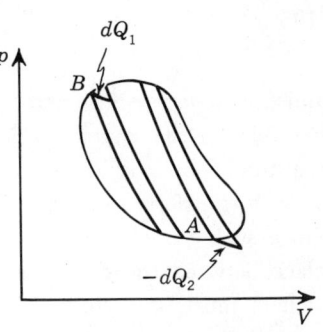

$$\text{(9)} \qquad \oint \frac{dQ_{rev}}{T} = 0$$

where integration is carried out over the whole contour. The subscript of dQ explicitly stresses the reversible nature of the cycle under consideration. According to Sec. 1 eq. (9) is the necessary and sufficient condition for

Fig. 6.
Representation of an arbitrary reversible process as a sum of infinitely narrow Carnot cycles.

$$\text{(10)} \qquad dS = \frac{dQ}{T}$$

to be a *perfect differential*, provided that dQ is added reversibly (utilizing the available work in full). Reversibility is assured if we put $dQ = dU + dW$ according to the First Law, i. e. if instead of (10) we write (10 a)

$$\text{(10 a)} \qquad dS = \frac{dU + p\,dV}{T}$$

for the simple working fluid now being considered. The absolute temperature, see eq. (7 a), defined in the above sense is seen to be the *integrating denominator* of the incomplete differential which appears in the numerator of (10 a).

Equation (9) has been shown to be true for a general path of integration but it still applies only to a very special thermodynamic system (homogeneous fluid). It is, however, true for any system, composed of different substances, appearing in different phases, and possessing any number of degrees of freedom (e. g. electrical or magnetic), *provided that the system does not perform any irreversible processes, such as friction, Joule heat, etc.*

If we first consider a single, say the i-th homogeneous component of the system possessing two degrees of freedom, we find according to (10) that

$$\text{(10 b)} \qquad dS_i = \frac{dQ_i}{T_i}$$

is a perfect differential; here T_i denotes the absolute temperature of this i-th component and dQ_i denotes the total quantity of heat added reversibly to it, whether externally or by the other components of the system.[1]

We now form the sum

(10 c) $$dS' = \sum_i dS_i = \sum_i \frac{dQ_i}{T_i},$$

and find that it is also a total differential independently of the choice of the variables of the system to describe the processes. This sum is simpler than the sum of the individual expressions in (10 b), because in (10 c) the quantities of heat transferred between the individual components need not be included. Since such transfers of heat have been assumed reversible they must take place between the components at equal temperatures (ordinary conduction of heat must be excluded!). Denoting two such sub-systems by i and i' we have $T_i = T_{i'}$, and $dQ_i = -dQ_{i'}$, (the heat added to i is rejected by i'). *It is thus seen that the terms which are due to such transfers cancel each other.* The same is true of the quantities of heat transferred at phase equilibrium which, as will be stressed in Sec. 8 B, stipulate equal temperatures for the two phases. *Consequently we may restrict the meaning of dQ_i to denote only the heat added externally to the i-th component.*

In principle an inequality between T_i and $T_{i'}$ is not excluded if the two sub-systems are separated from each other by an adiabatic wall. This would, however, necessitate fairly artificial combinations of the sub-systems. Normally we shall find that there is thermal equilibrium not only throughout a single component (T_i constant within i) but throughout the whole system, ($T_i = T$). In such cases (10 c) reduces to

(10 d) $$dS = \frac{1}{T} \sum_i dQ_i = \frac{dQ}{T}$$

which is identical with (10); dQ denotes here, as in eq. (10), the quantity of heat added reversibly to the whole system from the external surroundings. Equations (10), (10 a), as well as the slightly more general eq. (10 c), directly demonstrate *the existence of entropy S as a property of the system* and thus gives the proof *of the first part of the Second Law.*

---

[1] When more than two degrees of freedom are present it is possible to fix all degrees of freedom except two, using different combinations, and to apply eq. (10) to each partial process.

The difference in the entropy between two arbitrary states A and B is calculated with the aid of the equation

$$(11) \qquad S_B - S_A = \int_A^B \frac{dQ_{rev}}{T}.$$

We wish to emphasize the fact that the path of integration bears no relation to the way in which the system reaches B from A in actual fact. Real processes are always at least partly irreversible. Our rule (11) demands, however, the selection of an imaginary *reversible* path. The actual path selected is immaterial, because S is a property, and as such it is independent of the path.

The simplest example of such a calculation is afforded by the porous plug experiment described in Sec. 5 C, Fig. 4, where the points 2 and 3 in the sketch correspond to states A and B in eq. (11). Since the process is adiabatic, we have $dQ = 0$, so that for the *real* process

$$\int_2^3 \frac{dQ}{T} = 0$$

independently of how much the instantaneous temperature which prevails during the turbulent transition departs from the isotherm (shown dotted in the sketch, Fig. 3). On the other hand for the imaginary *reversible* process, which we may select along the isotherm, we have $dU = 0$, $dQ_{rev} = dU + p\,dV = p\,dV$, so that per mol of gas flowing:

$$\Delta S = S_3 - S_2 = \int_2^3 \frac{p\,dV}{T} = R \int_2^3 \frac{dv}{v} = R \log \frac{v_3}{v_2}.$$

We shall, obviously, find the same value if we integrate along 23 + 31 in Fig. 3 instead of the isotherm, as can easily be verified. Attention may be drawn to the fact that the preliminary evaluation of entropy in eq. (3.10) made use of the heat added reversibly in the sense of the preceding eq. (10 a), and the same is true of the van der Waals gas in Sec. 9 B.

Our example shows clearly that the existence and the value of entropy at the final state depend only on the state itself and not on whether it has been reached reversibly or irreversibly. Its value, denoted here by S_3, is determined except for a constant, denoted here by S_2.

In connection with the transition from eq. (10 b) to (10 c) we wish to remark that it implies that *partial entropies are additive*. This is usually assumed in classical thermodynamics, but from the higher point of view of statistical mechanics it is not necessarily so, see Sec. 31 A.

A system will be termed *isolated* when it does not interact with the surroundings, i. e. when no heat is transferred and no work is performed. The *energy* of such a system is constant, because $dQ = 0$ and $dW = 0$. According to (11) the *entropy* of such a system would also be constant:

(12) $$S_B = S_A.$$

This is a paradox and it seems to contradict the second part of the Second Law. The reason for it lies in the too narrow interpretation which we gave to the concept of a "thermodynamic system" in eq. (10 b) and following, because we have excluded all irreversible interactions between the components of the system thus implying thermodynamic equilibrium. It was in fact necessary to do so to calculate the entropy difference in (11). The proposition contained in (12), *namely that the entropy of an isolated system is constant, the system being in thermal equilibrium*, is true only under the above restrictive assumption.

C. The Second Part of the Second Law

We now assume that of the two engines **E** and **E′** considered in Section A one, say **E′**, is *not reversible*. In this case we can still achieve the mode of operation assumed in (3) when **E** driven by **E′** worked as a refrigerator between the same temperatures as **E′**, and can thus prove the impossibility of $\eta' > \eta$, but the reverse operation is not possible. Instead of eq. (4) we now have, therefore,

(13) $$\eta > \eta'$$

because $\eta = \eta'$ is also excluded by the assumption that **E′** is irreversible. The reversible Carnot cycle has a larger efficiency than an irreversible Carnot cycle which operates between the same temperatures and produces the same power. The latter is less economical than the former; it requires a greater expenditure of fuel for the same power: $Q_1' > Q_1$.

Retaining the definition of absolute temperature given in (8) and in accordance with (8 a) and (2) we conclude from $1 - \eta < 1 - \eta'$ that

$$T_2/T_1 = Q_2/Q_1 < Q_2'/Q_1'$$

and consequently

(13 a) $$\frac{Q_1'}{Q_2'} < \frac{T_1}{T_2}.$$

For an infinitely narrow Carnot diagram we have

$$\frac{dQ_1'}{T_1} < \frac{dQ_2'}{T_2}$$

instead of (8 b). Following the same reasoning as before (namely by subdividing into infinitely many cycles) we can prove that for an arbitrary cycle which is partly or wholly irreversible we must have

(14) $$\oint \frac{dQ'}{T} < 0,$$

if rejected heat quantities, dQ_2' etc., are considered negative. We now split this cycle into two segments, $A \to B$ and $B \to A$ and we assume that segment $B \to A$ consists of infinitesimal reversible processes only, whereas all irreversible processes take place along $A \to B$. Applying eq. (11) to the segment $B \to A$ we can rewrite eq. (14) to read

$$\int_A^B \frac{dQ'}{T} + S_A - S_B < 0$$

or

(15) $$S_B - S_A > \int_A^B \frac{dQ'}{T}.$$

This inequality applies to any kind of system. To be consistent we can now lift the restriction placed in connection with (10 c) and allow irreversible processes to take place within it. Consequently dQ' can be represented as the sum of dQ_e (heat introduced from the outside) and dQ_i (irreversible transfer of heat within the system). For an *isolated system* ($dQ_e = 0$)) we can write now

(15 a) $$S_B - S_A > \sum \int_A^B \frac{dQ_i}{T}.$$

For each individual process dQ_i the integral in (15 a) is positive because dQ_i appears twice, once as a positive quantity, and once as a negative quantity, the denominator being smaller in the former case (an example is afforded by the porous plug experiment or by the conduction of heat under a finite temperature difference). Thus *a fortiori*

(16) $$S_B > S_A.$$

The entropy of an isolated system can only increase. The Second Law ascribes a definite direction to natural phenomena which was absent from the mechanistic point of view.

In order to clarify the conflicting statements in eqs. (16) and (12) we shall introduce the concept of *retarded equilibrium*. We shall assume that state A is one of equilibrium both in (16) and in (12), but we shall suppose that it contains different components which can be prevented from interacting with each other by the application of artificial devices. In this connection we may think of an impermeable wall which separates two gaseous phases and prevents their mixing. If such a wall is removed, it being possible to do so with an arbitrarily small expenditure of work (e. g. the opening of a valve, the closing of an electric contact), an irreversible process will set in and it will continue until a new state of equilibrium, B, has been reached. We may also think of two substances which cannot combine chemically under the conditions prevailing at A but which can be prompted to react in the presence of a catalyzer. The latter takes no part in the transfer of energy but it makes an irreversible chemical process possible so that transition to a new state of equilibrium, B, takes place. The Second Law makes no statements about the details of such processes but it enables us to calculate the change in entropy on transition from equilibrium state A to equilibrium state B. In order to do this it suffices to think of any *reversible* process which leads from A to B and to form the integral

$$S_B - S_A = \int_A^B \frac{dQ_{rev}}{T};$$

its value is independent of the particular choise of the reversible path between A and B.

The conditions of equilibrium at state B in eq. (16) are now different from those at state A. In this manner an irreversible retarded process at A becomes possible and a new state of equilibrium B with $S_B > S_A$ can be reached.

D. Simplest numerical examples

According to (8 a), the efficiency of a perfect steam engine would be

(17) $$\eta = \frac{T_1 - T_2}{T_1} = \frac{80}{373} = 22\%,$$

if it operated on a Carnot cycle. We have assumed here that $T_1 = 100$ C and that $T_2 = 20$ C. The real indicator diagram of a steam engine is not identical with the Carnot cycle (Fig. 1 compared with Fig. 5); nevertheless, the high-pressure line in Fig. 1 is identical with the isotherm of boiling water and the low-pressure line is approximately that of atmospheric temperature. The indicated limit of 22% is reached approximately in modern designs but it can never be exceeded.

If we assume that T_2 is kept constant in (17), we find that η increases with an increasing temperature T_1.

Superheated steam (locomotive) is more effective than steam at a normal boiling temperature. However, there are certain limits to the magnitude of pressure used in engineering practice. For this reason the work of developing mercury vapor turbines was initiated in the USA many years ago. By coupling a mercury vapor turbine with a steam turbine it is possible to obtain a unit which operates between 535 C and 35 C. This corresponds to an ideal efficiency of

$$\eta = 62\%.\ \ ^1$$

The diesel engine (ignition temperature 400 C) operates at a larger temperature difference than the steam plant and has a considerably larger ideal efficiency. We remark here that the efficiency of a diesel cycle cannot be inferred directly from a Carnot cycle because its indicator diagram differs too much from the latter.

It is possible to state generally: Heat at a higher temperature is more precious than at a lower temperature. Work can be regarded as being equivalent to heat at an infinitely high temperature.

An efficiency of $\eta = 100\%$ could be attained in a prime mover if it were possible to maintain absolute zero in its cooler. We shall discover in Sec. 12 that, strictly speaking, such an efficiency cannot be achieved.

---

[1] These remarks are not quite relevant at the present time, as modern engineering practice has developed means of handling steam at very high (near-critical and super-critical) pressures and temperatures exceeding the 535 C quoted. (*Transl.*).

At this stage we shall interpose a remark of a historical nature. Written in terms of our original θ temperature scale, and assuming an infinitely small temperature difference $\theta_1 = \theta$ and $\theta_2 = \theta - d\theta$, the efficiency from eqs. (2) and (7) is given by

(17 a) $$\eta = 1 - \frac{\phi(\theta - d\theta)}{\phi(\theta)} = \frac{\phi'(\theta)}{\phi(\theta)} \, d\theta = C(\theta) \, d\theta.$$

The function $C(\theta) = \phi'(\theta)/\phi(\theta)$ is designated as "Carnot's function" in older references. The same designation in the absolute temperature scale belongs to the function

(17 b) $$C(T) = \frac{1}{T}.$$

A slightly unexpected application of the Second Law to the derivation of algebraic inequalities is contained in Problem 4.

We shall refrain here from discussing the application to the universe which was already given by Clausius and which predicts its "thermal death." The increase in entropy is supposed to equalize all temperature differences so that the performance of work will supposedly become impossible. We think that the nature of the universe, i. e. whether it is open or closed, expanding (even pulsating!) or stationary is still too much in doubt to enable us to discuss this problem.

Planck[1] opposes (and rightly so) the view of certain physicists that the essence of the Second Law consists in the statement that energy tends to degrade. Evidently an increase in entropy causes in many cases a decrease in the available temperature difference and hence also in the availability of work. Planck quotes the obvious example in which heat is transformed into work completely, namely the example of an isothermal expansion of a perfect gas with heat transfer from a reservoir of higher temperature and with complete utilization of the pressure of the gas to perform work. In this process energy is not degraded but quite to the contrary, it is ennobled (heat completely transformed into work).

In our and in Planck's opinion, the essence of the Second Law consists in the existence of entropy and in the impossibility of its decreasing under well defined conditions.

---

[1] "Theorie der Wärme", Vol. V of Einführung in die theoretische Physik, Sec. 36, or "Thermodynamik", Sec. 108. See also engl. transl. "Theory of Heat", Vol. V of "Introduction to Theoretical Physics", p. 50, or "Treatise on Thermodynamics", 3rd. ed. p. 81.

E. Remarks on the Literature of the Second Law

Our proof of the Second Law was based on that due to Clausius.[1] The proof given by Planck (*l. c.*) is perhaps simpler and certainly more precise; it is, however, more abstract and less instructive than ours. The proof given by Carathéodory[2] is even more abstract and, at the same time, simpler if the simplicity of a proof is judged by the small number of assumptions required. In fact, using a system of two fluids which can be separated by a heat conducting or a heat insulating wall, as required, Carathéodory needs only the assumption: *In the neighborhood of every state which can be reached reversibly there exist states which cannot be reached along a reversible adiabatic path, or, in other words, which can only be reached irreversibly or which cannot be reached at all.* This exceedingly economical postulate suffices to provide a mathematical proof of the existence of the property known as entropy.

We shall now quote the point of view which was adopted by Carathéodory in his inaugural address to the Prussian Academy:[3] "It is possible to ask the question as to how to construct the phenomenological science of thermodynamics when it is desired to include only directly measurable quantities, that is volumes, pressures, and the chemical composition of systems. The resulting theory is logically unassailable and satisfactory for the mathematician because, starting solely with observed facts, it succeeds with a minimum of hypotheses. And yet, precisely these merits impede its usefulness to the student of nature, because, on the one hand, temperature appears as a derived quantity, and on the other, and above all, it is impossible to establish a connection between the world of visible and tangible matter and the world of atoms through the smooth walls of the all too artificial structure."

In connection with the last question Planck[4] makes the casual remark: It is true that the First Law applies to 10 molecules enclosed in a fixed volume, but with the aid of such a system it is impossible to build a heat engine owing to excessive fluctuations. Applied to such a system the Second Law loses its sense. Carathéodory's proof does not exclude such systems in advance; it requires additional restrictive assumptions in order to adapt itself to reality.

---

[1] R. Clausius: Mechanische Wärmetheorie, 1876. 2nd. ed. of "Abhandlung über mechanische Wärmetheorie".

[2] C. Carathéodory, Math. Ann. 67, 1909 and Prussian Academy, Jan. 1925. Reference should also be made to M. Born: Natural Philosophy of Cause and Chance, Oxford 1949, who, in Caratheodory's own judgement, has given a particularly clear presentation of his method.

[3] Sitzungsberichte of 3rd. July 1919, No. XXXIII.

[4] *Ibid.* 1921, p. 453.

In order to obtain at least an approximate idea of Carathéodory's method let us consider the two fluid systems already mentioned, namely Σ_1 and Σ_2, whose states will be described by pressure, volume and an additional parameter θ. These properties are related through an equation of state each of which we can write in the form:

$$\theta_1 = F_1(p_1, V_1); \qquad \theta_2 = F_2(p_2, V_2).$$

Let the two systems be brought into thermal contact which may be defined by stipulating $\theta_1 = \theta_2$. The thermally coupled system $\Sigma = \Sigma_1 + \Sigma_2$ satisfies the equation

$$F_1(p_1, V_1) - F_2(p_2, V_2),$$

so that of the four variables only three are independent. We can choose them arbitrarily and we may denote them by x, y, and z. The First Law states then that the quantity of heat added to the system during a reversible process is given by Pfaff's differential expression (see Sec. 1):

(18) $$dQ = X\,dx + Y\,dy + Z\,dz,$$

whre X, Y, and Z denote functions of x, y, z. Generally speaking dQ is not a perfect differential, as seen from eq. (4 a) in Sec. 1. The corresponding Pfaff differentials for sub-systems Σ_1, Σ_2 containing two variables each can always be transformed into perfect differentials by the adoption of an integrating denominator, as also mentioned in Sec. 1. From this, together with Carathéodory's postulate, it is possible to conclude that the expression (18) also possesses an integrating denominator (in fact a family of them). This proves the existence of absolute temperature and of entropy and the fact that they are properties.

F. On the relative rank of energy and entropy

We quote here a note[1] by Robert Emden whose deep understanding of thermodynamics has withstood the test of time in fundamental papers on astrophysics (gaseous spheres!) and meteorology (grey atmosphere): "Why do we have Winter Heating? The layman will answer: 'To make the room warmer.' The student of thermodynamics will perhaps so express it: 'To impart the lacking (inner, thermal) energy.' If so, then the layman's answer is right, the scientist's wrong."

---

[1] Nature, Vol. 141, May 1938, p. 908 entitled: Why do we have Winter Heating?

"We suppose, to correspond to the actual state of affairs, that the pressure of the air in a room always equals that of the external air. In the usual notation, the (inner, thermal) energy is, per unit mass,

$$u = c_v T.$$

(An additive constant may be neglected). Then the energy content is, per unit of volume,

(19) $$u_1 = c_v \rho T$$

or, taking into account the equation of state, we have

(20) $$u_1 = c_v \mu p / R.$$

The energy content of the room is thus independent of the temperature, solely determined by the state of the barometer. The whole of the energy imparted by the heating escapes through the pores of the walls of the room to the outside air."

"I fetch a bottle of claret from the cold cellar and put it to be tempered in the warm room. It becomes warmer, but the increased energy content is not borrowed from the air of the room but is brought in from outside."

"Then why do we have heating? For the same reason that life on the earth needs the radiation of the sun. But this does not exist on the incident energy, for the latter, apart from a negligible amount, is re-radiated, just as a man, in spite of continual absorption of nourishment, maintains a constant body-weight. Our conditions of existence require a determinate degree of temperature, and for the maintenance of this there is needed not addition of energy but addition of entropy."

"As a student, I read with advantage a small book by F. Wald entitled 'The Mistress of the World and her Shadow'. These meant energy and entropy. In the course of advancing knowledge the two seem to me to have exchanged places. In the huge manufactory of natural processes, the principle of entropy occupies the position of manager, for it dictates the manner and method of the whole business, whilst the principle of energy merely does the book-keeping, balancing credits and debits."

Numerical examples and critical remarks are given in Problem 2.

7. The thermodynamic potentials and the reciprocity relations

We have at our disposal two pairs of variables

$$p, v \quad \text{and} \quad T, s$$

for each simple, homogeneous system (possessing one mechanical and one thermal degree of freedom, e. g. a gas, a vapor, or a liquid). The First Law expressed in terms of them, whether per mol or per unit mass, see Sec. 5 D, eq. (11), has the form

(1) $$du = T\,ds - p\,dv.$$

The two "extensive" quantities, s, v, are the independent variables, the two "intensive" quantities T, p being conjugate to them. The internal energy, u, is to be regarded as a function of the variables s, v:

$$u = u(s, v).$$

According to (1) the remaining two variables are given by

(2) $$T = \left(\frac{\partial u}{\partial s}\right)_v, \quad -p = \left(\frac{\partial u}{\partial v}\right)_s.$$

However the selection of independent variables is largely a matter of free choice. There are four possibilities of making such a choice with one mechanical and one thermal variable in the pair:

(3) $$s, v; \quad s, p; \quad T, v; \quad T, p.$$

At this point we recall the Legendre transformation whose great importance for analysis, for mechanics and for thermodynamics has already been stressed in Vol. I, Sec. 42. It gives us the rule: *If it is desired to replace one of the independent variables (e. g. s) in a Pfaff differential of the form (1) by its conjugate, it is necessary to subtract from the dependent variable (in our case u) the product of the two conjugate independent variables (in our case T s).*

A corresponding rule applies when it is desired to substitute both initial variables and when there are more variables than two. In this manner there are four expressions associated with the four possibilities (3), namely

(4) $$u(s,v); \quad h(s,p) = u + pv; \quad f(T,v) = u - Ts; \quad g(T,p) = u - Ts + pv.$$

     energy   enthalpy   free energy   free enthalpy

The expression for h, taken per mol or per unit mass, corresponds to the quantity H introduced previously in eq. (4.9). The remaining symbols and definitions are summarized in Table (6) below.

7.7a THE THERMODYNAMIC POTENTIALS

We shall now clarify the usefulness of the preceding definitions in a purely formal way by forming the corresponding differentials and by substituting du from (1):

(5) $$\begin{cases} dh = du + p\,dv + v\,dp = T\,ds + v\,dp, \\ df = du - T\,ds - s\,dT = -p\,dv - s\,dT, \\ dg = du - T\,ds - s\,dT + p\,dv + v\,dp = -s\,dT + v\,dp. \end{cases}$$

The last terms show that the differentials dh, df, dg have the same simple form when expressed in terms of the independent variables associated with them as du has when it is expressed in terms of the variables s and v. The expressions (4) are known as the *thermodynamic potentials* because the variables can be deduced from them by differentiation with respect to the independent variable associated with them in the same way as the components of a force are derived from a force potential. The same designation can be properly applied to energy as seen from eq. (2). The corresponding equations for the potentials h, f, and g will be found in the Table. We should like to stress here that the choice of a potential determines the choice of the associated variables. For example the free energy f possesses the properties of a potential with respect to the variables v, T; it loses this property with a different choice of variables.

From the representation in (2) and from the analogous expressions given in the Table there follow the most important and significant thermodynamic relations summarized in the fourth column of our Table.

We shall begin by considering the third of these relations:

(7) $$\left(\frac{\partial p}{\partial T}\right)_v = \left(\frac{\partial s}{\partial v}\right)_T.$$

Substituting $ds = dq/T$, we obtain

(7a) $$\left(\frac{\partial p}{\partial T}\right)_v = \frac{1}{T}\frac{dq}{dv}\bigg|_{T=const}$$

The left-hand side is the coefficient of tension, β, from eq. (1.5), except for the p in the denominator. The second factor on the right-hand side is the "isothermal heat of expansion"[1] which must be introduced during the expansion to maintain a constant temperature.

---

[1] The symbol M was used to denote it in older papers and it was measured in calories. Introducing the Carnot function $C(T)$ for $1/T$, eq. (6.17 a) and denoting the mechanical equivalent of heat by J, we find that the right-hand side reads J.C.M. This is the reason for which James Clerk Maxwell used dp/dt as his pen name. (Mnemonic rule for examinees who were asked for Clapeyron's equation.)

(6) Table

Potential	Independent variables	Conjugate variables	Thermodynamic relations	Definitions and notation
U, u $du = T\,ds - p\,dv$	v, s	$T = \left(\dfrac{\partial u}{\partial s}\right)_v$ $p = -\left(\dfrac{\partial u}{\partial v}\right)_s$	$\left(\dfrac{\partial T}{\partial v}\right)_s = -\left(\dfrac{\partial p}{\partial s}\right)_v$ $= \dfrac{\partial^2 u}{\partial v\,\partial s}$	Energy u (Clausius) ε (Gibbs)
H, h $h = u + pv$ $dh = T\,ds + v\,dp$	p, s	$T = \left(\dfrac{\partial h}{\partial s}\right)_p$ $v = \left(\dfrac{\partial h}{\partial p}\right)_s$	$\left(\dfrac{\partial T}{\partial p}\right)_s = \left(\dfrac{\partial v}{\partial s}\right)_p$ $= \dfrac{\partial^2 h}{\partial p\,\partial s}$	Enthalpy H (Lewis and Randall) X (Gibbs) J (Heat engineering in certain countries)
F, f $f = u - Ts$ $df = -s\,dT - p\,dv$	v, T	$s = -\left(\dfrac{\partial f}{\partial T}\right)_v$ $p = -\left(\dfrac{\partial f}{\partial v}\right)_T$	$\left(\dfrac{\partial s}{\partial v}\right)_T = \left(\dfrac{\partial p}{\partial T}\right)_v$ $= -\dfrac{\partial^2 f}{\partial v\,\partial T}$	Free energy F (Helmholtz) ψ (Gibbs)
G, g $g = h - Ts$ $= f + pv$ $= u - Ts + pv$ $dg = -s\,dT + v\,dp$	p, T	$s = -\left(\dfrac{\partial g}{\partial T}\right)_p$ $v = \left(\dfrac{\partial g}{\partial p}\right)_T$	$\left(\dfrac{\partial s}{\partial p}\right)_T = -\left(\dfrac{\partial v}{\partial T}\right)_p$ $= -\dfrac{\partial^2 g}{\partial p\,\partial T}$	Free enthalpy ζ (Gibbs), also called thermodynamic potential

It is remarkable that the relation (7 a) which we have justified for homogeneous systems contains a statement which is true for the transition between *two* homogeneous systems, namely two different phases. Equilibrium between water and steam is of particular interest. Equation (7 a) becomes identical with the famous Clapeyron equation which played such an important role in the development of the steam engine (see Sec. 1·6) if we interpret p as the vapor pressure at temperature T and replace $(dq/dv)_T$ by $\Delta q/\Delta v$, where now Δq denotes the heat of evaporation per mol (or per unit mass).

At present eq. (7) is used to obtain a remarkable formula for $c_p - c_v$ of general validity. Assuming that T is kept constant we can obtain from the First Law written in form (1) the expression

$$\left(\frac{\partial u}{\partial v}\right)_T = T\left(\frac{\partial s}{\partial v}\right)_T - p.$$

In view of (7) we have

(8) $$\left(\frac{\partial u}{\partial v}\right)_T + p = T\left(\frac{\partial p}{\partial T}\right)_v.$$

On the other hand the First Law can be written as:

$$dq = du + p\,dv = \left\{\left(\frac{\partial u}{\partial v}\right)_T + p\right\}dv + \left(\frac{\partial u}{\partial T}\right)_v dT,$$

and at constant v we have

(8 a) $$c_v = \frac{dq}{dT}\bigg|_{v=const} = \left(\frac{\partial u}{\partial T}\right)_v$$

whereas at constant p we obtain

(8 b) $$c_p = \frac{dq}{dT}\bigg|_{p=const} = \left\{\left(\frac{\partial u}{\partial v}\right)_T + p\right\}\left(\frac{\partial v}{\partial T}\right)_p + \left(\frac{\partial u}{\partial T}\right)_v.$$

Subtracting (8 b) and (8 a) we find

(8 c) $$c_p - c_v = \left\{\left(\frac{\partial u}{\partial v}\right)_T + p\right\}\left(\frac{\partial v}{\partial T}\right)_p$$

and from (8)

(9) $$c_p - c_v = T\left(\frac{\partial p}{\partial T}\right)_v\left(\frac{\partial v}{\partial T}\right)_p.$$

The last two factors represent the "coefficient of tension β" and the "coefficient of thermal expansion α," respectively, provided that the factors p and v appearing in the denominator of eq. (1.) are taken into account. Thus eq. (9) can also be written in the form

(9 a) $$c_p - c_v = \alpha\,\beta\,v\,p\,T$$

For a perfect gas we had $\alpha = \beta = 1/T$ and eq. (9 a) takes on the form

(9 b) $$c_p - c_v = \frac{pv}{T} = \begin{cases} R \text{ for one mol} \\ R/\mu \text{ for a unit of mass,} \end{cases}$$

as it should, in accordance with Sec. 3.

It will be noted that in deducing eq. (9 b) for the perfect gas there was no need to make use of the additional caloric condition of Sec. 4 A which stated that the internal energy of a perfect gas depended only on its temperature. This is due to the fact that the caloric condition does not really constitute a new requirement imposed on the gas but represents a property of the perfect gas which is a consequence of the Second Law. In order to see this it is sufficient to express the pressure on the right-hand side of eq. (8) in terms of the equation of state of a perfect gas; we then have $(\partial p/\partial T)_v = p/T$; hence eq. (8) leads to $(\partial u/\partial v)_T = 0$ which means that the internal energy is independent of the volume, being a function of temperature alone.

We now propose to examine more closely the last thermodynamic relation in our Table. Replacing once more ds by dq/T we obtain

(10)
$$\left(\frac{\partial v}{\partial T}\right)_p = -\frac{1}{T}\frac{\partial q}{\partial p}\bigg|_{T=const}$$

The left-hand side is the product of the coefficient of thermal expansion and volume. The last term on the right-hand side is known as the "isothermal heat of compression". It is, generally speaking, *negative*, which means that heat must be rejected if the system is to be maintained at the same temperature at a higher pressure. If this were not so it would become heated on compression. Correspondingly α is, generally speaking, *positive* (two negative signs in (10) cancel each other). There are, however, exceptions. The best known exception is water between 0 C and 4 C. Equation (10) shows that in this interval the heat of compression is *positive*: it is necessary to add heat in order to prevent the water from cooling on compression. (See also Problem I.6) The same is true of *raw rubber* and *silver iodide* in certain temperature intervals. The anomaly of water has led Roentgen to suppose that water tends to polymerize in the neighborhood of its freezing point, the supposition having been confirmed later by others. Thus the process of crystallization which takes place at 0 C occurs to a certain extent before it.

All four thermodynamic relations were deduced by Maxwell in his "Theory of Heat," London, 1883, from an elementary geometrical figure; he also stated them in words. It is evident that he felt himself that in this case the differential representation is much simpler than the elementary treatment given in his text-book; for this reason he appended the analytical formulation contained in our Table in a remark to Chap. IX. An intuitive understanding of the signs in these reciprocity relations is contained in the *principle due to Braun and Le Chatelier* somewhat in the manner of Lenz's rule in electrodynamics;

it does not, however, attain the same degree of precision as that possessed by the statements in our Table.[1]

The impressive regularity of our Table is due to the great student of thermodynamics and statistical mechanics – Willard Gibbs. His papers, which were at first buried in the Transactions of the Connecticut Academy of 1876 and 1878, became generally known only after Ostwald published them in German in 1902 under the title "Thermodynamische Studien." Adopting Gibbs' point of view we consider that the "four potentials" u, h, f, g, or U, H, F, G, are equivalent, the choice between them depending on the choice of independent variables, eq. (3). We have already stressed in eq. (5.7) that the simplest formulation of the theory of the Joule-Kelvin *porous plug experiment* is obtained in terms of *enthalpy*, H, which is equal on both sides. In relation to phase equilibria the same simplification is achieved by the use of the *free enthalpy*, G. The *free energy*, F, is the principal potential in physical chemistry and in electrochemistry. It furnishes a measure of *chemical affinity*. Planck prefers, as a rule, to use the *"potential* function"

$$\Phi = -\frac{G}{T} = S - \frac{U + pV}{T},$$

which is in fact convenient in problems involving statistical questions; it does not, however, fit Gibbs' beautiful system.

8. Thermodynamic equilibria

A. Unconstrained thermodynamic equilibrium and maximum of entropy

We have found in Sec. 6 C that the entropy of an isolated system cannot decrease. A system was called isolated when it absorbed no heat and performed no work. These conditions are equivalent to stating that the internal energy U and the volume V are kept constant ($dU = 0, dV = 0$). An isolated system will tend to a final state at which the entropy has a maximum if all constraints within the system are removed. We shall call this a state of unconstrained thermodynamic equilibrium.

---

[1] In this connection see: P. Ehrenfest, Z. Phys. Chem. **77**, 1911; Planck, Ann. d. Phys. **19**, 1934 with an Appendix *ibid* **20**, 1935; further Mrs. Tatiana Ehrenfest and Mrs. de Haas-Lorentz, Physica **2**, 1935 with reply from Planck, *ibid*. The main problem during this discussion was the distinction between intensive and extensive properties which is of paramount importance for an unambiguous formulation of the principle of Braun-Le-Chatelier.

A process which starts spontaneously from a state of unconstrained equilibrium is impossible; if this were not so the entropy would have to increase again in contradiction to our assumption that the entropy already has a maximum value. We can, however, consider virtual processes δ which are compatible with the restrictions $dU = 0, dV = 0$ and which cannot, evidently, occur spontaneously. (Example: let a vessel be filled with a gas whose pressure and temperature are constant; we now let half of the gas be heated to a temperature $T + \delta T$, the other half being cooled to $T - \delta T$.) Such virtual changes of state from unconstrained thermodynamic equilibrium satisfy the relations

(1) $$\delta S \leqslant 0 \quad \text{when} \quad \delta U = 0; \quad \delta V = 0,$$

or, in another form

(1 a) $$S = S_{max} \quad \text{when} \quad U = \text{const}, \quad V = \text{const}.$$

If it were possible to indicate a process with $\delta U = 0$, $\delta V = 0$ for which $\delta S > 0$, we would conclude that the initial state was not one of unconstrained equilibrium. We would conclude that there existed constraints whose removal caused the entropy to increase further. Equation (1), or eq. (1 a), constitutes one of the two conditions of equilibrium established by Gibbs. The second, which is less important to us, has the form

(2) $$\delta U \geqslant 0 \quad \text{when} \quad \delta S = 0, \quad \delta V = 0.$$

The condition

(2 a) $$U = U_{min} \quad \text{when} \quad S = \text{const}, \quad V = \text{const}.$$

is equivalent to it. *At a state of equilibrium the internal energy assumes a minimum value.* This last proposition is reminiscent of the criterion for equilibrium in general mechanics which requires the potential energy to assume a minimum value (see e. g. Vol VI Sec. 25).

B. AN ISOTHERMAL AND ISOBARIC SYSTEM IN UNCONSTRAINED THERMODYNAMIC EQUILIBRIUM

It follows from the characteristics of the state of an unconstrained equilibrium in eq. (1) that the pressure and temperature throughout the system are independent of space coordinates. If this were not so we could select two elements of space in which the temperatures were T_1 and T_2, say, the corresponding pressures being p_1 and p_2. We could now assume a virtual process during which the energy of the first element changed by δU_1 its volume changing by δV_1. The corresponding variations for the second element of space would then be $\delta U_2 = -\delta U_1$, $\delta V_2 = -\delta V_1$ if the conditions in (1) were to be satisfied. Changes in concentration or of the masses contained in the individual phases present will now be excluded. According to (1) we must then have

$$0 \geqslant \delta S = \delta S_1 + \delta S_2 = \frac{1}{T_1}(\delta U_1 + p_1 \delta V_1) + \frac{1}{T_2}(\delta U_2 + p_2 \delta V_2) =$$
$$= \left(\frac{1}{T_1} - \frac{1}{T_2}\right)\delta U_1 + \left(\frac{p_1}{T_1} - \frac{p_2}{T_2}\right)\delta V_1.$$

The virtual changes δU_1 and δV_1 are arbitrary and independent of each other. The above inequality cannot, therefore, be satisfied for $T_1 \neq T_2$, $p_1 \neq p_2$, if δU_1, δV_1 are to have any values, contrary to our assumption.

C. Additional degrees of freedom in retarded equilibrium

The state of a system which is in unconstrained thermodynamic equilibrium is often specified by indicating the internal energy, U, the volume, V, and the masses of the independent components (*cf.* Sec. 14). We shall now consider a system Σ which is not yet in equilibrium. Its state can only be determined if in addition to U, V and the masses of the independent components we specify further quantities x_i; these may denote, for example, the distribution of the independent components over the phases and the concentrations of the individual components which can interact chemically; they may, furthermore, describe local differences if the system is subdivided into sufficiently small elements of volume, and if the preceding quantities are specified for each element. We shall consider only states of disequilibrium of a kind, and this is an essential assumption for Sec. 21, which can be interpreted as states of constrained equilibrium, in which the x_i are kept constant, so that the *entropy of the system* may be taken to be *the sum of the entropies of all volume elements for such a state of constrained equilibrium*. We shall restrict ourselves to the consideration of isothermal and isobaric systems. On transition from constrained equilibrium U, V, x_i to the constrained equilibrium $U + dU$, $V + dV$, $x_i + dx_i$ the change in the entropy must be calculated from

(3) $$T \, dS = dU + p \, dV + \sum_i X_i \, dx_i,$$

which is a generalization of eq. (7.1). The kind of process taking place, whether reversible or irreversible, is here of no importance, because dS denotes the difference between the entropy of the final and initial state, both (only infinitesimally different) being constrained equilibria. The coefficients X_i will be called forces associated with the additional degrees of freedom x_i.

D. Extremum properties of the thermodynamic potentials

We shall now add to Σ a "surroundings" which we may imagine in the form of a very large heat reservoir, Σ_0. All quantities referring to Σ_0 will be denoted by the subscript o; the combined system consisting of Σ and Σ_0 will be assumed to be thermally insulated; under these assumptions the total entropy cannot decrease:

$$(4) \qquad dS + dS_0 > 0.$$

The sign of equality will be excluded by assuming that the *changes* of state which take place in Σ are irreversible like all real processes. As already mentioned, we assume that the system Σ is isobaric and isothermal which means that it is in mechanical and thermal equilibrium but not necessarily in chemical or phase equilibrium. The transfer of heat between Σ and Σ_0 will be allowed only on condition that Σ has the temperature T_0 of the heat reservoir. We assume that the system Σ_0 is so large that it can exchange *reversibly*, i. e. absorb or reject, a quantity of heat with Σ without markedly changing its own temperature. Thus

$$(5) \qquad dS_0 = \frac{dQ_0}{T_0}.$$

The changes in volume in Σ are assumed to take place at a pressure p which is always equal to that outside. Applying the First Law to Σ and to the quantity of heat $dQ = -dQ_0$ transferred to Σ from Σ_0, we obtain

$$(6) \qquad dU + p\,dV = -dQ_0.$$

Consequently

$$(7) \qquad dS > \frac{1}{T}(dU + p\,dV)$$

because, according to our assumption, $T = T_0$ when there is a flow of heat, and eq. (7) follows from (4), (5) and (6). If, however, $T \neq T_0$ then there is no exchange of heat, $dU + p\,dV$ vanishes according to the First Law, and (7) states simply that $dS > 0$, as already derived in (6.16) for an isolated system. Hence eq. (7) is true for any process involving the transfer of heat and the performance of work by our system.

From (3) and (7) we infer that for any process dU, dV, dx_i which can take place spontaneously we must have

(8) $$\sum_i X_i\,dx_i > 0.$$

In the case of *reversible* processes of the combined system $\Sigma + \Sigma_0$ and of the system Σ alone we must have a sign of equality in eq. (4), which leads to a sign of equality in (8). We can now make the following statement:

The necessary and sufficient conditions for a process in an isothermal and isobaric system Σ to be reversible, are: heat must be exchanged with the surroundings in a reversible way (i. e. at a temperature T of Σ equal to that of the surroundings), the internal pressure p must be equal to the external pressure and, in addition, we must have

(9) $$\sum_i X_i\,dx_i = 0$$

during the whole process.

The latter condition is satisfied, for example, when all x_i are kept constant. It is also satisfied when non-vanishing x_i's are associated with vanishing X_i' s. Any transition from state *1* with U_1, V_1, x_{i1} to a state *2* with U_2, V_2, x_{i2} can be conducted in many different ways and always so that condition (9) is satisfied during the whole process. An application of this rule to an actual example is given in Section *E*.

Let us now consider an isothermal system of fixed temperature and volume (e. g. a system immersed in a bath of constant temperature T); the differential of free energy is then $dF = dU - T\,dS$. If the changes in F, U, S refer to a spontaneous process we can apply eq. (7) with $dV = 0$ and we obtain

(10) $$dF \leqslant 0.$$

The free energy decreases.[1] There exists a minimum of free energy beyond which no spontaneous changes of state are possible. The condition of equilibrium of the present system is

(11) $$F = F_{min} \quad \text{when} \quad T = \text{const}, \quad V = \text{const}.$$

---

[1] The condition that the pressure is constant throughout the system (see C *infra*) is not required for the validity of this result because the existence of any differences in pressure between different parts of the system has no influence on the result in (7) owing to $dV = 0$.

The condition that for any virtual process

(11 a) $\qquad \delta F \geqslant 0$ when $\delta T = 0, \quad \delta V = 0$

is equivalent to that in (11).

In the case of an isothermal-isobaric system (for example one immersed in a heat bath of temperature T and pressure p in a way that ensures thermal and static equilibrium with the system) we make use of the free enthalpy $G = U - TS + pV$. Its differential is $dG = dU - T\,dS + p\,dV$ because $dT = 0, dp = 0$. If the changes in G, U, S, V, refer to a spontaneous process we can apply eq. (7) once more, and we obtain

(12) $\qquad dG \leqslant 0.$

The free enthalpy decreases. There is a minimum of *free enthalpy* beyond which spontaneous changes are no longer possible. The present condition of equilibrium is

(13) $\qquad G = G_{min}$ when $T = \text{const}, \quad p = \text{const}.$

The condition

(13 a) $\qquad \delta G \geqslant 0$ when $\delta T = 0, \quad \delta p = 0$

is equivalent to it.

This last statement is most important in the theory of processes which involve transformation of substances into different forms. It will be seen later that of the four potentials introduced in Sec. 7, the free enthalpy will prove, generally speaking, to be first in importance.

The proof of eqs. (2) and (2 a), as well as the derivation of an extremum property for the enthalpy H of an isobaric system under the conditions $\delta S = 0, \delta p = 0$ are left to the reader.

E. The theorem on maximum work

We now propose to calculate the quantity of work which a system Σ can perform on its surroundings when it passes reversibly from a state *1* of given temperature T_0 to another state *2* of equal temperature. We assume that the transition need not be isothermal but we postulate that Σ may exchange heat with its surroundings only at that temperature T_0; in other words, in ranges where $T \neq T_0$ the processes must be adiabatic and reversible.

Making use of the definition of free energy $F = U - TS$ we find that according to (3) we may write

$$dF = dU - T\,dS - S\,dT = -S\,dT - p\,dV - \sum_i X_i\,dx_i \tag{14}$$

for an elementary process. The requirement of reversibility means that the sum $\sum X_i\,dx_i = 0$ during the whole process of transition $1 \to 2$. Thus the quantity of work performed by the system becomes

$$\int_1^2 p\,dV = F_1 - F_2 - \int_1^2 S\,dT.$$

We now assert that

$$\int_1^2 p\,dV = F_1 - F_2 \tag{15}$$

i. e. that the integral of $S\,dT$ vanishes under our assumptions. This results from the following argument: The transition $1 \to 2$ may contain isothermal components at temperature T_0; along these we have $dT = 0$. In addition it contains one or more adiabatic and reversible components each of which corresponds to a constant entropy and to a temperature T_0 which is the same at the beginning as at the end. Hence such an adiabatic segment satisfies the relation

$$\int_{T_0}^{T_0} S\,dT = S \int_{T_0}^{T_0} dT = 0.$$

If the process is irreversible the quantity of work performed is smaller than $F_1 - F_2$. If this were not the case we could bring the system to its initial state in a reversible way performing a quantity of work which is $\geqslant 0$ at the expense of an equivalent quantity of heat from the surroundings at T_0. This, however, contradicts the Second Law. For this reason the change in the *free energy* is also described as the *maximum work* which the system Σ can perform on its surroundings when it undergoes a process during which its initial and final temperature is equal to that of the surroundings, T_0, on condition that it exchanges heat with the surroundings only at that temperature.

In many text-books on thermodynamics the above proposition on maximum work is stated on the additional restrictive assumption that the transition $1 \to 2$ is isothermal. Our formulation goes further because for some systems it is impossible to find a *reversible isothermal* process $1 \to 2$. In spite of that the maximum of work can be produced. Helmholtz's term "free energy" derives from eq. (15); its difference between two states 1 and 2 of equal temperature T_0 represents that portion of the change in energy, $U_1 - U_2$, which can be converted into external work (which is "available") during a reversible process $1 \to 2$, or that which it is necessary to perform externally on the system during the reverse process on condition that heat is exchanged with the surroundings only reversibly at a temperature T_0. Consistently we can call $U - F = TS$ the "bounded", or "unavailable" energy.

We shall now show with the aid of a simple example how it is possible to perform the process $1 \to 2$ in a way which will ensure that $\sum_i X_i \, dx_i = 0$ along the whole path; the specific example has been chosen so that it is representative of the general case. For this purpose we shall consider a dissociating gas whose velocity of dissociation we can impede at will. Let x denote its degree of dissociation; its value at the state of unrestrained equilibrium will be denoted by $\bar{x}\,(T, V)$. Instead of (3) we can write $T\,dS = dU + p\,dV + X\,dx$. The initial state is given by T_0, V_1, x_1, and the final state is denoted by T_0, V_2, x_2. Of the two degrees of dissociation x_1 and x_2 at least one will be assumed to correspond to a deviation from equilibrium; otherwise the succeeding argument would become trivial. The first process is assumed to be adiabatic and reversible from T_0, V_1, x_1 at constant x_1 (i. e. $dx_1 = 0$) to such values T', V' for which $x_1 = \bar{x}\,(T', V')$. In other words we are asking for those values of T and V for which the prescribed value x_1 denotes the degree of dissociation in unconstrained equilibrium. We can now remove the restraint which kept x_1 constant without causing any further changes in the system. Next we perform an adiabatic reversible change from T', V' to such values T'', V'' as to effect a change from x_1 to $x_2 = \bar{x}\,(T'', V'')$. All along the path of change the equilibria are unconstrained so that according to (9) $X\,dx$ vanishes at all points (it is seen that $X = 0$ is the condition of dissociation equilibrium owing to $dx \neq 0$). We now maintain $x_2 = $ const and perform a further adiabatic and reversible change from T'', V'' until the initial temperature T_0 and a volume V''' are reached. Finally we perform an isothermal reversible process at constant x_2 during which V''' is changed to the desired volume V_2. In the preceding example the whole process could have been performed isothermally because dissociation equilibrium depends on pressure. In order to achieve this we can perform a reversible isothermal process changing the pressure p_0 at constant x_1 until $x_1 = \bar{x}(T_0, p')$ or, in other words, until dissociation equilibrium is reached at the prescribed value of x_1. At this stage we again remove the constraint and change p' to p'', where p'' is so chosen that $x_2 = \bar{x}(T_0, p'')$ and $X = 0$ during the process. Finally we again keep $x_2 = $ const and vary the pressure until the prescribed final volume V_2 has been reached.

In conclusion we shall make the following remark. The *maximum work* which can be made available during an elementary process with equal initial and final temperature T_0 is $p\,dV + \sum_i X_i\,dx_i$. For this reason the preceding expression is known as the generalized differential of work. It is a perfect differential and there exists a property, our free energy F at constant temperature T_0, whose difference is equal to its integral.

9. The van der Waals equation

We shall now consider real gases, having so far devoted our attention to perfect gases, and we shall base our description on a dissertation entitled "The continuity of the gaseous and liquid states" published by van der Waals in Leiden in the year 1873.

In fact van der Waals succeeded in establishing an equation of state which reproduces qualitatively the process of liquefaction (condensation) of a gas and which introduces a quantitative correction into the equation of state of a perfect gas. Boltzmann[1] described van der Waals as the Newton of real gases.

Written for one mol, the equation established by van der Waals has the form:

(1) $$p = \frac{RT}{v-b} - \frac{a}{v^2}.$$

The constant b introduced here is due to the volume of the molecules; the constant a is a measure for the forces of cohesion between the gaseous molecules and is connected with the capillarity of the free liquid surface. (The atomic significance of a and b is discussed in Sec. 26.) For $a = b = 0$, or, which amounts to the same, for sufficiently large v eq. (1) transforms into the perfect-gas equation, as it should. Instead of (1) we can also write

(1 a) $$(p + p_a)(v - b) = RT, \quad p_a = a/v^2.$$

---

[1] Enzykl. der Mathem. Wiss., Vol. V. 1, p. 550.

The quantity p_a denotes the "cohesion pressure" which must be added to the "kinetic pressure" p. We shall begin by calculating the coefficient of thermal expansion from (1). Putting $dp = 0$ and differentiating (1), we obtain

$$0 = \frac{dT}{v-b} - \left(\frac{T}{(v-b)^2} - \frac{2a}{Rv^3}\right) dv,$$

and consequently

(2) $$\alpha = \frac{1}{v}\left(\frac{\partial v}{\partial T}\right)_p = \frac{v-b}{vT - \frac{2a}{R}\left(\frac{v-b}{v}\right)^2}$$

which is a generalization of the value $\alpha = 1/T$ for a perfect gas and which can be obtained from (2) by substituting $a = b = 0$.

We shall also calculate the difference

(2 a) $$\alpha - \frac{1}{T} = \left\{\frac{2a}{RT}\left(\frac{v-b}{v}\right)^2 - b\right\} \Big/ \left\{vT - \frac{2a}{R}\left(\frac{v-b}{v}\right)^2\right\}.$$

This can be simplified by retaining only the first powers of the parameters a, b implying $v \gg b$, $RTv \gg a$:

(2 b) $$\alpha - \frac{1}{T} = \left(\frac{2a}{RT} - b\right) \Big/ vT.$$

For most gases, e. g. O_2, N_2, the right-hand side of (2) is positive in the range of ordinary temperatures, meaning that the coefficient of expansion is larger than that for the perfect gas. Hydrogen H_2 and the noble gases are the only exceptions. The forces of cohesion of these gases, as defined by a, are so small that the right-hand side of (2 a) becomes negative at ordinary temperatures. For this reason H_2 was designated in the past as a "*gaz plus que parfait*."

A. Course of isotherms

Figure 7 shows the course of the van der Waals isotherms. The v-axis and the straight line $v = b$ parallel to the p-axis are their asymptotes. Equation (1) has no physical meaning for $v < b$. According to (1) the points of intersection of an isobar $b = $ const with an isotherm $T = $ const are determined by a cubic equation. This has either one or three real roots. The limit between these two cases lies along the critical isotherm $T = T_{cr}$ on which the three points of intersection coalesce into one point of inflection with a horizontal tangent – the *critical point* $v = v_{cr}$, $p = p_{cr}$.

In order to determine v_{cr} and T_{cr} we calculate from (2):

$$\frac{\partial p}{\partial v} = 0 \quad \text{i. e.} \quad \frac{RT}{(v-b)^2} = \frac{2a}{v^3}, \quad \frac{1}{2}\frac{\partial^2 p}{\partial v^2} = 0 \quad \text{i. e.} \quad \frac{RT}{(v-b)^3} = \frac{3a}{v^4}.$$

It follows that

(3) $$\begin{cases} v = v_{cr} = 3b, \\ RT = RT_{cr} = \frac{8}{27}\frac{a}{b}. \end{cases}$$

The corresponding value of p is found from (1)

(4) $$p = p_{cr} = \frac{1}{27}\frac{a}{b^2}.$$

Since the constants a, b can be expressed in terms of the critical parameters eq. (1) can be rewritten to contain only the ratios

$$\mathsf{v} = \frac{v}{v_{cr}} \;;\; \mathsf{p} = \frac{p}{p_{cr}} \;;\; \mathsf{t} = \frac{T}{T_{cr}}.$$

We then obtain

(5) $$\left(\mathsf{p} + \frac{3}{\mathsf{v}^2}\right)(3\mathsf{v} - 1) = 8\mathsf{t}.$$

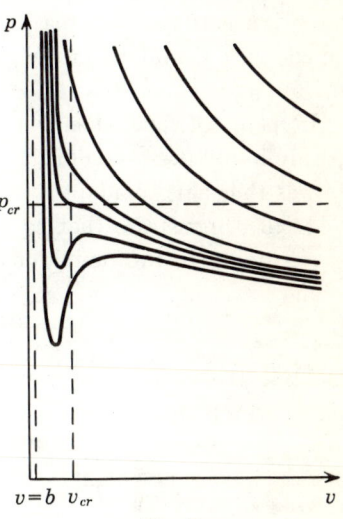

Fig. 7.
The van der Waals isotherms in the p,v plane

The preceding equation expresses the *law of corresponding states* due to van der Waals and establishes a universal law of similarity; its accuracy is the same as that of eq. (1). Incidentally, we may note that an analogous law of similarity can be established for any equation which contains only three individual constants.[1] It suffices to eliminate these three constants by introducing new dimensionless properties v, p, t.

B. Entropy and the caloric behavior of the van der Waals gas

In discussing the behavior of a perfect gas in Sec. 5 D we have proved the existence of entropy from our additional caloric condition, but in the case of the van der Waals gas we shall proceed in a reverse manner in that

---

[1] J. de Boer and collaborators, Physica **14**, 139, 149, 320 (1948).

we shall deduce its caloric behavior from the proposition on the existence of entropy. In the case of a perfect gas this was defined by eq. (5.7 a), i. e. by $\partial u/\partial v = 0$; in the present case we shall be led to the more general condition

$$(6) \qquad \left(\frac{\partial u}{\partial v}\right)_T = \frac{a}{v^2}.$$

This condition is physically revealing: the internal energy of a gas is now seen to consist not only of the kinetic energy of the molecules but also of the potential energy of their forces of cohesion which is associated with the constant a. As is the case with a gravitating system of mass points, this energy is negative and approaches zero with increasing expansion. Consequently the energy u contained in the gas must increase with v, as borne out by our eq. (6).

In point of fact when van der Waals established his equation he was already in full possession of the fundamental propositions of thermodynamics. He was thus able to adapt the form of his equation to the entropy principle. We shall demonstrate that eq. (6) can be deduced from that principle.

According to the definition in eq. (6.10 a), we write

$$(7) \qquad ds = \frac{du + p\,dV}{T}.$$

Inserting the value for p from (1) and considering u to be a function of T and v, we obtain at first

$$(8) \qquad \begin{aligned} ds &= \frac{1}{T}\left(\frac{\partial u}{\partial T}dT + \frac{\partial u}{\partial v}dv\right) + \left(\frac{R}{v-b} - \frac{a}{v^2\,T}\right)dv \\ &= \frac{1}{T}\frac{\partial u}{\partial T}dT + \frac{1}{T}\left(\frac{\partial u}{\partial v} + \frac{R}{v-b} - \frac{a}{v^2\,T}\right)dv. \end{aligned}$$

The necessary and sufficient condition for this expression to be a perfect differential is:

$$(9) \qquad \frac{1}{T}\frac{\partial^2 u}{\partial v\,\partial T} = \frac{\partial}{\partial T}\left(\frac{1}{T}\frac{\partial u}{\partial v} + \frac{R}{v-b} - \frac{a}{v^2\,T}\right).$$

In carrying out the differentiation indicated on the right-hand side it is noticed that the middle term vanishes and that the last term gives $a/v^2\,T^2$; the first term gives rise to two, one which cancels the left-hand side, the other being equal to $-\dfrac{\partial u}{\partial v}\Big/T^2$. Thus eq. (9) becomes

$$(9\text{ a}) \qquad 0 = -\frac{1}{T^2}\frac{\partial u}{\partial v} + \frac{1}{T^2}\frac{a}{v_2}.$$

This is identical with eq. (6) which we seek to prove.

By partial differentiation we can deduce from (6) that

$$\frac{\partial c_v}{\partial v} = \frac{\partial^2 u}{\partial T\, \partial v} = \frac{\partial}{\partial T} \frac{a}{v^2} = 0, \tag{10}$$

so that c_v is a pure function of T, as for a perfect gas.

We shall now proceed to calculate the difference between the molar heats $c_p - c_v$ which for a perfect gas was equal to the characteristic value R. With reference to the general eq. (7.8 c), we obtain

$$c_p - c_v = \left\{ \left(\frac{\partial u}{\partial v}\right)_T + p \right\} \left(\frac{\partial v}{\partial T}\right)_p. \tag{11}$$

The value of $(\partial v/\partial T)_p$ can be taken from (2), that of $(\partial u/\partial v)_T$ follows from (6). Hence for a van der Waals gas we find that

$$c_p - c_v = R \left/ \left(1 - \frac{2a}{RT} \frac{(v-b)^2}{v^3}\right)\right.. \tag{12}$$

In this equation a may be considered small and the product of the small quantities a, b may be omitted. Equation (12) now becomes

$$c_p - c_v = R \left/ \left(1 - \frac{2a}{RTv}\right)\right.. \tag{13}$$

Reverting to the expression for entropy in (8) we can simplify it to

$$ds = \frac{c_v\, dT}{T} + R \frac{dv}{v-b}, \tag{14}$$

in view of (6) and (10). Integrating on the assumption that the molar specific heat c_v is almost a constant, as for a perfect gas, we obtain

$$\int_{T_0, v_0}^{T, v} ds = s - s_0 = c_v \log \frac{T}{T_0} + R \log \frac{v-b}{v_0-b}. \tag{15}$$

10. Remarks on the liquefaction of gases according to van der Waals

A. The integral and the differential Joule – Thomson effect

The condition that the *enthalpy H is constant* is valid for any equation of state and not only for perfect gases; this has already been stressed in eq. (5.7). From the Table in Sec. 7 we obtain

(1) $$\Delta h = T \Delta s + v \Delta p$$

if the infinitesimal quantities ds, dp are extrapolated to small finite differences $\Delta s, \Delta p$ by the way of an approximation. Changing from the variables s, p to the variables T, p we may put

$$\Delta s = \left(\frac{\partial s}{\partial T}\right)_p \Delta T + \left(\frac{\partial s}{\partial p}\right)_T \Delta p.$$

Recalling the significance of c_p we have

$$\left(\frac{\partial s}{\partial T}\right)_p = \frac{1}{T}\left(\frac{\partial q}{\partial T}\right)_p = \frac{c_p}{T}.$$

Making use of the relation

$$\left(\frac{\partial s}{\partial p}\right)_T = -\left(\frac{\partial v}{\partial T}\right)_p$$

from our Table in Sec. 7 we can transform eq. (1) to read:

(2) $$\Delta h = c_p \Delta T + \left[v - T\left(\frac{\partial v}{\partial T}\right)_p\right]\Delta p.$$

Thus the fact that the enthalpy is constant gives

(3) $$\frac{\Delta T}{\Delta p} = \frac{1}{c_p}\left[T\left(\frac{\partial v}{\partial T}\right)_p - v\right] = \frac{vT}{c_p}\left(\alpha - \frac{1}{T}\right),$$

where α denotes the coefficient of thermal expansion. We restrict our attention to the van der Waals gas in that we substitute the special value for α from eq. (9.2). Taking into account eq. (9.2 b) we obtain

(4) $$\frac{\Delta T}{\Delta p} = \left(\frac{2a}{RT} - b\right)\bigg/c_p,$$

from which we conclude: When the gas is expanded, $\Delta p < 0$, there will be *cooling*, $\Delta T < 0$, if

(4 a) $$\frac{2a}{RT} > b.$$

This is the case with air and with most other gases. *Air can be cooled at will by repeated expansion and can finally be liquefied.*

The industrial production of liquid air (and its separation from, Ne, A, ...) in Linde-type plants need not, of course, be restricted to an exact realization of the Joule-Thomson effect in all its details. Instead of "porous plugs" throttling valves are used and the performance is improved by Linde's regenerative counter-flow heat exchanger.

We shall encounter eq. (3) once more in Sec. 11. It represents the *finite Joule-Thomson effect*. It originated by extrapolation from the *differential Joule-Thomson effect* which has been rigorously defined earlier. From eq. (3) we find that for the latter

(5) $$\left(\frac{\partial T}{\partial p}\right)_h = \frac{vT}{c_p}\left(\alpha - \frac{1}{T}\right).$$

B. The inversion curve and its practical utilization

We shall now try to determine quite generally the region in the p, T plane in which expansion ($\Delta p < 0$) is associated with a decrease in temperature ($\Delta T < 0$), as in eq. (4 a) or, in other words, where $(\partial T/\partial p)_h > 0$. This, from the point of view of practical applications, desirable region will be called positive. It is bounded by the *inversion curve* on which $(\partial T/\partial p)_h = 0$, so that from (5) we see that it is given by

(5 a) $$\alpha(p, T) = \frac{1}{T}.$$

It separates the desirable from the undesirable, *negative* region. As mentioned previously the states of air and of most other gases which correspond to ordinary conditions of pressure and temperature always lie within the inversion curve. This is confirmed by Fig. 8.

Making use of the accurate van der Waals equation we obtain the following expression for the inversion curve from eq. (9.2 a):

(5 b) $$\frac{2a}{RT}\left(\frac{v-b}{v}\right)^2 = b,$$

where it is only necessary to express v in terms of p and T. Introducing at the same time the reduced coordinates p and t we obtain after some rearrangement:

(5 c) $\qquad \mathsf{p} = 24\sqrt{3\mathsf{t}} - 12\mathsf{t} - 27.$

Figure 8 contains an inversion curve determined experimentally for H_2 (W. Meissner) in addition to the inversion curve from eq. (5 c). The conversion data for hydrogen and air are given in the caption to the figure. It is seen that $(\partial T/\partial p)_h$ for air is positive at room temperature up to 450 at; on the other hand for H_2 at room temperature it is always negative. This circumstance was responsible for many accidents due to the fact that highly compressed hydrogen ignited spontaneously when leaking from damaged pipes. Hydrogen can be cooled on sudden expansion ("throttling") only after its temperature had been reduced below -80 C.

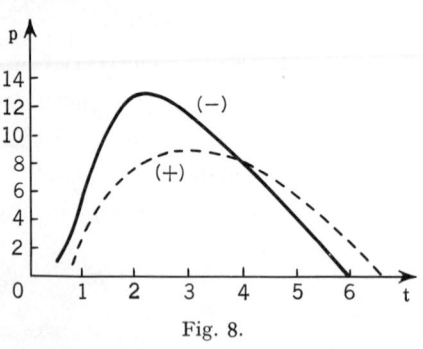

Fig. 8.

Inversion curve for the differential Joule-Thomson effect in reduced coordinates:

——— determined experimentally for H_2 by W. Meissner

------ calculated for a van der Waals gas from eq. (10.5 c)

For air: $T = 132.5 \times \mathsf{t}$ K; $p = 34.5 \times$ p at.
For H_2: $T = 33.2 \times \mathsf{t}$ K; $p = 13.2 \times$ p at.

Returning to the performance of a machine operating on the Joule-Thomson process we find that it depends on the *integral Joule-Thomson effect*

(5 d) $\qquad (T_2 - T_1) = \int\limits_{p_1}^{p_2} \left(\frac{\partial T}{\partial p}\right)_h dp.$

In practical applications the pressure p_2 after expansion is, in most cases, approximately equal to atmospheric. The temperature T_1 is determined essentially by the choice of the fluid for pre-cooling (for the liquefaction of air cooling water is used, whereas liquid nitrogen is used for the liquefaction of H_2). The pressure p_1 at which the gas enters the liquefaction chamber is the only variable which may be adjusted freely within certain limits. We shall now try to determine that value for p_1 which leads to a maximum cooling effect. Differentiating the integral in (5 d) with respect to the lower limit and equating to zero we have

$$\left(\frac{\partial T}{\partial p}\right)_h = 0 \quad \text{for} \quad p = p_1, \quad T = T_1.$$

This, however, is exactly the condition for the state p_1, T_1 to lie on the inversion curve of the differential Joule-Thomson effect. In designing a liquefaction plant it is necessary to satisfy this condition. For example, it is found that the most favorable temperature of pre-cooling for the liquefaction of H_2 is 64.5 K (it is generated by evaporating liquid nitrogen under reduced pressure). The corresponding favorable pressure is found from the inversion curve to be 160 at. In practice the values of 72 K and 140 at are often selected for operation. In the case of helium a temperature of pre-cooling of 14 K and a pressure $p_1 = 29$ at is used. The temperature of 14 K is produced either with liquid H_2 which is allowed to boil under a reduced pressure (Kammerlingh-Onnes), or, more recently, with helium gas which is cooled with the aid of a reversible adiabatic process (performance of work in an expansion engine). (Kapitza and Meissner; the latter gave a complete analysis of the required thermodynamic conditions.)

C. The boundary of the region of co-existing liquid-vapor phases in the p,v plane

The word *phase* has several meanings; generally speaking it denotes the "form of a phenomenon." In this connection we may think of the phase of an optical vibration, of the phases of the moon, of the various phases in a political development; later we shall speak of the multi-dimensional "phase space" in statistical mechanics. In thermodynamics the term "phase" is used to denote the different states of aggregation of a single substance including the various structural forms of the solid (crystal structures, amorphous structures). When applied to several substances the term includes the different chemical groupings of which they are capable.

In the theory evolved by van der Waals we shall study the equilibrium between the gaseous phase *2* and the liquid phase *1*, i. e. that between a saturated vapor over its liquid. *Equality of pressure* constitutes the mechanical condition for such equilibrium and *equality of temperature* is the thermal condition of equilibrium. Of the four potentials listed in the Table in Sec. 7 the *free enthalpy G* is the one most suitable for use when p and T are constant. According to eq. (8.13 a) the equality of p and T carries with it the equality of g in both phases: $g_1 = g_2$. If the masses of substance contained in the two phases, *1* and *2*, are denoted by m_1 and m_2 then according to eq. (8.13 a) we may write

$$\delta G = \delta(m_1 g_1 + m_2 g_2) = (g_1 - g_2) \cdot \delta m_1 = 0$$

because $\delta m_1 = - \delta m_2$; the total mass $m_1 + m_2$ remains constant during the variation.

With reference to Fig. 7 we consider an isotherm $T < T_{cr}$ and make it intersect with an isobar $p < p_{cr}$. Of the three points of intersection we shall denote the two external ones by A and B, Fig. 9, the corresponding values of g being denoted by g_A and g_B. These are also equal because they belong to the same p and T. The equality of g_A and g_B leads to

(6) $\quad u_B - u_A - T(s_B - s_A) + p(v_B - v_A) = 0,$

because $g = u - Ts + pv$. The difference $s_B - s_A$ can be taken from eq. (9.15) Taking into account the equality of T at both points of intersection A and B we find that

(7) $\quad s_B - s_A = R \log \dfrac{v_B - b}{v_A - b}.$

We now calculate $u_B - u_A$ integrating du along the isobar $p = \text{const}$ (any other path of integration would lead to the same result):

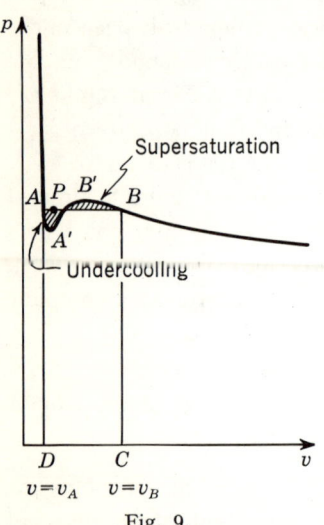

Fig. 9.
Definition of Maxwell's line.

$$u_B - u_A = \int_A^B du = \int_A^B \left\{ \left(\dfrac{\partial u}{\partial v}\right)_T dv + \left(\dfrac{\partial u}{\partial T}\right)_v dT \right\}.$$

Making use of eqs. (9.6), (9.10) we have:

$$u_B - u_A = \int_A^B \dfrac{a}{v^2} dv + \int_A^B c_v(T) \, dT,$$

where the second term on the right-hand side vanishes as $T_A = T_B$. The first term gives:

(8) $\quad u_B - u_A = -\dfrac{a}{v_B} + \dfrac{a}{v_A}.$

Substituting (7) and (8) into (6) we find:

(9) $\quad -\dfrac{a}{v_B} - RT \log(v_B - b) + \dfrac{a}{v_A} + RT \log(v_A - b) + p(v_B - v_A) = 0.$

We now calculate the area in the p,v plane bounded by the isotherm $T = \text{const}$, the axis of abscissae and the two straight lines $v = v_A$, $v = v_B$. According to the van der Waals equation this area is equal to

10.11 LIQUEFACTION OF GASES ACCORDING TO VAN DER WAALS

(10) $$\int_B^A p\,dv = RT\int_A^B \frac{dv}{v-b} - a\int_A^B \frac{dv}{v^2} = \left[RT\log(v-b) + \frac{a}{v}\right]_{v=v_A}^{v=v_B}.$$

The expression on the right-hand side is equal to the first four terms in eq. (9), except for the sign. Substituting (9) into (10) we have

(11) $$\int_A^B p\,dv = p\,(v_B - v_A).$$

Geometrically $p(v_B - v_A)$ represents the rectangle $ABCD$ in Fig. 9. According to (9) its area is equal to that between the isotherm and the axis of abscissae considered in (10). Hence the two crescents shaded in the figure must have equal areas. This rule gives a convenient graphical method of determining the boundary points A and B for phase equilibrium. It was first given by Maxwell (Nature 1875); the line AB is known as "Maxwell's line."

Carrying out the above construction for every isotherm below the critical in Fig. 7 we obtain the boundary enclosing that region in the p,v plane within which the two phases, *liquid* and *gaseous*, co-exist in equilibrium. The boundary is shown sketched in Fig. 10. To the right of this line only gas can exist, whereas to the left of it only the liquid will be present. The vertex of this region coincides with the critical point v_{cr}, p_{cr}. The two branches of the boundary

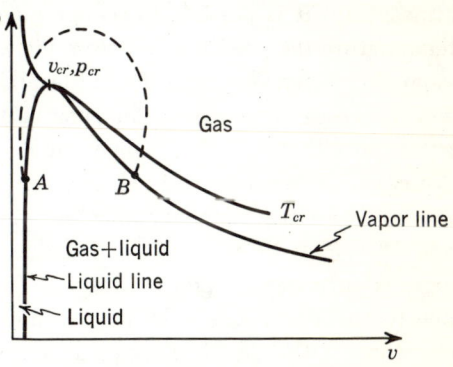

Fig. 10.

Boundary of the region of co-existing phases gas + liquid in the p,v plane.

curve meet at that point. The locus of points A is the water (or liquid) line, and that of points B is the steam (or vapor) line. Proceeding from the gaseous phase we find that the first droplets of liquid appear on the vapor line; proceeding from the liquid phase we shall notice the first bubbles of vapor on crossing the liquid line.

Let us return for a moment to Fig. 9 and let us inquire into the physical meaning of the points along Maxwell's line AB. They are points of varying volume but equal pressure and temperature. The variation in volume is produced by the varying proportions of the liquid and gaseous phase present.

At B we have pure saturated vapor of molar volume v_B, whereas at A we have pure liquid of molar volume v_A. Denoting the relative mass of the liquid by x and that of the vapor by $1-x$, we find that at A $x=1$ and at B $x=0$. At an intermediate point P we have

$$v_P = x v_A + (1-x) v_B.$$

It follows that

(11 a) $$\quad x = \frac{v_B - v_P}{v_B - v_A} = \frac{PB}{AB}\ ;\quad 1-x = \frac{v_P - v_A}{v_B - v_A} = \frac{AP}{AB}.$$

The parts x and $1-x$ can be inferred from the diagram as they are equal to the ratios PB/AB and AP/AB respectively.

We shall now explain the meaning of the captions "supersaturation" and "undercooling" in Fig. 9. They denote *unstable* states of equilibrium, which are *not* states of *true equilibrium* like those corresponding to the points on Maxwell's line. Under favorable conditions (for example steam in dust-free atmosphere) it is possible to obtain supersaturated steam when at constant temperature the pressure assumes a higher value than that at point B on the vapor line. Similarly the liquid state can be maintained at a temperature which exceeds slightly the boiling point if heated in a vessel which is completely free from vibration. At a temperature equal to that at point A the pressure decreases. It must be realized, however, that these unstable extensions over the limits on the boundary lines are very small and the states between A' and B' cannot be attained at all because they correspond to an isothermal increase in pressure upon expansion. It is very remarkable that the theory due to van der Waals is in a position to predict, at least qualitatively, the existence of the unstable states along the branches AA' or BB'.

We now return to Fig. 10 and directly deduce from it the following facts:

1. At temperatures $T > T_{cr}$ it is impossible to liquefy a gas however large the compression.

2. In order to liquefy a gas it is not enough to reduce the temperature to $T < T_i$; according to eqs. (5 b) and (9.3) the inversion temperature T_i is about seven times larger than T_{cr}. It is, however, possible to liquefy a gas, as was the case with air, by a succession of Joule-Thomson expansions and to reach the range of partial and, finally, total liquefaction, provided that $T < T_i$ has been attained.

3. It is possible to reach the liquid region at the lower left-hand corner of the diagram in Fig. 10 directly from the rarefied gaseous state at the lower

right-hand region without crossing the shaded two-phase region by following a path which traverses the region above the critical point. According to the van der Waals equation the states along such a path form a continuum of states of stable equilibrium. There is no discontinuity even on crossing the critical isotherm $T = T_{cr}$, either on its lower branch where $p < p_{cr}$ or on its upper branch for which $p > p_{cr}$. This behavior explains the title of van der Waals' paper "The continuity of the gaseous and liquid states."

4. In particular we can traverse from one end, B, of a Maxwell line to the other at A along the path shown by a broken line in Fig. 10. The analytic expression for g_B will vary continuously until the value g_A is attained. This has been anticipated in eq. (6). Strictly speaking we could have added to this equation a linear function of T owing to the fact that the zero levels of energy and entropy are undetermined. Our considerations of continuity do, however, demonstrate that this should, in fact, have been set equal to zero.

In the preceding description the concept of the critical point, as well as that of Maxwell's line, have been explained with the aid of the idealized model of a van der Waals gas.

In actual fact the course of the isotherms of a real gas shows qualitative agreement with the van der Waals model. In particular the existence of a critical point is always observed being determined by the condition from Sec. 9, namely

$$\left(\frac{\partial p}{\partial v}\right)_T = 0; \quad \left(\frac{\partial^2 p}{\partial v^2}\right)_T = 0$$

together with the equation of state $p = p(v, T)$. Moreover, for analytically formulated equations of state the relation (11) for Maxwell's line remains true in general, as it is easy to prove.

In order to do this let us consider an isotherm below the critical whose shape is qualitatively reproduced by that shown in Fig. 9. From the equality of free enthalpy at A and B we have that

(12) $$f_A + p v_A = f_B + p v_B$$

since $g = f + p v$. On the other hand

(13) $$f_A - f_B = -\int_A^B \left(\frac{\partial f}{\partial v}\right)_T dv.$$

According to our Table in Sec. 7 we find that $(\partial f/\partial v)_T = -p$ so that eqs. (12) and (13) yield

$$p(v_B - v_A) = \int_A^B p\, dv$$

where the integral extends over the extrapolated isotherm $AA'B'B$. This proves that (11) is true for any analytic form of the equation of state.

11. The Kelvin temperature scale

The absolute temperature has been defined in Sec. 6 as that function $\phi(\theta)$ of an arbitrary, conventionally measured temperature θ which satisfies Carnot's ratio

(1) $$\frac{Q_1}{Q_2} = \frac{\phi(\theta_1)}{\phi(\theta_2)} = \frac{T_1}{T_2}.$$

This definition was proposed by Thomson as early as 1848. It has also been pointed out that a temperature measured with the aid of a gas thermometer would satisfy condition (1) with a degree of precision which is directly related to the deviation of the thermometric substance within the range under consideration from that of a *perfect gas*. We now proceed to show how the temperature θ measured with the aid of a *real* gas thermometer can be reduced to the absolute temperature T.

Evidently eq. (1) is not very suitable for this purpose because it is not possible to measure Q calorimetrically with a high enough degree of precision. In its place, however, we may use any relation deduced from the Second Law which contains the absolute temperature. Lord Kelvin recognized that his analytical formulation of the Joule-Thomson effect given in Sec. 8 was particularly suitable. Clapeyron's equation given in Sec. 14 constitutes another practical starting point.

Equation (10.3) had the form:

(2) $$\frac{\Delta T}{\Delta p} = \frac{v\, T}{c_p}\left(\alpha - \frac{1}{T}\right).$$

The quantities c_p and α, which have been defined with the aid of derivatives with respect to T can be rewritten in terms of derivatives with respect to θ, because $T = \phi(\theta)$ is a pure function of θ:

$$\alpha = \frac{1}{v}\left(\frac{\partial v}{\partial T}\right)_p = \frac{1}{v}\left(\frac{\partial v}{\partial \theta}\right)_p \frac{d\theta}{dT} = \alpha' \frac{d\theta}{dT};$$

$$c_p = \left(\frac{\partial q}{\partial T}\right)_{p=\text{const}} = \left(\frac{\partial q}{\partial \theta}\right)_{p=\text{const}} \frac{d\theta}{dT} = c_p' \frac{d\theta}{dT}.$$

The newly defined quantities α' and c_p' are to be regarded as being empirically determined functions of θ. Dividing the numerator and denominator on the right-hand side of eq. (2) by $d\theta/dT$ we find that it is equal to

(2 a) $$\frac{vT}{c_p'}\left(\alpha' - \frac{1}{T}\frac{dT}{d\theta}\right).$$

The left-hand side can be rewritten as

(2 b) $$\frac{\Delta\theta}{\Delta p}\frac{dT}{d\theta}$$

where the quantity $\Delta\theta/\Delta p$ is an empirically given function of θ, namely that measured with the aid of the Joule-Thomson effect. Equating (2 a) and (2 b) we have

$$c_p'\frac{\Delta\theta}{\Delta p}\frac{1}{T}\frac{dT}{d\theta} = v\left(\alpha' - \frac{1}{T}\frac{dT}{d\theta}\right)$$

so that

(3) $$\frac{1}{T}\frac{dT}{d\theta} = \frac{v\alpha'}{v + c_p'\frac{\Delta\theta}{\Delta p}},$$

or

(4) $$\log\frac{T}{T_0} = \int_{\theta_0}^{\theta} \frac{v\alpha'}{v + c_p'\frac{\Delta\theta}{\Delta p}}\, d\theta.$$

In general the integration must, of course, be carried out numerically because it involves only empirically determined functions of θ (v also belongs to this group). If θ is measured in deg Celsius and if the unit on the scale of T is suitably chosen then we have the correspondence

$$\theta = 0 \quad \text{and} \quad T = T_0,$$
$$\theta = 100\,\text{C} \quad \text{and} \quad T = T_0 + 100\,\text{C}.$$

In order to determine T_0 we can use eq. (4), or

$$\text{(4 a)} \qquad \log\left(1 + \frac{100\,C}{T_0}\right) = \int_0^{100\,C} \frac{v\,\alpha'}{v + c_p'\,\dfrac{\Delta\theta}{\Delta p}}\,d\theta.$$

The scale of temperature defined above is now universally known as the *Kelvin temperature scale*, and we write $T = \ldots$ K. In the interval 0 to 100 C the differences between T and θ are very small for most thermometric substances. In accordance with the determinations carried out by the German Physikalisch-Technische Reichsanstalt the maximum deviation when θ is determined by an air thermometer is -0.0026 C. It is smaller for the more perfect hydrogen for which it reaches 0.0007 C.

These differences naturally increase as the temperature approaches the point of liquefaction of the respective substance, and the Joule-Thomson process fails below it. The most effective method for the generation of the lowest temperatures near the (unattainable) absolute zero is that based on the *magneto-caloric effect* (P. Debye 1926, W. J. Giauque 1937); the two authors suggested (independently of each other) to use the paramagnetic salt-gadolinium sulphate as the most suitable substance for the purpose. The process is as follows: The substance is first cooled in a liquid helium bath of very low temperature (~ 1.3 K) and placed in a powerful magnetic field until thermal equilibrium has been reached. In this way all magnetic dipoles become practically unidirectional, provided that the field is sufficiently powerful. The entropy at this ordered state is smaller than at the state of disorder which prevails in the absence of a field (*cf.* Sec. 19), and the result is an appreciable reduction in entropy. At this moment the paramagnetic salt is insulated thermally and the field is switched off. The total entropy must remain constant in the process, but as the external field is decreased a certain amount of disorder will set in, the magnetic contribution to entropy increasing from zero to an appreciable value. Owing to the presence of the thermal insulation the total entropy remains constant which causes the temperature to decrease in consequence. The specific heat c_p is very small at these very low temperatures and the magnetic contribution to entropy constitutes the leading term so that on adiabatic demagnetization the temperature can be made to decrease to a value of the order of several thousandths of one degree K (de Haas). We shall see in Sec. 12 that the point of absolute zero cannot be reached in this way. We shall only note here that the thermodynamic theory of the process of adiabatic demagnetization (*cf.* Sec. 19) supplies us with a method of extending the definition of the Kelvin temperature scale down to temperatures in the neighborhood of absolute zero.

The skeptical Mach[1] expressed the view that the zero point on the absolute temperature scale has a meaning only in relation to the special case of a gas. This view is flatly contradicted by the creation of the absolute Kelvin temperature scale. The whole structure of the science of thermodynamics would collapse without the existence of this (fixed, but unattainable) lower limit of temperature.

12. Nernst's Third Law of thermodynamics

Nernst's proposition is quite rightly termed the Third Law of Thermodynamics. It does not lead to a new property, like the First and Second Laws of thermodynamics but it makes the properties S, F, G, ... numerically determinate and hence usable.

The entropy is defined in terms of its differential dS but S itself is only defined short of a constant S_0. In itself this does not constitute a drawback because in applications we almost always deal with differences in entropy. The same is true of the energy U in whose expression a constant of integration remains undefined. However, in the expressions of the potentials F and G a linear function of the form $S_0 T +$ const remains undetermined in view of the term TS. Thus their usefulness for states at different temperatures and in the equations of chemical equilibrium becomes illusory.

Thus there arises the question of the absolute value of entropy: As is the case with all fundamental questions, nature provides the simplest conceivable and the mathematically most satisfactory answer: *As the temperature of a system tends to absolute zero its entropy tends to a constant value S_0 which is independent of pressure, state of aggregation, etc.* We may put it equal to zero so that the entropy of every substance becomes normalized in an absolute way. This also removes the indeterminateness in the potentials F, G because in the integral formulae for entropy (*cf.* (6.11)) we may assume the lower limit to be at $T = 0$.

This is the formulation which was given by Planck, but it is interesting to quote here Nernst's *own* first statement of the same fact. In connection with his researches on electrochemical phenomena Nernst was able to handle the Second Law of thermodynamics with greater success than anybody else but, remarkably enough, he disliked the concept of entropy and preferred to

---

[1] "Prinzipien der Wärmelehre", Leipzig 1896, p. 341. Gay Lussac's inaccessible account of his free-expansion experiment has been published in an Appendix to this book.

use that of "maximum (available) work" which he also used as a measure of chemical affinity. Introducing Nernst's symbol A for it we find that according to eq. (8.15) we have

(1) $$A = \Delta F \quad \text{with} \quad \Delta F = F_1 - F_2.$$

From $F = U - TS$ together with the relation

$$S = -\left(\frac{\partial F}{\partial T}\right)_V, \quad \text{see Table in Sec. 7}$$

we have

$$F = U + T\left(\frac{\partial F}{\partial T}\right)_V.$$

so that from (1) we find that

(2) $$A - \Delta U = T\left(\frac{\partial A}{\partial T}\right)_V \quad \text{with} \quad \Delta U = U_1 - U_2.$$

Nernst regarded (2) as the *fundamental equation of thermodynamics*[1] because it combines the First Law with the Second.

According to Berthelot and Nernst, ΔU denotes the heat tone of the process at constant V.[2] Berthelot thought that the left-hand side of (2) should be equal to zero which, however, is not the case. Nernst (*l. c.*) made the following comment: "It occurred to me that we have here a limiting law since often the difference between A and ΔU is very small. It seems that at absolute zero A and ΔU are not only equal but also asymptotically tangent to each other. Thus we should have

(3) $$\lim \frac{dA}{dT} = \lim \frac{d\Delta U}{dT} = 0 \quad \text{for} \quad T = 0."$$

This is the historical origin of the most far-reaching generalization of classical thermodynamics of our century.

Returning from A to

$$\Delta F = \Delta U - T\Delta S$$

we obtain at once from (3) that

$$\lim\left\{\frac{d\Delta U}{dT} - \Delta S - T\frac{d\Delta S}{dT}\right\} = \lim \frac{d\Delta U}{dT}$$

---

[1] "Theoretische Chemie", ed. 1926, p. 795. We have written ΔU instead of Nernst's U and we have added the subscript V in the partial derivative of A which was omitted by Nernst.

[2] In most cases it is more useful to consider the heat tone at constant pressure and to define it as the difference of enthalpies ΔH.

and consequently

(4) $\qquad \Delta S \to 0$ i. e. $S_2 \to S_1$ for $T \to 0$.

Thus we have deduced Planck's formulation $S \to S_0$ and the possibility of normalizing $S_0 = 0$ for any substance from Nernst's statement (3). Reciprocally, it is evident that eq. (3) follows from (4).

Originally Nernst limited the validity of his theorem to pure condensed substances (solids or liquids) and excluded the gaseous state. However, in a subsequent publication[1] he considered the question of degenerate gases which was beginning to occupy the attention of physicists at the time. This problem was formulated shortly afterwards in the Bose-Einstein and the Fermi-Dirac statistics. The two statistics show that gases degenerate at the lowest temperatures. In particular in his sketch of a theory of metals (1927) the present author demonstrated that, with respect to all its properties, the free electron gas obeys Nernst's Third Law. It is not at all astonishing that the expressions (5.10) and (9.15) for the entropy of a perfect gas and for that of a van der Waals gas respectively do not lead to $S \to 0$ as $T \to 0$ because they do not claim any validity in this region. Apparent contradictions to the Third Law observed in connection with deuterium compounds have been resolved by Clausius who showed that they represented frozen, i. e. metastable states of equilibrium.

The science of thermodynamics can give no information about the duration of such exceptional states; this problem must be left to be dealt with by the methods of quantum mechanics. In any case we did not find it necessary to restrict the validity of Nernst's Third Law to condensed substances and we take the view that its validity is universal.

We shall now summarize a series of the most simple consequences of this law as regards homogeneous substances. Since we are now referring to mols or units of mass we shall use lower case symbols s, v instead of the capital letters used so far.

1. *The coefficient of thermal expansion tends to zero as* $T \to 0$. According to the Table in Sec. 7, line G, we have

$$\left(\frac{\partial s}{\partial p}\right)_T = -\left(\frac{\partial v}{\partial T}\right)_p.$$

The left-hand side vanishes because the limit s_0 of s is independent of pressure. Hence the change in volume which appears on the right-hand side must also vanish. The same is true of the coefficient of thermal expansion, α.

---

[1] "Die theoretischen und experimentellen Grundlagen des neuen Wärmetheorems," Knapp, Halle, 1918.

2. *The coefficient of tension tends to zero as* $T \to 0$. According to the table in Sec. 7, line F, we have

$$\left(\frac{\partial s}{\partial v}\right)_T = \left(\frac{\partial p}{\partial T}\right)_v$$

Hence, again, the left-hand side vanishes because the limit s_0 is independent of v as well. The same is true of the right-hand side and of the coefficient of tension β, provided $p \neq 0$.

3. *The specific (or molar) heats* c_p *and* c_v *tend to zero as* $T \to 0$. From the definitions

$$c_v = T\left(\frac{\partial s}{\partial T}\right)_v, \qquad c_p = T\left(\frac{\partial s}{\partial T}\right)_p$$

we obtain by integration

(5) $$s(v, T) = \int_0^T \frac{c_v}{T} dT; \qquad s(p, T) = \int_0^T \frac{c_p}{T} dT.$$

The constants of integration would be functions of v only in the first, and of p only in the second integral. However, for sufficiently small values of T their existence would contradict Nernst's Third Law which stipulates that the limiting value s_0 is asymptotically independent of both v and p. We have normalized the value of s_0 in both eqs. (5) at $s_0 = 0$ by assuming the lower limit of integration to be equal to zero. Now these integrals show that c_v and c_p must vanish for $T = 0$; if this were not so the integrals would diverge in view of the lower limit of integration $T = 0$. The German Reichsanstalt carried out extensive measurements under Nernst's direction providing a full confirmation of the fact that the specific heats, and that c_p in particular, decrease as absolute zero is approached. The rate at which these quantities tend to zero is left undetermined by eqs. (5). According to Debye (*cf.* Vol. II, Sec. 44) the rate of decrease for an elastic body is proportional to T^3, for the electron gas to T (*cf.* this volume, Sec. 39). These theoretical results have also been fully confirmed by experiment.

4. *The absolute zero temperature cannot be reached by any finite process; it may only be reached asymptotically.* We have made use of the process of adiabatic expansion, see Sec. 5, to cool real gases. In the range of lowest temperatures its place, as the most efficient process, is taken by adiabatic demagnetization, see Sec. 11.

12. NERNST'S THIRD LAW OF THERMODYNAMICS

The T,S diagram in Fig. 11 represents the curve of constant field intensity, $H = \text{const}$,[1] for a salt in a field of strength H. The pressure is also considered to be constant (e. g. $p = 0$, vacuum; in the case of a salt in the solid state the volume is unimportant). This curve passes through the origin 0 and its slope increases with T because the ordering of dipoles forced on the salt by the field is increasingly disturbed as T increases. We now supplement the diagram with the curve $H = 0$ for an unmagnetized salt. It lies above the curve $H = \text{const}$ because in this case no magnetic ordering exists. According to Nernst's Third Law this curve must also pass through the origin at $T = 0$. If we now start with the magnetized state at the initial temperature T_1 and proceed along an adiabatic (isentropic) curve, i. e. along a horizontal line $S = \text{const}$ until the demagnetized state on the curve $H = 0$ [2] has been reached we attain a temperature T_2 which is, as is seen from the drawing, considerably lower than T_1; but absolute zero is certainly not attained.

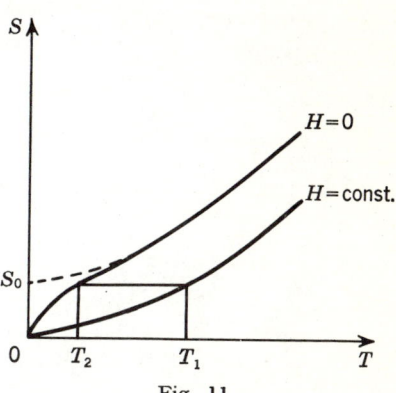

Fig. 11.
Illustrating the impossibility of reaching absolute zero by adiabatic demagnetization

This would not be so if the Third Law were not true. Suppose that the curve $H = 0$ had a limit $S_0 \neq 0$ for $T = 0$, as shown by the broken line in Fig. 11. With the present choice of the initial temperature T_1 we would attain absolute zero starting from $H = \text{const}$ in *one step*, namely we would reach point $S = S_0$ at $H = 0$.

It is quite natural to try to come nearer to absolute zero by magnetization at T_2 and repeated demagnetization. However, such a scheme would encounter practical difficulties. The only effective way of going one step further would be to demagnetize the nuclei, but this is still in the future.

In any case the following conclusion can be drawn from the Third Law: *The point of absolute zero temperature can only be attained asymptotically*. Consequently, the realization of a Carnot process with heat rejection at $T = 0$ and an efficiency of $\eta = 1$, *cf.* Sec. 6, is impossible.

---

[1] In the present Section and in Secs. 19 and 25 the letter H is used to denote the magnetic field strength and not enthalpy.

[2] The characteristic bulge in the curve is due to the fact at very low temperatures paramagnetic materials exhibit a kind of spontaneous magnetization (ordering of magnetic moments), similar to that exhibited by ferromagnetic substances.

CHAPTER II

THE APPLICATION OF THERMODYNAMICS TO SPECIAL SYSTEMS

13. Gaseous mixtures, Gibbs' paradox. The law due to Guldberg and Waage

The air which surrounds us is a mixture of

$$78\% \; N_2, \; 21\% \; O_2 \text{ and almost } 1\% \; A.$$

The surprisingly high argon content escaped the notice of earlier investigators because of its nobility; it was only discovered by Lord Rayleigh and W. Ramsay as late as the beginning of the century. The preceding figures represent molar percentages and are proportional to the number of molecules of the respective gas in the mixture. The ratio of percentages by weight is obtained by multiplying the above numbers with the molecular weights 28, 32, 40 respectively.

On the assumption that the gases are perfect, which is permissible at ordinary temperatures (point-molecules with zero cohesion; the van der Waals constants $b = 0$, $a = 0$), each component of the mixture behaves as if it alone occupied the volume V. It is, therefore, possible to define *partial pressures* p_i and to add them in order to obtain the total pressure p.

(1) $$p = \sum_i p_i, \quad \text{Dalton's law.}$$

Since each component obeys the perfect-gas law due to Gay-Lussac in the form (3.10), we also have

(1 a) $$V p_i = n_i R T.$$

Adding eqs. (1 a) for all components we obtain

(2) $$V p = n R T; \quad n = \sum_i n_i.$$

The energy of the mixture is also the sum of the molar energies of the components:

(3) $$U = \sum_i n_i u_i; \quad u_i = c_{vi} T.$$

It follows that the specific heats C_v and C_p of the mixture are given by

(3 a) $$C_v = \sum_i n_i c_{vi}; \quad C_p = \sum_i n_i c_{pi}; \quad C_p - C_v = n R.$$

Equation (3) can be understood by imagining that each component is compressed isothermally to the smaller volume

(3 b) $$V_i = V \frac{p_i}{p},$$

that all these volumes are placed side by side in compartments of equal cross-section A (similar to those in Fig. 12 below), and that they are separated by partitions, so that the sum of the volumes $V = \sum_i V_i$. If the partitions are withdrawn the i-th component will spread from the smaller volume V_i over the larger volume V and during this process its temperature T, its molar energy u_i, and its total energy $n_i u_i$ will remain constant. According to our preceding representation the presence of the remaining components may be disregarded. The physical nature of this process is better described by the term *diffusion*.

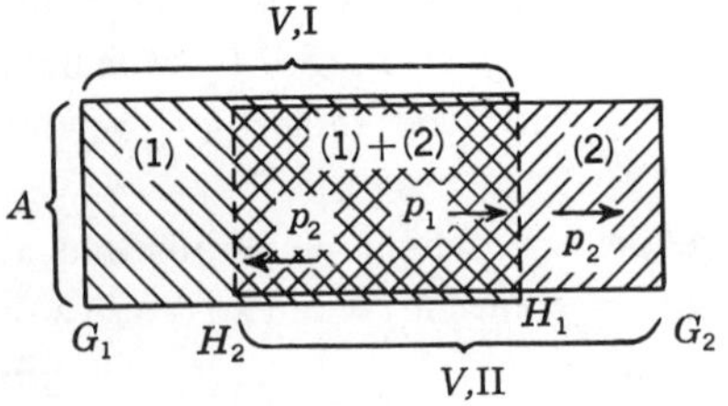

Fig. 12.
Reversible separation of two gases.

A. Reversible Separation of Gases

In the following argument we may restrict ourselves to the consideration of two gases: 1 and 2. In order to achieve separation we shall need *semipermeable membranes*. These occur in nature as walls of organic cells and can also be realized approximately with the aid of chemical means (copper ferrocyanide membrane). Their practical and theoretical application is due to the botanist Pfeffer, *cf.* Sec. 15. Figure 12 shows two cylinders, I and II of equal volumes V, which can slide into one another without friction. The side surfaces and the ends (G_1 of I and G_2 of II) are assumed *impermeable* to both gases 1 and 2. However, the end H_2 of II is assumed impermeable only to 2, allowing 1 *to pass through it*. The reverse is assumed about the end H_1 of I.

The two cylinders are assumed to have been pushed into each other initially, both gases being mixed. Now cylinder II is imagined drawn out infinitely slowly by a distance dx to the right. The compartment $G_1 H_2$ can only be entered by gas 1 which, so to say, ignores the presence of membrane H_2.

The end H_2 is acted upon only by the pressure p_2 which opposes the displacement dx. The work performed by it during the process is

(4) $$dW = -A\, p_2\, dx.$$

The membrane H_1, being at rest, does not perform any work and the same is true of wall G_1. However, there is work associated with wall G_2 which is moved by dx. Since compartment $H_1 G_2$ contains only gas 2 at a pressure equal to that in $H_2 H_1$ (H_1 is, so to say, ignored by the gas) the work done by p_2 on G_2 is

(4 a) $$dW = +A\, p_2\, dx,$$

i. e. equal and opposite to (4). The process of separation requires no work, and occurs without the exchange of heat, as we shall assume from now on; this means that the energy and the temperature of the two components remain constant. The final outcome of this experiment is that both gases 1 and 2 become separated, either of them in a volume V and at a pressure p_1 or p_2. Their entropies will be denoted by S_1 and S_2. Since this imaginary process of separation proved to be reversible, requiring neither work nor heat, we can say that the entropy S of the mixture:

(5) $$S = S_1 + S_2.$$

It is evident that the process represented in Fig. 12 can be reversed by sliding the cylinders slowly into each other. In this way the *mixing* of the gases also becomes *reversible*.

Equation (5) provides a sure basis for the calculation of the entropy of a mixture. Written out more fully it is

(5 a) $$S(T, V) = S_1(T, V) + S_2(T, V),$$

or (by changing the mathematical definition of the symbol S for the function)

(5 b) $$S(T, p) = S_1(T, p_1) + S_2(T, p_2).$$

In words: *In order to calculate the entropy of a mixture as the sum of the entropies of its components it is necessary to imagine each component occupying the same volume V as the mixture so that the pressure p is reduced to p_1 and p_2 respectively.*

Generalizing to more than two components we have, evidently:

(6) $$S(T, p, n_1, n_2, \ldots) = \sum_i n_i\, s_i(T, p_i),$$

where n_i is the number of mols and s_i is the molar entropy of the i-th component.

B. The Increase in Entropy during Diffusion and Gibbs' Paradox

During the process of diffusion considered in the preceding Section the initial volumes V_i of the components, imagined arranged side by side, have been assumed to be different from each other, each being smaller than V; their pressure was brought to the common value p by isothermal compression. Let S_0 denote the sum of the entropies of the components before diffusion. In analogy with (6) we can write

(6 a) $$S_0 = \sum_i n_i s_i(T, p).$$

From (6) and (6 a) we can calculate the change in entropy due to diffusion:

(6 b) $$S - S_0 = \sum_i n_i \{s_i(T, p_i) - s_i(T, p)\}.$$

Using our original definition of entropy for a perfect gas as given in (5.10) we can calculate this difference per mol and at constant temperature[1] and we obtain:

$$s_i(T, p_i) - s_i(T, p) = R \log \frac{v}{v_i} ;$$

owing to (3 b) we can replace the right-hand side by $R \log p/p_i$. Thus eq. (6 b) transforms into

(7) $$S - S_0 = R \sum_i n_i \log \frac{p}{p_i}.$$

Diffusion is an irreversible process in the same way as the conduction of heat. The increase in entropy found in (7) can be reversed only by the performance of work.

Taking into account the preceding equations of state (2) and (1 a) we have $p/p_i = n/n_i$. The increase in entropy becomes a pure function of the numbers of mols n_i and of their sum $n = \sum_i n_i$:

(8) $$S - S_0 = R \sum_i n_i \log \frac{n}{n_i} = R \left[n \log n - \sum_i n_i \log n_i \right].$$

---

[1] It is true that Clausius and Waldmann have shown that diffusion is accompanied by measurable temperature differences. However, we are here only interested in the final state which sets in after the temperatures have become equalized.

The right-hand side of (8) may be called the *mixing term*. It depends solely on the number of molecules and not on their nature. This leads to the *paradox enunciated by Gibbs*: On going over to the limit of identical molecules eq. (8) would, apparently, remain unchanged. This is absurd, because when the partitions are removed from compartments enclosing completely identical molecules there is no diffusion. *The process of going over to the limit is inadmissible*. It contradicts the *atomistic nature of matter* and it is inconsistent with the fact that there is no continuous transition between different kinds of molecules (e. g. the atoms H and He).

In order to explain it in greater detail we consider the case of very similar molecules, e. g. the isotopes of a noble gas. Equation (8) applies to this case without any correction; the same is true of a mixture of ortho- and para-hydrogen whose components differ only by their spin as well as of a mixture of molecules some of which are in the ground state, the remainder being in an excited state of energy. On the other hand eq. (8) fails when the molecules of the components are completely indistinguishable.

C. The law of mass action due to Guldberg and Waage

So far we have assumed that the components of the mixture under consideration are chemically inert with respect to each other. When chemical reactions are possible in the system a state of equilibrium between the reactants and products will be reached eventually so that we now propose to study the precise nature of such an equilibrium. We shall assume that the chemical reaction takes place at constant pressure, p, (e. g. at constant atmospheric pressure) and at constant temperature, T. Such conditions can almost always be achieved if the experiment is suitably arranged. In accordance with Sec. 8 we can write down the conditions of equilibrium as

(9) $\qquad \delta G = 0 \quad \text{with} \quad G = U - TS + pV = H - TS.$

The dissociation of steam into hydrogen and oxygen affords a simple example of such a process during which a stoichiometric mixture of oxyhydrogen gas remains in equilibrium with water vapor. (The fact that steam cannot be regarded as a perfect gas at ordinary temperatures need not concern us here because no appreciable dissociation occurs unless the temperature has become high.) The chemical formula of the reaction, namely

$$2 H_2O \rightleftarrows 2 H_2 + O_2$$

asserts that 2 mols of H_2O must disappear for every two mols of H_2 and every mol of O_2 appearing in the system. If we denote the number of mols of H_2,

O_2 and H_2O present in the system by n_1, n_2 and n_3, respectively, and if we denote the integers associated with the chemical equation by ν_1, ν_2 and ν_3 (reckoned positive for the substances on the right-hand side of the chemical equation, and negative for the other side) then we have for the present example

(9 a) $$\begin{cases} \nu_1 = 2, \quad \nu_2 = 1, \quad \nu_3 = -2 \\ \text{for } H_2, \quad O_2, \quad \text{and} \quad H_2O \text{ respectively.} \end{cases}$$

Further, considering a real or virtual change in the number of mols in the system we can establish the following proportions

(10) $$\delta n_1 : \delta n_2 : \delta n_3 = \nu_1 : \nu_2 : \nu_3.$$

We shall now proceed to express the condition of equilibrium in eq. (9) in terms of the variables p, T, n_i and the parameters ν_i. The molar energies, as shown previously, are functions of T only. Hence

$$U = \sum_i n_i u_i(T).$$

The same is true of enthalpy, as seen from eqs. (1) and (1 a), or, of the molar enthalpy which will be denoted by h_i. Thus we have

(11) $$H = \sum_i n_i h_i(T).$$

The rule concerning entropy is, however, different. In this connection it is necessary to recall that the molar entropies, $s_i(T, p)$, must be augmented by the term due to mixing and given in eq. (8), i. e.

(12) $$S = \sum_i n_i \{s_i(T, p) + R \log n/n_i\}.$$

From eqs. (9), (11) and (12) we can deduce that

$$G = \sum_i n_i \{h_i(T) - T s_i(T, p) - RT \log n/n_i\}$$

which can also be written as

(13) $$G = \sum_i n_i \{g_i(T, p) - RT \log n/n_i\}$$

where g_i denotes the free enthalpy (Gibbs' function) of one mol of each individual component.

Considering that T and p are constant, we can obtain from eq. (13) that

$$\delta G = \sum_i \delta n_i \{g_i(T, p) - RT \log n/n_i\} - RT \sum_i n_i \, \delta \log n/n_i.$$

Since, however, $n = \sum_i n_i$, the last term in the above equation vanishes. The condition of equilibrium $\delta G = 0$ can thus be rewritten as

(14) $$\sum_i \nu_i \{g_i(T, p) - RT \log n/n_i\} = 0$$

where ν_i has been substituted for δn_i from the condition of proportionality, eq. (10), suitably modified. Taking anti-logarithms we obtain the *law of mass action* in the form

(15) $$\prod_i (n_i/n)^{\nu_i} = K, \text{ where } \log K = -\frac{1}{RT} \sum_i \nu_i g_i(p, T).$$

This law was first discovered in 1867 by the Norwegian scientists Guldberg and Waage who used a line of argument based on statistical mechanics (probability of molecular encounters). Shortly afterwards, Gibbs demonstrated the validity of this law for perfect gases with the aid of purely thermodynamic considerations. He further extended the scope of this law by actually computing the value of the constant K. With certain limitations, the law of mass action can also be applied to vapors[1] and this constitutes one of the foundations of physical chemistry whose early development took place at the period under consideration.

If we now introduce molar concentrations, i. e. if we introduce the molar fractions $c_i = n_i/n$, we obtain from eq. (15)

(15 a) $$\prod_i c_i^{\nu_i} = K.$$

The quantity K is known as the "constant" of the equation of mass action or the *equilibrium constant*. In the example of the dissociation of water vapor we see from eq. (15 a) that

(16) $$\frac{c_1^2 c_2}{c_3^2} = K.$$

---

[1] The processes of polymerization and dissociation which may occur with vapors can also be included in the preceding argument if they are described with the aid of additional equations, and if the law is applied by setting up the conditions of equilibrium for several simultaneous reactions.

In order to determine the individual values of the three unknowns c_1, c_2 and c_3, we can utilize the additional condition:

(16 a) $$c_1 + c_2 + c_3 = 1$$

which follows from $\Sigma n_i = n$, as well as the known (directly or by measurement) ratio of the number of atoms of hydrogen to that of oxygen which exist in the system:

(16 b) $$\frac{N_H}{N_O} = \frac{2 c_1 + 2 c_3}{2 c_2 + c_3}.$$

In the more general case when more than three components take part in the reaction and when more than two atomic species are involved, the number of available equations is still sufficient to compute the individual values of c_i.

It is possible to eliminate the molar fractions, c_i, from eq. (15 a) and to use partial pressures, p_i, instead. We then have

(17) $$\prod_i p_i^{\nu_i} = K_p \quad \text{where} \quad K_p = p^{\sum_i \nu_i} \times K.$$

This form of the law of mass action is useful because K_p is independent of p being a function of T alone. This fact can be proved as follows. Differentiating K in eq. (15) with respect to pressure we obtain

(18) $$\frac{\partial \log K}{\partial p} = - \frac{1}{RT} \sum_i \left\{ \nu_i \frac{\partial g_i(p, T)}{\partial p} \right\}.$$

From the Table in Sec. 7 we find that

(18 a) $$\frac{\partial g_i}{\partial p} = v_i(p, T) \quad \text{and} \quad \left(\frac{\partial g_i}{\partial T} \right) = - s_i(p, T).$$

Here v_i denotes the molar volume of the i-th component under the pressure p and is equal to RT/p. Hence eq. (18) may be transformed into

$$\frac{\partial \log K}{\partial p} = - \frac{\sum_i \nu_i}{p} = \frac{\partial}{\partial p} \log p^{-\sum_i \nu_i}.$$

Furthermore, integrating with respect to p, we have

(18 b) $$K = C \times p^{-\sum_i \nu_i},$$

where C is independent of p. However, as seen from eq. (17), C is identical with K_p which proves our proposition.

We can obtain the relation between K and T in a similar manner. Differentiating eq. (15) with respect to T, we obtain

(19) $$\frac{\partial \log K}{\partial T} = \frac{1}{RT^2} \sum_i \nu_i g_i - \frac{1}{RT} \sum_i \nu_i \frac{\partial g_i}{\partial T}.$$

In view of eq. (18 a) we can write

(19 a) $$\frac{1}{RT^2} \sum_i \nu_i (g_i + T s_i) = \frac{1}{RT^2} \sum_i \nu_i h_i$$

for the right-hand side of eq. (19).

Having found the derivatives of $\log K$ with respect to p and T we can calculate the value of K itself and we find that it is determined except for a constant factor which depends only on the nature of the system under consideration. We shall obtain the latter in Sec. 14 C.

A discussion of the numerous applications of the law of mass action to problems of chemistry and engineering is beyond the scope of these lectures. It is only possible to sketch some general consequences of the equations just deduced. We shall repeat the most important ones here in a somewhat more complete form for the sake of future reference:

(15 a) $$\Pi c_i^{\nu_i} = K(p, T)$$

(18) with (18 a) $$\frac{\partial \log K}{\partial p} = -\frac{\sum_i \nu_i v_i}{RT} = -\frac{\Delta v}{RT},$$

(19 b) with (19 a) $$\frac{\partial \log K}{\partial T} = \frac{\sum_i \nu_i h_i}{RT^2} = \frac{\Delta h}{RT^2}.$$

Here the symbol Δv denotes the change in the molar volume of the whole system and Δh denotes the change in the total molar enthalpy during a process which involves the completion of the chemical reaction from left to right on the assumption that both the pressure and the temperature remain constant. It is seen from the first equation that an increase in K causes an increase in that c_i which is associated with a positive value of ν_i, i. e. with the substances which appear on the right-hand side of the chemical equation in accordance with the previously adopted convention (cf. eq. (9 a)).

The second of the above equations shows that an increase in pressure at constant temperature shifts the equilibrium in favor of that side of the chemical equation which corresponds to the *smaller volume*. Since it has been assumed that only perfect gases take part in the reaction the change in volume, Δv, follows from Gay-Lussac's law of integral volume ratios discussed in Sec. 3 C and can be evaluated directly from the chemical equation. Thus in the case of the dissociation of steam considered previously in eq. (10) steam has the smaller volume, 2, compared with that of the sum of the volumes of the products of dissociation 2 H_2 and O_2 which is equal $2 + 1 = 3$. Hence an increase in pressure causes the concentration of steam to increase.

In the case of the explosive mixture of chlorine and hydrogen

$$H_2 + Cl_2 \rightleftarrows 2\,HCl$$

$\Delta v = 0$ and equilibrium depends on temperature alone.

The last of the foregoing equations demonstrates that an increase in temperature at constant pressure shifts the position of equilibrium in the direction of that side in the equation which is associated with the higher enthalpy. At lower temperatures the gases which correspond to higher enthalpy are, practically, non-existent. Thus in the case of steam the degree of dissociation at the boiling point at 100 C is negligible.

In this connection it should be borne in mind that the science of thermodynamic equilibrium is only concerned with final stable states and makes no statements about the speed of reaction with which a certain state of equilibrium is reached. These speeds may be so low that "metastable" states become possible, but the latter cannot be dealt with on the ground of classical thermodynamics. This explains why a mixture composed of two parts hydrogen and one part oxygen by volume can exist for any length of time even though it is not in a thermodynamically stable state of equilibrium. Such metastable states can be included under the heading of "constrained equilibria" as defined in Sec. 6 C.

The *synthesis of ammonia* from hydrogen and atmospheric nitrogen is of great practical importance in industry. It occurs in accordance with the equation

$$N_2 + 3\,H_2 \rightleftarrows 2\,NH_3,$$

which shows that the molar volume decreases from 4 to 2. In accordance with the second of the foregoing equations the concentration of ammonia increases with pressure. The extraordinary success with which this synthesis is now carried out in industry is due to the complete understanding of the conditions for thermodynamic equilibrium (Haber), to the mastery of the engineering problems connected with high pressure (Bosch) and, finally, to the successful selection of catalyzers which promote high reaction rates (Mittasch).

14. Chemical potentials and chemical constants

In the preceding Sections we encountered the parameters n_i, i. e. the numbers of mols of the individual components, in addition to the properties p, T or v, T. They are properties, being typical *extensive quantities*.

In eq. (8.3) we have already admitted the possibility of having any number of properties (always, naturally, a finite number of them). Starting with this equation we can combine the First Law with the Second and we may write

$$(1) \qquad T\,dS = dU + p\,dV + \sum_i X_i\,dx_i.$$

We shall now identify the quantities x_i with our n_i' s. Following Gibbs, the *intensive quantities* which are canonically conjugate with them will be denoted by $-\mu_i$, so that eq. (1) now becomes

$$(2) \qquad T\,dS = dU + p\,dV - \sum_i \mu_i\,dn_i,$$

$$(2\,a) \qquad dU = T\,dS - p\,dV + \sum_i \mu_i\,dn_i.$$

Instead of (2) we can also write

$$(2\,b) \qquad dH = dU + d(pV) = T\,dS + V\,dp + \sum_i{}' \mu_i\,dn_i,$$

$$(2\,c) \qquad dF = dU - d(TS) = -S\,dT - p\,dV + \sum_i \mu_i\,dn_i,$$

$$(2\,d) \qquad dG = dH - d(TS) = -S\,dT + V\,dp + \sum_i \mu_i\,dn_i.$$

A. The chemical potentials μ_i

First we notice that no changes have to be introduced to our Table in Sec. 7, which was originally limited to two independent variables, provided that the new additional variables n_i are kept constant. If, however, their variation is permitted it is necessary to add the following to the preceding differential relations, depending on whether we use (2 a, b, c, d)[1]:

---

[1] The subscript n_j denotes that all n_j' s with respect to which we do not differentiate are kept constant.

(3 a) $$\mu_i = \left(\frac{\partial U}{\partial n_i}\right)_{S, V, n_j},$$

(3 b) $$\mu_i = \left(\frac{\partial H}{\partial n_i}\right)_{S, p, n_j},$$

(3 c) $$\mu_i = \left(\frac{\partial F}{\partial n_i}\right)_{T, v, n_j},$$

(3 d) $$\mu_i = \left(\frac{\partial G}{\partial n_i}\right)_{T, p, n_j}.$$

It is evident that it is possible to deduce in the same way relations which correspond to Maxwell's equations in Sec. 7. For example:

(4) $$\frac{\partial V}{\partial n_i} = \frac{\partial \mu_i}{\partial p}$$

(it is implied that on the right-hand and on the left-hand sides $S, p, n_j \neq n_i$ and S, n_i, respectively, are kept constant). Of the eqs. (3 a, b, c, d) the last one is the most important relation. In order to examine it more closely let us imagine that our system has been increased by a given factor, say γ. All extensive quantities, i. e. the numbers of mols n_i, the volume V, and the entropy S, will become multiplied by γ, whereas all intensive quantities, pressure, temperature, and the potentials μ_i, will remain unchanged. We conclude from (3 d) that G becomes multiplied by γ in the same way as the n_i' s. Thus we see that G must be a *homogeneous function of the first degree in the n_i' s*:

(5) $$G(p, T, \gamma n_i) = \gamma G(p, T, n_i).$$

We can make the same assertion with regard to the quantities U, H, F, when we consider them to be functions of T, p, n_i; but, for example, the free energy, regarded as a potential and expressed with the aid of the variables T, V, n_i which correspond to it, must satisfy the relation (*cf.* Sec. 7) $F(T, \gamma V, \gamma n_i) = \gamma F(T, V, n_i)$.

Applying Euler's rule for homogeneous functions (differentiation with respect to γ, with $\gamma = 1$) we obtain

(6) $$G = \sum_i n_i \left(\frac{\partial G}{\partial n_i}\right)_{T, p, n_j}.$$

Combining this with (3 d), we have

(7) $$G = \sum_i \mu_i n_i.$$

14. 11b GASEOUS MIXTURES, GIBBS' PARADOX

The μ_i' s depend on p, T, and the n_i' s. The dependence on the latter must be such as to yield a homogeneous function of order zero, i. e. pure functions of the ratios of numbers of mols, or of the so-called "molar concentrations." It follows, further, from eq. (3 d) that the condition of equilibrium $\delta G = 0$ can be written

$$(8) \qquad \delta G = \sum_i \left(\frac{\partial G}{\partial n_i}\right)_{T, p, n_j} \delta n_i = \sum_i \mu_i \, \delta n_i = 0, \quad \delta p = 0, \quad \delta T = 0.$$

This must be supplemented with the auxiliary conditions which express the fact that the number of chemical atoms is preserved. Applying this to a single equation for the chemical reaction in the same way as in the case of the law of mass action in eq. (13.10) we have:

$$(9) \qquad \delta n_1 : \delta n_2 : \delta n_3 : \ldots = \nu_1 : \nu_2 : \nu_3 : \ldots .$$

Combining this with eq. (8), we obtain

$$(10) \qquad \sum_i \mu_i \nu_i = 0.$$

In this equation we have not stipulated, as was the case with the law of mass action, that the reactants are perfect gases. Thus we may regard eq. (10) as a *universally valid* formulation of the *law of mass action*.

The real difficulty consists now in the determination of the μ_i' s; it constitutes the main problem in the science of physical chemistry and can be solved only on the basis of measurements. Fortunately the extent to which this is required is reduced by the existence of certain identities, and we shall now proceed to deduce them. From (7) we have quite generally:

$$(11) \qquad dG = \sum_i n_i \, d\mu_i + \sum_i \mu_i \, dn_i.$$

Since G is a property we may write:

$$(11\,a) \qquad dG = \left(\frac{\partial G}{\partial p}\right)_{n_j, T} dp + \left(\frac{\partial G}{\partial T}\right)_{n_j, p} dT + \sum_i \left(\frac{\partial G}{\partial n_i}\right)_{T, p, n_j} dn_i.$$

According to (2 d) the factors of dp and dT are equal to V and $-S$ respectively, and it is seen from eq. (3 d) that the factors of dn_i are exactly equal to our μ_i' s. On comparing (11) with (11 a) we conclude that

$$(11\,b) \qquad \sum_i n_i \, d\mu_i = V \, dp - S \, dT.$$

We shall now apply this general equation to the case when, of the independent variables p, T, n_i in eq. (11 a), p, T are kept constant and only the n_i' s are varied. Taking into account that the μ_i' s are properties, we obtain

(11 c) $$\sum_i \sum_k n_i \left(\frac{\partial \mu_i}{\partial n_k}\right)_{T, p, n_j} dn_k = 0.$$

Now in this equation the n_i' s are independent variables (the auxiliary conditions which had to be taken into consideration, for example in connection with the law of mass action, are unimportant, because (11 b) is true not only at chemical equilibrium, being also valid in the case when the numbers of atoms involved vary), and eq. (11 c) must be satisfied for each dn_k. Hence

(12) $$\sum_i n_i \left(\frac{\partial \mu_i}{\partial n_k}\right)_{T, p, n_j} = 0 \quad (k = 1, 2, \ldots, K).$$

Since the μ_i' s depend only on the ratios of the n_k, i. e. on the concentrations c_k, the system of equations (12) can be written as

(12 a) $$\sum_i c_i \left(\frac{\partial \mu_i}{\partial c_k}\right)_{T, p, c_j} = 0 \quad (k = 1, 2, \ldots, K-1),$$

assuming K substances, or $K-1$ concentrations. The μ_i' s are to be regarded as functions of the $K-1$ variables $c_1, c_2, \ldots, c_{K-1}$. Equations (12 a) occur already in Gibbs' writings; they are, however, usually described as the *Duhem-Margule conditions*.

B. Relation between the μ_i' s and the g_i' s for ideal mixtures

We have expressed in Sec. 13 the free enthalpy for the special case when perfect gases are involved in terms of all the variables p, T, n_i in giving eq. (13.13). This equation has the form of our present eq. (7) so that on comparing the latter with (13.13) we can at once write down an equation for the chemical potentials:

(13) $$\mu_i = g_i(T, p) - RT \log \frac{n}{n_i}; \quad n = \sum_i n_i.$$

It has already been stressed in Sec. 13 that the symbol g_i denotes the molar free enthalpy of the pure component i. As we can see now, the chemical potential differs from it. What is the origin of the term $RT \log n/n_i$? The answer is: The increase in entropy on mixing (c/. (13.8)). Mixtures which obey eq. (13) will be called *ideal mixtures* even when no perfect gases are involved. It is easy to verify that the chemical potentials given in (13) satisfy the Duhem-Margule conditions (see Problems to II). In the case of non-ideal mixtures the interaction of components may give rise to heat tones, changes in volume, etc.

C. The chemical constant of a perfect gas

We now revert to the question of how far we can predict the equilibrium constant $K(p, T)$ in the law of mass action, already posed at the end of Sec. 13. According to (13.15) it is necessary to know completely the quantities g_i which we shall now write in the form

$$(14) \qquad g_i = h_i - T s_i.$$

We know that for a perfect gas and for temperatures which are not too low we have

$$(15) \qquad h_i = c_{pi} T + h_{io};$$
$$(16) \qquad s_i = c_{pi} \cdot \log T - R \log p + s_{io}.$$

Here h_{io} and s_{io} denote integration constants whose values cannot be determined with the aid of thermodynamics alone. The Third Law does not help here directly because the laws under consideration must not be extrapolated to $T \to 0$. We shall stress here that quantum mechanics confirms eq. (16) for sufficiently large temperatures, T, and thus leads to a definite value of s_0. The exact value of c_p is also given by quantum mechanics so that the quantity

$$(17) \qquad i_j = \frac{s_{jo} - c_{pj}}{R}$$

can be calculated for each component j. The quantities i_j are called *chemical constants* and their interpretation will become clear on substituting our present eqs. (14), (15), (16) into eq. (13.15). We thus obtain

$$\log K = \sum_j v_j \left[\frac{c_{pj}}{R} \log T - \log p - \frac{h_{jo}}{RT} + \frac{s_{jo} - c_{pj}}{R} \right],$$

where the last term in the square bracket is identical with i_j. Hence the law of mass action can be written

$$(18) \qquad \prod_i c_i^{v_i} = p^{-\sum_j v_j} \, T^{\sum_j v_j c_{pj}/R} \times e^{\sum_j v_j i_j - r_0/RT}$$

The only quantity which must be determined experimentally, and which we may formally denote as "the heat of reaction at absolute zero" is

$$r_0 = \sum_j \nu_j h_{jo}. \tag{19}$$

This is the constant which was mentioned in connection with eq. (13.19 a).

15. Dilute solutions

A solution is called dilute when the quantity of the solvent (e. g. water) is much larger than that of the solute (e. g. sugar). Dilute solutions differ from concentrated solutions by the simplicity of their behavior in a way similar to the simplicity of the behavior of perfect gases, as compared with real gases, with the exception of strong electrolytes.

A. General and historical remarks

When the solute is introduced into the solvent it will diffuse *uniformly* over the solvent irrespective of the initial state, in the same way as a gas will diffuse over the volume at its disposal. This behavior, in the same way as with a gas, is ascribed to the action of a pressure acting on the solute. It is called *osmotic pressure* and it is denoted by P. Its presence can be shown by the application of a semipermeable membrane which is permeable to the solvent but not to the solute, *cf.* Sec. 13 B. This membrane experiences only the osmotic pressure P being insensitive to the pressure in the solvent.

If the membrane is placed between the solution and the solvent and if it is free to move then work will have to be performed in order to move it towards the side on which there is the solution. The solvent will cross the membrane and the solution will become more concentrated. Conversely, work can be obtained by moving the membrane in the direction of the solvent thus allowing it to penetrate into the solution so that the concentration decreases. There is, consequently, a bias towards dilution which manifests itself by the availability of this positive work. We can say that the membrane exerts a suction on the solvent which is opposed to the solute's tendency to spread out and which is proportional to the latter's osmotic pressure.

This corresponds to the arrangement which was first used by Pfeffer (investigations into osmosis, Leipzig 1877) to measure osmotic pressures. A long tube is inserted into a beaker filled with water, the tube being closed at the bottom with the aid of a semipermeable membrane (copper ferrocyanide, *cf.* Sec. 13 A). Since the membrane is permeable to water the levels in the tube and beaker will at first be equal. If now sugar is added to the water in the tube, the water will begin to rise in it in proportion to the quantity of sugar added. Equilibrium is reached when the hydrostatic pressure at the lower end of the tube becomes equal to the osmotic pressure P of the solution. When the solutions are concentrated the differences in level may be equal to several meters of water and Pfeffer was forced to use a closed mercury manometer in his later experiments.

Osmotic pressure and semipermeable membranes play a most important part in nature's economy. The riddle of how the juices can penetrate to the tops of tall trees can only be solved by the recognition of the existence of osmotic pressure. The walls of organic cells both in animals and plants are all semipermeable. The protoplasm in the walls of cells must have the same osmotic pressure as the external fluid in which the cell is immersed: both must be "isotonic" (iso-osmotic). If the external osmotic pressure is larger, the cell will contract, and it will burst if the reverse is true. In the field of medicine, for example in illnesses involving blood corpuscules, both possibilities play a remarkable role.

B. Van 't Hoff's equation of state for dilute solutions

If we now wish to obtain quantitative results in addition to the preceding general statements we must consider the special case of reversible processes in dilute solutions. Van 't Hoff followed this line of thought in 1885 and discovered a certain similarity between dilute solutions and perfect gases.

In order to prove the analogy it is customary to consider *cycles* involving a moving piston which is assumed to be semipermeable during one stroke and impervious during the next and to compare the amounts of work performed by or on the system. We shall, however, base the argument on the consideration of our general *equilibrium conditions*, because in this way we shall achieve our goals much faster.

The system will be assumed to consist of two parts — the pure solvent and the solution, both interacting across a semipermeable wall. All quantities relating to the substance of the "solvent" will be denoted by the subscript 1, and those relating to the "solute" will obtain the subscript 2. The sub-system "solution" will be denoted by the superscript 1 and the sub-system "pure solvent" on the other side will be denoted by the superscript 2. Thus we shall distinguish between the mol numbers n_1^1, n_1^2 and n_2^1. Since the sub-system 2 contains no solute, we have

(1) $$n_2^2 = 0.$$

We stipulate that the semipermeable wall shall not prevent an exchange of heat so that the temperature is assumed uniform throughout the system. We must, however, admit the possibility of the pressures p^1 and p^2 on the two sides of the wall being different.

When deducing the equilibrium condition in Sec. 8 we could have considered the case $T = \text{const}$, $p^1 = \text{const}$ and not equal to $p^2 = \text{const}$. It is, however, easy to see that we would have obtained nothing new and that the old condition

(2) $$\delta G = 0$$

would turn out to apply to the system $p^1 \neq p^2$.

The auxiliary conditions are

(3) $$\delta n_1^1 + \delta n_1^2 = 0,$$
(4) $$\delta n_2^1 = \delta n_2^2 = 0.$$

Equation (3) expresses the conservation of mass for the solvent, and eq. (4) is a mathematical expression of the properties of the semipermeable wall. Introducing the chemical potentials from Sec. 14 A into (2) and taking into account (4), we obtain

(5) $$\mu_1^1 \delta n_1^1 + \mu_1^2 \delta n_1^2 = 0.$$

In view of (3) we have further

(6) $$\mu_1^2(p^2, T) = \mu_1^1\left(p^1, T, \frac{n_2^1}{n_1^1}\right).$$

The respective variables have been shown in the brackets in order to make the relations clearer. We have already established in Sec. 14 that the μ's depend only on the ratios of the numbers of mols — the "concentrations." The argument n_2^2/n_1^2 could be omitted from μ_1^2 because it refers to the homogeneous sub-system consisting of pure solvent in which, according to (1), the concentration n_2^2/n_1^2 is equal to zero. We can now see quite clearly that a non-zero value of n_2^1 must necessarily imply a difference $p^2 - p^1 \neq 0$ because otherwise the condition of equilibrium (6) could not be satisfied. The reason lies in the relation $\delta n_1^1 + \delta n_1^1 = 0$ which is so characteristic in the consideration of equilibrium and which expresses a conservation law. The difference

(7) $$P = p^1 - p^2$$

is called *osmotic pressure*.

Equation (6) is the exact equation of state for any solution; in order to be in a position to apply it, it would be necessary to know the chemical potentials. Since, generally speaking, this is not the case, we are forced to use experimental results and semi-empirical formulae, just as was the case with real gases. In analogy with gases, where the perfect gas constitutes a limiting case, we can consider a limiting case in this connection too; it is, namely, possible to treat the case of highly dilute solutions with purely theoretical means. Hence we now imply that

$$n_1^1 \gg n_2^1.$$

When the substance 2 penetrates into the solvent 1 the only change *with respect to substance* 1 is an increase in molecular disorder, i. e. the generation of entropy due to mixing. The assumption implied here is borne out by experiment to a high degree of accuracy; it can be further reinforced by considerations of a thermodynamic or of a statistical nature.

Denoting the molar free enthalpy of the solvent by g_1 we have to assume that

(8) $$\begin{cases} \mu_1^1 = g_1(p^1, T) - RT \log \frac{n_1^1 + n_2^1}{n_1^1} \\ \mu_1^2 = g_1(p^2, T) \end{cases}$$

in accordance with Sec. 14 B. In view of the assumption that $n_2^1 \ll n_1^1$ we have $\log(1 + n_2^1/n_1^1) \approx n_2^1/n_1^1$. Hence it follows from (6) and (8) that

(9) $$g_1(p^2, T) = g_1(p^1, T) - RT \frac{n_2^1}{n_1^1}.$$

We now expand the left-hand side into a Taylor series and retain the first term only:

(10) $$g_1(p^2, T) = g_1(p^1, T) + (p^2 - p^1) \left[\frac{\partial g_1(p^1, T)}{\partial p} \right]_T + \ldots$$

According to our fundamental Table in Sec. 7 we have

(11) $$\frac{\partial g_1}{\partial p} = v_1,$$

where v_1 is the molar volume of the pure solvent. Substituting (10), (11) and (7) into (9), we obtain

(12) $$P v_1 = RT \frac{n_2^1}{n_1^1}.$$

Since the partial volume of the solute is small compared with the volume V of the solution we may put

$$V = n_1^1 v_1,$$

with the same degree of approximation as that in all preceding formulae. From now on we shall use the symbol n to denote the number of mols n_2^1 of the solute. Hence

(13) $$PV = nRT.$$

This is van 't Hoff's proposition: *The osmotic pressure of a dilute solution of n mols of solute is equal to the pressure of a perfect gas which would be measured if n mols of the gas exerted pressure on the walls of a vessel of total volume V equal to that of solvent and solute.*

When more solutes exist in the solution, their quantities being $n_1, n_2, \ldots$, we would find in the same way that

(14) $$PV = (n_1 + n_2 + \ldots)RT,$$

which is the equation of state of a mixture of perfect gases. It is, therefore, possible to define the partial pressure P_i of a single solvent

(14 a) $$P_i V = n_i RT$$

and the total osmotic pressure

$$P = \sum_i P_i,$$

in the same way as for perfect gases. The unexpected form of eq. (13) was difficult to grasp when first discovered. Any doubts have now been removed by the results of an enormous volume of experimental material and by considerations based on kinetic theory (due to H. A. Lorentz among others).

16. The different phases of water. Remarks on the theory of the steam engine

In the present Section we shall study the equilibrium between the different phases of water. On this occasion we wish to make some remarks on the origins of thermodynamics and, in particular, on its connection with the development of the steam engine; the reader is also reminded of the paragraph on Carnot in the introduction to Sec. 6.

A. The Vapor-Pressure Curve and Clapeyron's Equation

We begin by considering the following well-known facts: Imagine a cylindrical vessel filled with water and a piston which fits snugly over its upper surface. There is no air between the water and the piston. We now pull the piston out of the cylinder, the temperature being kept constant all the time, and notice that some water evaporates. The quantity of steam formed between the water level and the piston is just sufficient to maintain a constant pressure which is independent of that volume and is a function of temperature alone. The steam is said to be "saturated." On reversing the piston the steam is not compressed but water is formed, its quantity being again just sufficient to insure that the steam remains saturated. It is thus seen that there exists an equation:

(1) $$\phi(p, T) = 0$$

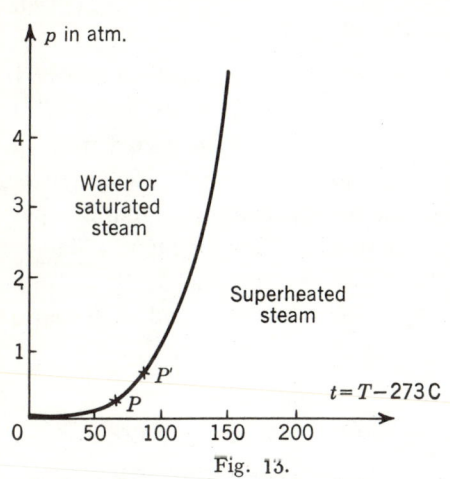

Fig. 13.
Phase equilibrium between water and steam.

which is independent of volume, and which connects the prescribed temperature T with the pressure p of saturated steam. The plot of eq. (1) in p, T coordinates yields the vapor-pressure curve, as shown by the curve in Fig. 13. Some pairs of values are given as follows:

$t =$	0 C	50 C	100 C	120 C	200 C	
$p =$	6×10^{-3}	0.126	1.03	2.02	15.9	kp/cm².

This independence of volume is consistent with the van der Waals *model of the process of liquefaction*. Reverting to Fig. 10 we can imagine that the temperature is the third coordinate and that it is plotted at right angles to the p,V plane, and we can fix our attention on the resulting "state surface." Looking at the surface from the direction of the V-axis, i. e. projecting it onto the p,T plane we shall notice that each Maxwell line AB will yield one point in the p,T plane. The fact that along every such straight line the volume varies according to the ratio of the mass of liquid to that of vapor, as given by eq. (10.12), cannot be deduced from Fig. 13. In particular, the two curves denoted as "liquid line" and "vapor line" in the p,V plane give one single projection in the p,T plane, namely our vapor-pressure curve (1). This curve ends abruptly at the point which corresponds to the critical point on the state surface. It is seen that the van der Waals equation takes into account the fact expressed in eq. (1) (not only in the case of steam, but for all condensing gases).

We shall now endeavor to find an analytic expression for eq. (1). Since the variables p, T are being considered, use will be made of the free enthalpy G and of the equilibrium condition $\delta G = 0$, in accordance with Sec. 8. We put

(2) $$G = n\, g_2 + (N-n)\, g_1,$$

where the subscript 2 refers to the "higher" phase (steam) and the subscript 1 to the "lower" phase (water). A phase will be called calorically higher than another if the transition from the latter to the former is connected with an *addition* of heat. Later we shall use the subscript 0 to denote the solid phase. The symbols g_1 and g_2 denote the free enthalpies per mol of liquid and vapor respectively, n denotes the number of mols of steam, N that for vapor + liquid; g_1 and g_2 are pure functions of p and T.

The condition $\delta G = 0$ applied to (2) at constant p, T and N leads to $(g_2 - g_1)\, \delta n = 0$, so that

(3) $$g_2 = g_1.$$

This is the analytical expression for eq. (1). We shall now proceed to evaluate it, but before doing so we shall deduce from it a useful differential relationship.

Consider two neighboring points P and P', Fig. 13, whose coordinates are p, T and $p + dp$, $T + dT$. With reference to the Table in Sec. 7 we calculate

(4) $$g(p+dp, T+dT) = g(p,T) + dp\left(\frac{\partial g}{\partial p}\right)_T + dT\left(\frac{\partial g}{\partial T}\right)_p =$$
$$= g(p,T) + dp \times v - dT \times s.$$

We now form the difference for phases 1 and 2, denoting it by Δ, for example

$$\Delta v = v_2 - v_1; \quad \Delta s = s_2 - s_1; \quad \Delta g = g_2 - g_1,$$

and conclude from (4) that

(4 a) $$\Delta g(p+dp, T+dT) = \Delta g(p,T) + dp \times \Delta v + dT \times \Delta s.$$

The left-hand side must vanish as well as the first term on the right-hand side in view of the fact that eq. (3) applies to both points P and P'. Hence we have

(5) $$\frac{dp}{dT} = \frac{\Delta s}{\Delta v}.$$

It is convenient to express Δs in terms of enthalpy h. Remembering the relation $g = h - T s$, cf. Sec. 7, and noting that $\Delta g = 0$ and $\Delta T = 0$, we obtain

(5 a) $$\Delta h = T \Delta s.$$

$\Delta h = h_2 - h_1$ represents the quantity of heat per mol required for the phase transition $1 \rightarrow 2$ at constant pressure. It is called the "latent heat of evaporation" and it will be denoted by r. Substituting (5 a) into (5), we obtain

(6) $$\frac{dp}{dT} = \frac{r}{T \Delta v}.$$

This is the famous equation due to Clapeyron; it was proved thermodynamically for the first time by Clausius. If Δv denotes the difference in the specific volume as is usual, and not that in the molar volumes, the meaning of the symbol r must be adjusted accordingly (specific and not molar latent heat of evaporation).

On comparing (6) with the preceding formula (7.7) we notice that the total derivative dp/dT and the ratio $\Delta s/\Delta v$ replace the partial derivatives $(\partial p/\partial T)_v$ and $(\partial s/\partial v)_T$ in the expression for a single-phase system.

Clapeyron's equation provides the means for the point by point calculation of the vapor-pressure curve from its tangents if r and v are known from measurement. Instead of performing such a step-by-step process of integration we revert to eq. (3) which must, evidently, contain the result of such an integration. We shall assume that the pressure is so low that the vapor can be treated like a perfect gas. According to Sec. 14, eqs. (14) to (17), we obtain

(7) $$g_2 = R T \left\{ \log p - \frac{c_{p2}}{R} \log T + \frac{h_{20}}{R T} - i_2 \right\}.$$

In order to establish the corresponding expression for g_1, we neglect the changes in volume, as is usual for liquids and solids. It is now superfluous to differentiate between c_p and c_v and it suffices to consider one specific heat, c_{liq}. Hence we obtain the following expressions for entropy and enthalpy:

(8) $$h_1 = h_{1m} + \int_{T_m}^{T} c_{liq} \, dT,$$

(9) $$s_1 = s_{1m} + \int_{T_m}^{T} \frac{c_{liq}}{T} \, dT.$$

The symbols h_{1m} and s_{1m} denote the values of enthalpy and entropy at a provisionally arbitrary temperature T_m which may be chosen as that at the melting point. The constants h_{1m} and s_{1m} can thus be determined with the aid of the Third Law and the caloric properties of the solid phase.

It follows that the free enthalpy of the liquid is given by

$$(10) \quad g_1 = \int_{T_m}^{T} c_{liq}\, dT - T \int_{T_m}^{T} \frac{c_{liq}}{T}\, dT + h_{1m} - T s_{1m}$$

and our conditions of equilibrium (3) together with (7) and (10) yield:

$$(11) \quad \log p = \frac{c_{p2}}{R} \log T + \frac{1}{RT} \int_{T_m}^{T} c_{liq}\, dT - \frac{1}{R} \int_{T_m}^{T} \frac{c_{liq}}{T}\, dT - \frac{h_{20} - h_{1m}}{RT} + i_2 - \frac{s_{1m}}{R}.$$

In principle all quantities in the above equation, with the exception of the chemical constant i_2, must be obtained from measurements on the solid phase. If it is considered that the value of i_2 given by quantum mechanics is not reliable enough it is possible to check it with the aid of a single measurement of vapor pressure. The vapor-pressure curve for water obtained in the above way, and fully confirmed by experiment, is shown plotted in Fig. 13. It ascends monotonically as could have been already inferred from the Clapeyron equation. In fact r is always positive (this is in agreement with our previous definition of a "higher phase"); in addition, we must naturally have $\Delta v > 0$ (since $v_2 \gg v_1$).

The highest temperature which need be considered (see above) is the critical temperature. This is the physically natural end point of the curve (it is not shown in Fig. 13 because it lies outside its range). The curve also has a natural initial point which lies near the origin of our diagram in the case of water, cf. Sec. 17 A; it can be used to make a direct check on the value of i_2.

B. Phase Equilibrium Between Ice and Water

Ice is the lower phase with respect to water because the melting of, for example, one gram of ice requires the introduction of the *latent heat of fusion* (melting), r. Since we have agreed to denote the solid phase by the subscript 0, the symbol Δv will now denote the difference $v_{water} - v_{ice}$. Applying the proper interpretation of the symbol Δ to g, h, s we again obtain formally Clapeyron's equation (6) from eqs. (4) and (5), *with the important difference that Δv is now negative*:

$$\Delta v = v_1 - v_0 = (1.00 - 1.091) \text{ cm}^3/\text{g} = -0.091 \text{ cm}^3/\text{g}.$$

This fact is fundamental for the existence of life on earth. *Ice floats on water.* If this were not so, all fishes would die in winter and no life could develop at our latitudes. (As is well known, land animals have evolved from water animals.) *Water expands on freezing*. The erosion of mountains which allows fertile soil to reach the valleys is a consequence of this fact (bursting of rock when water freezes in fissures).

Clapeyron's equation shows that: The *melting curve* descends as T increases, unlike the *vapor-pressure curve*. In the neighborhood of 0 C the numerical value of the latent heat is

$$r = 80 \frac{\text{cal}}{\text{g}} = 80 \times 42.7 \frac{\text{at cm}^3}{\text{g}}.$$

The last value follows from (4.6), (3.2) and (3.2 a); hence at $T \sim 273$ K and $t = T - 273$ C = Celsius temperature, we have

(12) $$\frac{dp}{dt} = -\frac{80 \times 42.7}{273 \times 0.091} \frac{\text{at}}{\text{deg}} = -138 \frac{\text{at}}{\text{deg}}.$$

Accordingly the melting curve $\phi(p, t) = 0$ starts at $p \approx 0$ at $t \approx 0$ C and passes through the second quadrant of Fig. 13; it is a very steep, nearly straight, line, it being necessary to reach $p = 138$ at in order to depress the melting temperature to $t = -1$ C. This "depression of the melting point" plays an important part in the motion of glaciers, notwithstanding the fact that it is so small. The deeper parts of a glacier begin to move under the pressure of the masses above them but freeze again when the pressure decreases (regelation). The ease with which a skater moves on ice also depends on this fact; the ice which melts under the pressure of the skate acts as a lubricant.

C. The specific heat of saturated steam

So far we have only discussed the specific heats c_p and c_v. It is, however, possible to define a specific heat for any process, i. e. for any path in the p, v plane. It is immediately clear that on progressing along an isentrope ($dq = 0$) we have $c_s = 0$; on the other hand we can assert that: $c_T \approx \infty$ because on progressing along an isotherm there is no change in temperature no matter how large dq is. We can also define a specific heat along any path in the p, T plane.

We are particularly interested in the specific heat of steam, c_ϕ, on progressing along the vapor-pressure curve $\phi(p, T) = 0$. Applying the definitions of the latent heat of evaporation $r = \Delta h$ from eq. (5 a,) and taking into account that $h = u + p v$, we obtain

$$\text{(13)} \qquad \frac{dr}{dT} = \frac{d\Delta u}{dT} + p \frac{d\Delta v}{dT} + \frac{dp}{dT} \cdot \Delta v.$$

According to the First Law

$$\text{(13 a)} \qquad \frac{dq}{dT} = \frac{du}{dT} + p \frac{dv}{dT}$$

along any path, so that in the case of steam, along the vapor-pressure curve we must have in particular:

$$\text{(13 b)} \qquad c_\phi = \frac{dq_\phi}{dT} = \frac{du_2}{dT} + p \frac{dv_2}{dT}.$$

The corresponding equation for the liquid phase is

$$\text{(13 c)} \qquad c_{liq} = \frac{du_1}{dT} + p \frac{dv_1}{dT}$$

because the difference between c_p and c_v for the liquid phase can be neglected, as already remarked in connection with eq. (2.4). Hence we may put $c_p \approx c_v = c_{liq}$ and it follows from (13 b, c) that

$$c_\phi - c_{liq} = \frac{d\Delta u}{dT} + p \frac{d\Delta v}{dT}.$$

Substituting this into (13), we find:

$$\text{(13 d)} \qquad \frac{dr}{dT} = c_\phi - c_{liq} + \frac{dp}{dT} \cdot \Delta v.$$

Taking, finally, into account the Clapeyron equation (6), we have:

$$\text{(14)} \qquad c_\phi = \frac{dr}{dT} + c_{liq} - \frac{r}{T}.$$

According to the very precise measurements carried out by engineers[1] we have at $T \sim 373$ K:

$$\text{(14 a)} \qquad \frac{dr}{dT} = -0.64 \, \frac{\text{cal}}{\text{deg g}},$$

---

[1] In this connection *cf.* Problem II. 2.

so that with the value $r = 539$ cal/g for $T = 373$ K and $c_{liq} = 1$ cal/deg $\times$ g

(14 b) $$c_\phi = (1 - 0.64 - 1.44) \frac{\text{cal}}{\text{deg g}} = -1.07 \frac{\text{cal}}{\text{deg g}}.$$

Saturated steam requires no heat when its state is changed along the vapor-pressure curve; it is in a position to reject heat. On the other hand, if saturated steam is expanded without the addition of heat it will, as seen from Fig. 13, enter the region below the vapor-pressure curve, marked "water or supersaturated steam." It tends to condense in this region.

We shall now quote two examples, one trivial, the other fundamental for modern physics: In a bottle containing mineral water the atmosphere between the free surface of the liquid and the plug is one of saturated steam. When the bottle is opened suddenly the speed of the process insures its being adiabatic. The steam condenses into drops. This phenomenon finds a beautiful application in C. T. R. Wilson's (1912) *cloud chamber*. The chamber contains saturated steam and is expanded very suddenly. If just prior to the expansion ionizing material particles have been allowed to penetrate into it, the resulting ions will act as nuclei of condensation and thus the paths of the particles will be made visible. The importance of this method of research (cosmic rays, discovery of positrons, mesons, etc.) is very well known.

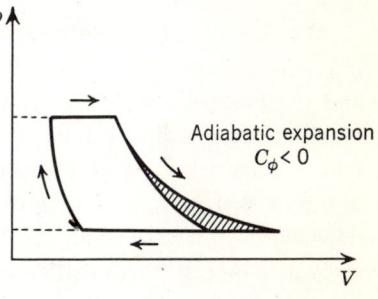

Fig. 14.

The indicator diagram, corrected for negative specific heat along the vapor-pressure line.

The fact that c_ϕ is negative is of some importance in steam engineering. When steam is expanded isentropically, the corresponding curve is less steep than the isentrope $p v^\gamma = $ const for a perfect gas. Owing to this fact the indicator diagram, Fig. 14, increases by the area shown shaded in the sketch. This has some advantages from the point of view of the design of a reciprocating steam engine.

17. General remarks on the theory of phase equilibria

The study of the different phases of water is only a fortuitous example which belongs to the much more general problem of the co-existence of the phases of substances of arbitrary chemical composition. Even in the case of water the preceding analysis was incomplete. In the solid state, in addition to the ordinary hexagonal ice whose structure is so beautifully exhibited by microscopic pictures of snow flakes, there exist, according to Tammann, numerous allotropic modifications; these show a preference for other regions in the p,T plane. Furthermore, the complete study of the phases of water would have to include its dissociation into oxy-hydrogen gas, *cf.* Sec. 13, when the system ceases to be *homogeneous*, (H_2O) and becomes *heterogeneous* (H_2, O_2).

A. The triple point of water

We now revert to the state diagram for water in the p,T plane. With reference to Fig. 13 we now draw a diagram in which the p-axis is horizontal and the t-axis is vertical (t denotes the Celsius temperature), both to a very much enlarged scale. The vapor-pressure curve has been drawn schematically and it is seen that it now is convex upwards. Further, the melting line is also seen sketched; according to Sec. 16 B it is nearly a straight line inclined downwards at a very small angle of 1/138 with respect to the p-axis. At this stage it is necessary to settle the question as to how to draw the boundary between the ice and steam regions in the p,t plane. It is known that ice can be transformed directly into steam and not only through the intermediate stage of being melted first. This process is called *sublimation*. (The term did not originate with water but with mercury.) The process can be observed in the spring during a light frost accompanied by brilliant sunshine when the snow seems to disappear quickly, without melting. In actual fact the resulting water vapor escapes into the atmosphere. This condition is again described by the equation $\phi(p, T) = 0$ which defines the *sublimation curve* in the p,t plane. We now propose to prove that it passes through the point of intersection of the melting curve with the vapor-pressure curve. In order to do this consider the analytic representation of the three curves

(1)
melting curve	$\phi_{01} = 0$	$g_0 = g_1$,
vapor-pressure curve	$\phi_{12} = 0$	$g_1 = g_2$,
sublimation curve	$\phi_{02} = 0$	$g_0 = g_2$.

It is seen that the equation $g_0 = g_2$ is satisfied when the two equations, $g_0 = g_1$ and $g_1 = g_2$, are satisfied simultaneously. The common point of intersection of all three curves is termed the *triple point*. It is defined by

(1 a) $$g_0 = g_1 = g_2.$$

The latent heats which correspond to the three transitions will be denoted by r_{01}, r_{12}, r_{02} so that r_{02} is the *latent heat of sublimation*. According to the First Law they must satisfy the relation

$$(2) \qquad r_{01} + r_{12} + r_{20} = 0.$$

(A path around the triple point along which the energy reverts to its initial value.) It follows from (2) that

$$(2\text{ a}) \qquad r_{02} = -r_{20} = r_{01} + r_{12}.$$

Making use of the preceding values $r_{01} = 80$ cal/g from Sec. 16 B and[1] $r_{12} = 603$ cal/g, we find that

$$(2\text{ b}) \qquad r_{02} = 683 \frac{\text{cal}}{\text{g}}.$$

Since the analytic expressions for g_0 and g_1 are not known (that for g_2 is known only approximately from the perfect gas equation), we are not in a position to solve eq. (1) and we must resort to experiment to find the *thermodynamic coordinates of the triple point*. The respective experimental values, Fig. 15, are

$$(3) \qquad t = 0.0074 \text{ C}, \qquad p = 0.0061 \text{ atm}.$$

The Clapeyron equation is again satisfied along the sublimation curve:

$$(4) \qquad \frac{dp}{dT} = \frac{r_{02}}{T \Delta v},$$

where Δv at the triple point can be found as follows:

$$(4\text{ a}) \qquad \Delta v = \Delta v_{20} = v_2 - v_0 = v_2 - v_1 + v_1 - v_0 = \Delta v_{21} + \Delta v_{10}.$$

The three equilibrium curves have been sketched in Fig. 15. They have been extended beyond the triple point with the aid of broken lines in order to stress the fact that they correspond to states of unstable equilibrium in the respective regions of the p,t plane.

Summing up we can state: The *three* phases of water under consideration can coexist only at a single *point* in the p,t plane; the coexistence of two of the three phases is possible only along one specific *curve;* each phase taken *singly* can exist only in a well-defined *area*.

---

[1] Extrapolated from the values for $t = 100$ C: $r = 539$ cal/g and $dr/dt = -0.64$ cal/g deg, eq. (16.14 a).

Figure 16 shows an isometric projection of a three-dimensional model for water. The temperature axis, T, is drawn upwards, the axis of volumes, v, is drawn horizontally inwards and the pressure axis, p, is drawn to the left and inwards. The hyperboloidal surface rising steeply corresponds to the gaseous phase, its equation being $T = pv/R$; the isobars and isochores which are straight lines are seen marked on the model. Its lower edge is the *vapor line*. The *solid phase* appears in the form of a bulge at the bottom on the

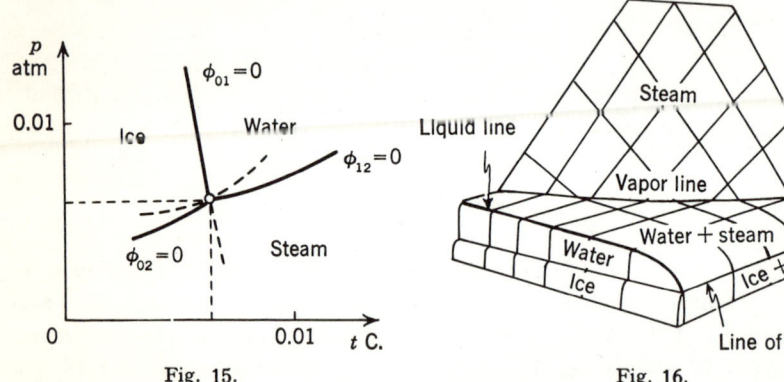

Fig. 15.

The neighborhood of the triple point in the p, t plane.

Fig. 16.

Model illustrating the three phases of water. In order to facilitate comparison with Fig. 15 it is necessary to imagine that the p-axis has been drawn to the left.

left. It borders on the *liquid phase* along the *melting curve*. The upper boundary of the latter, in turn, lies along the *liquid line*. The developable surface between the liquid line and the vapor line corresponds to the region of *water + steam*. The projection of this surface onto the p,T plane in the direction of v from left inwards lies along the vapor-pressure curve as already described in connection with eq. (16.1). The bulge at the right-hand bottom edge represents the coexistence of *ice + steam*; it is also developable. It borders on the region *water + steam* along the *triple – point edge*. The most forwards point on the *triple – point edge* is the *triple point* itself. The *sublimation curve* appears on projecting this part of the model onto the p,T plane.

B. Gibbs' phase rule

In the preceding Sections we have restricted ourselves to the consideration of a *single component*, H_2O, and we now propose to extend our description to *any number* of substances (molecular or atomic species). We shall number them as follows:

$$1, 2, \ldots, k, \ldots, K.$$

Furthermore we shall now admit the existence of *any number* of physical states and chemical groupings of the K components, instead of the *two* or *three* phases considered so far. We shall again denote them as phases and we shall number them consecutively

$$1, 2, \ldots, i, \ldots, J.$$

The symbol J denotes the largest number of phases which can coexist, and K denotes the total number of substances which react with one another.

The number of mols of the k-th component present in the i-th phase will be denoted by n_{ik}. The free enthalpy, G, of the whole system is the sum of the contributions from the individual phases and components. The condition $\delta G = 0$ leads to a system of simultaneous equations of the form

(5) $$\sum_{i=1}^{J} \sum_{k=1}^{K} g_{ik}\, \delta n_{ik} = 0.$$

The number of conditions included in (5) would equal $J \times K$ if the δn_{ik} were independent of each other. They are, however, linked by the condition that the total mass of each component (the sum of the numbers of mols over all the phases) must be preserved. Consequently, the following additional conditions must be satisfied:

(5 a) $$\sum_{i=1}^{J} \delta n_{ik} = 0 \quad \text{for} \quad k = 1, 2, \ldots, K.$$

The number of independent equations does not equal $J \times K$; it is only

(6) $$J \times K - K.$$

We now inquire into the number of variables which we have at our disposal and which must satisfy these equations. These are given by the numbers of mols n_{ik} whose number is, again, $J \times K$. Since, however, G is homogeneous in the n_{ik}'s, *cf.* Sec. 14 A, the conditions of equilibrium (5), (5 a) written for each phase will contain only ratios of the n_{ik}'s. The number of such ratios per phase is equal to $K - 1$ so that in the sum for all phases we shall have

$$J(K-1) = J \times K - J$$

ratios. In addition there are the two variables p, T which determine a point in the p, T plane. *The total number of independent variables at our disposal is thus*

(7) $$JK - J + 2.$$

When this number is smaller than the number of equations to be satisfied, eq. (6), the system cannot, generally speaking, be solved. It is, therefore, necessary to stipulate

$$JK - J + 2 \geqslant JK - K.$$

It follows that

(8) $$J \leqslant K + 2.$$

This is the famous *phase rule* discovered by Willard Gibbs. Instead of (8) we can also write

$$J + F = K + 2,$$

where F denotes the *number of degrees of freedom* possessed by the system of K components forming J coexisting phases.

When $F = 0$ the corresponding state of the system is called *invariant* which is the case, for example, for water, $K = 1$, at the triple point: we have here $J = 3$, ice, water, and steam coexisting.

When $F = 1$ the corresponding system is called *monovariant*. In the particular case when $K = 1$ we have $J = 2$, i. e. the coexistence of two phases along the vapor-pressure, melting, or sublimation curve.

When $F = 2$ the system is *bivariant*. For $K = 1$ we have also $J = 1$, no phases coexisting; instead, there is the possibility of the existence of a single phase in a two-dimensional region of the p,T plane.

If we now consider one of Tammann's allotropic modifications of ice in addition to the ordinary (hexagonal) phase we find that the phase rule allows their coexistence with only *one* more phase, *either* the liquid, *or* the vapor. There can be no "quadruple point" in cases when $K = 1$. Such a point can exist in at least a "binary" system composed of two substances: $K = 2$ gives $J = 4$ for $F = 0$.

The study of the phase rule in application to ternary or to multi-component systems leads to the consideration of multi-dimensional spaces describing phase equilibria. We must, however, leave this topic to physical chemistry.

C. Raoult's laws for dilute solutions

A particularly simple case of phase equilibrium occurs when dilute solutions are evaporated, on condition that the solute is not volatile. We shall refer here to our derivation of van 't Hoff's law in Sec. 15 B, excluding, from our considerations the case of electrolytes, as before. We wish to make the following changes in notation: Instead of denoting phases by subscripts as heretofore, we shall now denote the liquid and gaseous phase of the solvent by superscripts 1 and 2 respectively; these have been used in Sec. 15 to distinguish between the two subsystems "solution" and "pure solvent." The subscripts 1 and 2 are now required to denote the substances "solvent" and "solute," as was done in Sec. 14. The place of the "pure solvent" of Sec. 15 is now taken by the gaseous phase (number of mols n_1^2). Instead of the semipermeable wall we now have to consider the free surface of the fluid which is impermeable to the solute. The numbers of mols in the dilute solution are now, as before, denoted by n_1^1 and n_2^1.

The difference between the two cases consists in the fact that the pressures of the two phases are now equal. The former condition of equilibrium (15.5)

$$(9) \qquad \mu_1^1 \, \delta n_1^1 + \mu_1^2 \, \delta n_1^2 = 0$$

and the former auxiliary condition (15.3)

$$(10) \qquad \delta n_1^1 + \delta n_1^2 = 0$$

remain unaltered. Consequently, the equation for our present "vapor-pressure line" is determined by the preceding eq. (15.6)

$$(11) \qquad \mu_1^2(p, T) = \mu_1^1\left(p, T, \frac{n_2^1}{n_1^1}\right)$$

but with the proviso that $p^1 = p^2 = p$. In analogy with the previous eq. (15.8) we must assume now

$$(12) \qquad \mu_1^1 = g_1^1(p, T) - R\,T \log \frac{n_1^1 + n_2^1}{n_1^1}$$

$$\mu_1^2 = g_1^2(p, T),$$

where, unlike in Sec. 15, the symbol g_1^2 denotes the free enthalpy of the pure vapor, g_1^1 denotes that for the pure solvent. Thus the condition of equilibrium becomes:

$$(13) \qquad g_1^2(p, T) = g_1^1(p, T) - R\,T \log \frac{n_1^1 + n_2^1}{n_1^1}.$$

If the solution is dilute to a sufficient degree we have $n_2^1 \ll n_1^1$ and we may replace the logarithm by n_2^1/n_1^1 or by n_2/n_1 if we omit the superscript which has now become superfluous. Hence eq. (13) changes to

$$(14) \qquad g_1^2(p, T) = g_1^1(p, T) + R\,T \frac{n_2}{n_1}.$$

This is the equation for the *vapor-pressure curve of the solvent above the solution* in the p, T plane.

For the purpose of comparison we now write the equation for the *vapor-pressure curve for the pure solvent*. In order to make a distinction we shall denote its coordinates by p^*, T^*:

(14 a) $$g_1^2(p^*, T^*) = g_1^1(p^*, T^*).$$

Subtracting (14) from (14 a) we have

(15) $$g_1^2(p^*, T^*) - g_1^2(p, T) = g_1^1(p^*, T^*) - g_1^1(p, T) + RT \frac{n_2}{n_1}.$$

We now expand both sides into Taylor series and retain the first term of each only. Hence

(16) $$(p^* - p) \frac{\partial g_1^2}{\partial p} + (T - T^*) \left[-\frac{\partial g_1^2}{\partial T} \right] =$$
$$= (p^* - p) \frac{\partial g_1^1}{\partial p} + (T - T^*) \left[-\frac{\partial g_1^1}{\partial T} \right] + RT \frac{n_2}{n_1}.$$

The derivatives can be taken from the Table in Sec. 7. Thus eq. (16) becomes

(17) $$(p^* - p)\, [v_1^2(p, T) - v_1^1(p, T)] + (T - T^*)\, [s_1^2(p, T) - s_1^1(p, T)] = RT \frac{n_2}{n_1}.$$

This equation can be analyzed from two points of view. (a) First we shall inquire into the change in the vapor pressure at constant T on adding n_2 mols of substance 2 to the solvent. This means that we have to put $T = T^*$ in (17) and so we determine the *decrease in vapor pressure* $\Delta p = p^* - p$:

(18) $$\Delta p = \frac{RT}{v_1^2 - v_1^1} \cdot \frac{n_2}{n_1}.$$

Equation (18) applies to any dilute solution, that is, whenever $n_2 \ll n_1$. In particular if the vapor may be treated as a perfect gas

$$v_1^2 = \frac{RT}{p}$$

and if the molar volume v_1^1 of the liquid may be neglected with respect to that of the vapor, we can simplify (18) to read

(19) $$\frac{\Delta p}{p} = \frac{n_2}{n_1}.$$

This is the *law of vapor pressures*. It is independent of the solvent as well as of the nature of the solute and is given directly by the ratio of the number of mols of the two. This surprisingly simple law was discovered empirically by Raoult in 1886. It was proved thermodynamically shortly afterwards by van 't Hoff.

(b) Secondly we shall inquire into the change in the boiling point $T - T^* = \Delta T$ at constant pressure on adding substance 2 to the solvent 1. We now put $p = p^*$ in (17) and take into account that

$$(s_1^2 - s_1^1) T = r \qquad (r = \text{latent heat of evaporation}).$$

It follows from eq. (17) for $\Delta T = T - T^*$ that

(20) $$\Delta T = \frac{R T^2}{r} \cdot \frac{n_2}{n_1}.$$

Raoult's law of *decrease in vapor pressure* is seen to be associated with the law of *rise in boiling point*. Concerning the validity and the historical origin of this law the same remarks can be made as in connection with the law of decrease in vapor pressure.

To the *rise in boiling point* there corresponds a *freezing point depression* on transition from phase 1 to 0 (freezing). Equation (20) remains valid for this case as well except that $r = r_{12}$ must be replaced by the heat of reaction $r_{10} = - r_{01}$ which is equal to the negative of the heat of solidification.

In all these considerations the assumption was made that all molar masses are preserved during the phase transition which implies the exclusion of polymerization and dissociation. There is no difficulty in including such phenomena. It is only necessary to include the mass ratio m_2/m_1 associated with the respective process where necessary, e. g. on the right-hand side of eq. (11). Planck[1] stressed the importance of this circumstance for the determination of molecular weights.

D. Henry's law of absorption (1803)

The condition of equilibrium $\delta G = 0$ finds a simple application in the study of the solubility of a gas in a liquid whose vapor pressure is negligible compared with that in the gas chamber. In such cases only the chemical potential of the gas in the gaseous phase, μ^2, and in the solution, μ^1, are

---

[1] Thermodynamics, Sec. 269 and 270.

important (the subscript 1 denoting "gas" can be dropped here and subsequently). The auxiliary condition

$$\delta n^1 + \delta n^2 = 0$$

leads to

(21) $$\mu^1 = \mu^2$$

in the same way as in (11). We shall again assume that the gaseous phase may be regarded as a perfect gas, so that

(22) $$\mu^2 = g^2 = RT \log p + \psi(T).$$

The function $\psi(T)$ which includes the chemical constant of the gas need not be described in more detail. Regarding the solution we make the assumption

(23) $$\mu^1 = g^1(p, T) + RT \log c,$$

which is analogous to eq. (12). Here c denotes the molar concentration of the dissolved gas (represented as n_1^1/n with $n = n_1^1 + n_2^1$ in the above mentioned eq. (12)). With certain reservations the free enthalpy of a liquid of fixed chemical nature is practically independent of pressure, so that (23) may be written

$$\mu^1 = RT \log c + \chi(T).$$

It follows from (21), (22) and (23) that

$$\log c = \log p - \frac{\chi(T) - \psi(T)}{RT}$$

and also

(24) $$c = p \times f(T) \quad \text{with} \quad \log f(T) = \frac{\psi(T) - \chi(T)}{RT}.$$

This is the very well known law due to Henry: *The quantity of gas absorbed by a liquid is proportional to the partial pressure of the gas remaining above it.* The coefficient of proportionality depends only on the temperature (equal for the gas and the liquid); it is unaffected, for example, by the presence of any additional gases in the chamber.

18. The electromotive force of galvanic cells

We have until now considered only thermodynamic systems which were composed of electrically neutral particles (atoms, molecules). We now propose to investigate the changes to be introduced into our equations when charged particles (electrons, ions) are included. This field includes problems in the thermal and caloric equations of state for electrons in metals, electrolytes, ionized gases, etc. We shall continue to restrict ourselves to problems of equilibrium thermodynamics which implies the exclusion of such problems as the flow of an electric current through metals or electrolytes, etc.

From among the rich collection of remaining problems we shall now fix our attention on the question of the electromotive force (emf for short) of open galvanic cells because we can formulate several general statements without undue labor. We have purposely added the qualification "open" in the preceding sentence in order to emphasize the fact that we are interested in the static cases only. It is assumed that equilibrium has been attained and that it is not disturbed by any irreversible processes, such as the generation of Joule heat during the passage of a current. It is, therefore, necessary to imagine that the emf of the cell is measured with the aid of an electrostatic voltmeter.

When two phases which are capable of exchanging charged particles are permanently separated by a boundary there will appear between them a difference in potential in the same way as the existence of equilibrium between two solutions of different concentrations, separated by a semipermeable wall, implies that an osmotic pressure difference is permanently maintained between them.

A. Electrochemical potentials

We shall now consider a thermodynamic system Σ and its "surroundings" Σ' as was done in Sec. 8. The system Σ will be described by specifying the properties V, T, n_i. According to (14.1) we can write for Σ:

(1) $$T\,dS = dU + p\,dV - \sum_i \mu_i dn_i.$$

Let us fix our attention on an infinitesimal process

$$dV = 0, \quad dT = 0, \quad dn_i = 0 \quad \text{for} \quad i \neq j$$
$$dn_j \neq 0;$$

the product $\mu_j\, dn_j$ represents that quantity of work which must be performed on system Σ in order to change the number of mols n_i by an increment dn_j, as already mentioned.

We shall suppose that the component j is no longer electrically neutral as was the case before, but that it is charged positively, in other words, that it consists of molecules or atoms which have lost a certain number of electrons each, that number being denoted by z. Consequently, one mol of that j-th component carries a charge zF where F denotes Faraday's equivalent charge of 96494 coulombs and is equal to the product of the elementary charge × Avogadro's number per mol. The system Σ is assumed to have an electrical potential Φ with respect to Σ'. Without loss of generality it is possible to assume that Σ' is grounded so that its potential is zero. Since we are now performing only an imaginary experiment we need not be concerned with the manner in which the potential Φ is maintained. If we now introduce into the system dn_j mols of charged particles we shall have to perform the work

$$\Phi z_j F\, dn_j$$

in addition to the "chemical" work $\mu_j\, dn_j$. The total amount of work to be performed on Σ is

$$+ (\mu_j + z_j F \Phi)\, dn_j.$$

The transfer of the charge must occur infinitely slowly, and must not generate Joule heat if the process is to be reversible.

Accordingly, eq. (1) must be replaced by

(2) $$T\, dS = dU + p\, dV - \sum \eta_i\, dn_i$$

where

(3) $$\eta_i = \mu_i + z_i F\, \Phi.$$

The η_i's are known as the *electrochemical potentials* as distinct from the chemical potentials μ_i. For negatively charged particles z_i is to be taken to be negative.

Generally speaking, however, a galvanic cell will be described by more than one potential. It consists of a whole *chain* of phases and each phase has its own electrical potential and is in equilibrium only with its two neighboring phases. We can imagine, at least during our imaginary experiment, that the individual phases are separated from each other by semipermeable membranes which allow only certain ions to pass. These remarks will be made clear on the example of the Daniell cell.

B. The Daniell Cell, 1836

This is represented schematically in Fig. 17. The individual phases have been marked by Roman numerals. The wall M (made of clay) allows only the SO_4^{--} ions to pass. We imagine that a copper terminal V is connected to the zinc electrode IV in order to insure that the emf $\Phi_I - \Phi_V$ of the open cell is measured between identical metals (copper) thus excluding any contact emf's when measured electrostatically. The electrochemical potentials, η, and the numbers of mols, n, will be distinguished from each other respectively by writing the phase in the upper left superscript and the kind of particle in the lower right subscript, e. g. $^{II}\eta_{Cu^{++}}$. The symbol $\ominus$ will be used to denote electrons. The condition of equilibrium $\delta G = 0$, e. g. between the phases I and II and the auxiliary condition which expresses the preservation of molar masses are

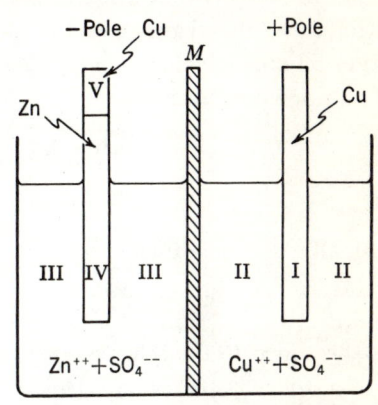

Fig. 17.
Schematic diagram of Daniell cell.

Zn	$ZnSO_4 + H_2O$	$CuSO_4 + H_2O$	Cu
solid	liquid	liquid	solid
		M	

(4) $\qquad ^I\eta_{Cu^{++}} \delta^I n_{Cu^{++}} + {}^{II}\eta_{Cu^{++}} \delta^{II} n_{Cu^{++}} = 0,$

(5) $\qquad \delta^I n_{Cu^{++}} + \delta^{II} n_{Cu^{++}} = 0.$

These equations are similar to those in Sec. 17, eqs. (9) and (10) because the virtual changes are performed at constant p and T and the chemical potentials must be replaced by the electro-chemical potentials. From (4) and (5) we obtain

(6) $\qquad ^I\eta_{Cu^{++}} = {}^{II}\eta_{Cu^{++}}.$

According to (3) with $z = +2$ we have:

(7) Phases I/II: $\qquad ^I\mu_{Cu^{++}} + 2F\Phi_I = {}^{II}\mu_{Cu^{++}} + 2F\Phi_{II}.$

The conditions of equilibrium between the remaining pairs of phases are analogous and can be written

(8) Phases II/III: $\qquad ^{II}\mu_{SO_4^{--}} - 2F\Phi_{II} = {}^{III}\mu_{SO_4^{--}} - 2F\Phi_{III},$

(9) Phases III/IV: $\qquad ^{III}\mu_{Zn^{++}} + 2F\Phi_{III} = {}^{IV}\mu_{Zn^{++}} + 2F\Phi_{IV},$

(10) Phases IV/V: $\qquad ^{IV}\mu_{\ominus} - F\Phi_{IV} = {}^V\mu_{\ominus} - F\Phi_V.$

In addition there are the equations describing the chemical reactions at the copper and the zinc electrode

$$Cu \rightleftarrows Cu^{++} + 2\ominus, \quad Zn \rightleftarrows Zn^{++} + 2\ominus.$$

This means the neutralization of the Cu^{++} ions on deposition on the copper electrode and the charging of the atoms of zinc with two positive elementary charges on crossing the boundary $Zn-ZnSO_4$. This occurs according to the two additional conditions:

$$(11) \qquad {}^{I}\mu_{Cu} - 2\,{}^{I}\mu_{\ominus} = {}^{I}\mu_{Cu^{++}},$$

$$(12) \qquad {}^{IV}\mu_{Zn} - 2\,{}^{IV}\mu_{\ominus} = {}^{IV}\mu_{Zn^{++}}.$$

Finally, the condition

$$(13) \qquad {}^{V}\mu_{\ominus} = {}^{I}\mu_{\ominus}$$

must be added to the scheme owing to the fact that the material (Cu) of the two electrodes is the same. This completes the description of the reactions taking place in our chain at equilibrium in all its details.

C. Contraction of individual reactions into a simplified overall reaction

It is possible successively to eliminate the potentials Φ_{II}, Φ_{III}, Φ_{IV} from eqs. (7), (8), (9), (10). In this way we find that

$$(14) \quad 2F(\Phi_I - \Phi_V) = ({}^{II}\mu - {}^{I}\mu)_{Cu^{++}} + ({}^{IV}\mu - {}^{III}\mu)_{Zn^{++}} + ({}^{II}\mu - {}^{III}\mu)_{SO_4^{--}} + \\ + 2({}^{IV}\mu - {}^{V}\mu)_{\ominus}.$$

In view of (11) the first term on the right-hand side can be transformed to

$$(15) \qquad {}^{II}\mu_{Cu^{++}} - {}^{I}\mu_{Cu} + 2\,{}^{I}\mu_{\ominus},$$

and, similarly, in view of (12) the second term can be transformed to

$$(16) \qquad {}^{IV}\mu_{Zn} - {}^{III}\mu_{Zn^{++}} + 2\,{}^{IV}\mu_{\ominus}.$$

Taking into account condition (13) we see that the sum of the terms in the last two equations, together with the fourth term on the right-hand side, is equal to zero. Hence it follows from (14) that

$$(17) \quad \Phi_I - \Phi_V = \frac{1}{2F}\left\{ {}^{IV}\mu_{Zn} - {}^{III}\mu_{Zn^{++}} + {}^{II}\mu_{Cu^{++}} - {}^{I}\mu_{Cu} + ({}^{II}\mu - {}^{III}\mu)_{SO_4^{--}} \right\}.$$

$\Phi_I - \Phi_V$ is the emf of the open cell, and we shall denote it by E.

It is permissible to assume that $^{II}\mu_{SO_4^{--}} = {}^{III}\mu_{SO_4^{--}}$ which implies $\Phi_{II} = \Phi_{III}$. This can be justified with the aid of certain artifices. Instead of (17) we then have

(17 a) $$E \cdot 2F = {}^{IV}\mu_{Zn} + {}^{II}\mu_{Cu^{++}} - {}^{III}\mu_{Zn^{++}} - {}^{I}\mu_{Cu}.$$

The physico-chemical interpretation of this relation is as follows:

Imagine that a charge $2F$ is transferred in a reversible (electrostatic) manner from the positive pole to the negative pole outside the cell so that an equal charge is transferred from the negative to the positive pole within the Daniell cell causing the following reaction to take place:

(17 b)
$$\frac{\begin{array}{c} Zn \to Zn^{++} \\ Cu^{++} \to Cu \end{array}}{Zn + Cu^{++} \to Zn^{++} + Cu}$$

for one mol at constant pressure and temperature. The product $2F \times E$ represents the (electrical) work performed by the cell. Hence the right-hand side of (17 a) can be interpreted as a decrease ΔG in the free enthalpy during the occurrence of the process described by (17 b). This can be inferred from the representation

(18) $$\Delta G = \sum_i \mu_i(-\Delta n_i)$$

which follows from (14.7) for $\Delta n_{Zn} = -1$, $\Delta n_{Zn^{++}} = +1$, $\Delta n_{Cu} = +1$ and $\Delta n_{Cu^{++}} = -1$ and for constant chemical potentials. Furthermore, to be precise it would be necessary to conduct the preceding reaction with an infinitesimally small fraction of one mol if the latter condition is to be satisfied.

The reaction (17 b) represents the *overall* reaction in the cell. It results from the individual reactions (7) – (13). Equations (7) – (13) result from *virtual* changes, (18) corresponds to a *real* transfer of charge. It will, however, be seen from eqs. (17 a, b) that the individual reactions are unimportant. Their function was to make it clear that the emf is determined by the *equilibrium of the individual sub-systems*.

During the transfer of the charge $2F$ as described by eq. (17 b), i. e. when one mol undergoes the overall reaction, the system will liberate a certain heat of reaction which can be measured directly. We now propose to deduce a relation between this heat and the emf.

D. The Gibbs-Helmholtz fundamental equation

We shall begin by generalizing from the special case of a Daniell cell to any galvanic cell. It is natural to suppose that eqs. (17 a) and (18) have general validity:

(19) $$E = \frac{1}{zF} \Delta G.$$

Here ΔG denotes the decrease in free enthalpy per one mol at $p = $ const, $T = $ const and z is the valency of the anions or kations or their least common multiple.

By forming differences we find from our Table in Sec. 7 that (e. g. $G = H - TS$; $S = -(\partial G/\partial T)_p$)

(20) $$\Delta G - T \left(\frac{\partial \Delta G}{\partial T}\right)_p = \Delta H.$$

If the reaction took place at $V = $ const, $T = $ const eqs. (19) and (20) would have been replaced[1] by

(21) $$E = \frac{1}{zF} \Delta F,$$

(22) $$\Delta F - T \left(\frac{\partial \Delta F}{\partial T}\right)_v = \Delta U.$$

Here ΔF, ΔH, and ΔU denote the decrease in F, H, and U per one mol undergoing the overall reaction and ΔH and ΔU denote the heats of reaction at $p = $ const and $V = $ const respectively. It follows at once from (19) – (22) that

(23) $$E - T \left(\frac{\partial E}{\partial T}\right)_p = \frac{\Delta H}{zF} \quad \text{(Gibbs)},$$

(24) $$E - T \left(\frac{\partial E}{\partial T}\right)_v = \frac{\Delta U}{zF} \quad \text{(Helmholtz)}.$$

These two equations played a very important part when Nernst formulated the Third Law, because experimental results showed that even at moderately low temperatures the "naive" formulae

(25) $$E \sim \frac{\Delta H}{zF} \quad \text{or} \quad E \sim \frac{\Delta U}{zF}$$

appear to be correct. From this fact Nernst concluded that the curves for U and F must osculate, and not only meet, at absolute zero (*cf.* Sec. 12).

---

[1] In the following we shall provisionally use the symbol F to denote free energy to avoid confusion with the Faraday constant F.

E. Numerical example

We now return to the Daniell cell. Measurement of the emf gives

$$E = 1.0999 \text{ volt at the ice point}$$

$$\frac{dE}{dT} = -4.3 \times 10^{-4} \frac{\text{volt}}{\text{deg}} \text{ at the ice point.}$$

It is seen from the way the Daniell cell is made that no difference between $(\partial E/\partial T)_p$ and $(\partial E/\partial T)_v$ is to be expected and, in fact, none is observed. Consequently, the two heats of reaction ΔH and ΔU given in (23) and (24) respectively do not differ from each other. Denoting their common value by q we obtain from (23) or (24), with $z = 2$:

$$q = 2 \times 96494 \text{ coulomb} \times (1.0999 + 273 \times 4.3 \times 10^{-4}) \text{ volt} =$$

$$= 192988 \text{ coulomb} \times (1.0999 + 0.1173) \text{ volt} = 2.35 \times 10^5 \text{ coulomb} \times \text{volt.}$$

The unit coulomb $\times$ volt = Joule = Erg. Since 1 Erg = 0.239 cal, we have

$$q = 56100 \text{ cal}$$

compared with the measured value $q = 55200$ cal.

If we calculated "naively" from (25) we would obtain

$$E = \frac{55200}{2 \times 96494 \times 0.239} \text{ volt} = 1.195 \text{ volt}$$

instead of 1.0999 volt. The primitive equation is seen to be fairly good and this can easily be understood if it is noticed that $T \cdot dE/dT = 0.1173$ is relatively small compared with $E = 1.0999$. The same is not true for the cell

$$Hg \,|\, Hg\,Cl_2 + KCl \,|\, KOH + Hg_2 \,|\, Hg.$$

Here we have

$$E = 0.1483 \text{ volt at the ice point}$$

$$\frac{dE}{dT} = 8.37 \times 10^{-4} \frac{\text{volt}}{\text{deg}} \text{ at the ice point}$$

The observed heat of reaction is

$$-3280 \text{ cal}$$

so that on applying the naive formula we would obtain a negative emf. In actual fact at 0 C the term $T \cdot dE/dT = 0.227$ volt exceeds E itself in value which is consistent with the negative value of the heat of reaction.

It is sometimes stated in text-books that the emf of a galvanic cell, particularly that of a Daniell cell, can be deduced from the First Law and that, generally speaking, the Second Law introduces only a correction term. The last example shows that the statement is erroneous. It is necessary, in principle, to base the calculation on the Gibbs–Helmholtz equations, which have been deduced with the aid of the First *and* the Second Law.

F. Remarks on the integration of the fundamental equation

In view of the identity

$$(26) \qquad E - T \frac{dE}{dT} = - T^2 \frac{d}{dT}\left(\frac{E}{T}\right)$$

eq. (23) can also be written

$$(27) \qquad \frac{d}{dT}\left(\frac{E}{T}\right) = - \frac{\Delta H}{zF \cdot T^2}.$$

We can split ΔH into a term ΔH_0 which is independent of temperature and into a term ΔH_T which depends on it. We shall prove that the latter vanishes very rapidly as $T \to 0$. We can write that

$$(28) \qquad \Delta C_p = \frac{\partial \Delta H}{\partial T} = \frac{\partial \Delta H_T}{\partial T}.$$

According to the Third Law ΔC_p vanishes for $T \to 0$, consequently ΔH_T must tend to zero *more rapidly* than T (like T^4 in most cases; no constant term in ΔH_T which would vanish on differentiating (28) need be considered because it would have had to be included in ΔH_0). On integrating it follows at once from (27) that

$$(29) \qquad E = \frac{\Delta H_0}{zF} - \frac{T}{zF} \int_0^T \frac{\Delta H_T}{T^2} dT.$$

By choosing the lower limit of integration at $T = 0$ we have given the right value to the constant of integration, namely, the value is such that it corresponds to our original definition of E in eq. (19). In fact, because $TS \to 0$ as $T \to 0$, as required by the Third Law, we have

$$\varDelta G_0 = \varDelta H_0, \quad \text{that is} \quad E = \frac{\varDelta H_0}{zF}.$$

This "limiting condition" which must be satisfied at absolute zero is seen to be satisfied by (29) which justifies our choice of the lower limit of integration.

Equation (29) enables us to predict the emf together with its variation with temperature from measured values of the variation of the heat of reaction with temperature. The emf of many cells was determined by this method, excellent agreement with direct measurement having been found. Fundamentally this agreement is equivalent to an additional verification of the Second and Third Laws.

19. Ferro- and paramagnetism

Diamagnetic phenomena are independent of temperature but para- and ferromagnetism depends very strongly on it. Both increase with decreasing temperature. Above a certain limit, known as the Curie point, ferromagnetic substances behave like paramagnetic solids. It is now our purpose to apply the principles of thermodynamics to such phenomena. As usual we can only expect to obtain a general framework within which such phenomena take place. Their details must be obtained with the aid of statistical methods and supplemented with statements from the field of atomic physics (magnetic moment of an electron, *cf.* Vol. III Sec. 14 B). Diamagnetic processes belong entirely to the realm of atomic physics.

A. Work of magnetization and magnetic equation of state

According to Vol. III., eqs. (5.66) and (12.2) the differential of the magnetic energy density is given by $(\mathbf{H}, d\mathbf{B})$ with $\mathbf{B} = \mu_0(\mathbf{H} + \mathbf{M})$ where $\mathbf{M}$ is the magnetization (magnetic moment per unit volume) and μ_0 is a constant for the vacuum which must be added on dimensional grounds. We are not interested in the term

$$\mu_0(\mathbf{H}, d\mathbf{H})$$

of this differential because it is present even in the absence of magnetism. Disregarding the vectorial nature of the process of magnetization (inadmissible in the case of single crystals) we find that the contribution to the energy density which is due to the second term and which is the only important one in the present considerations can be written as (cf. first footnote on p. 75).

(1) $$\mu_0 \, H \, dM.$$

It is, furthermore, convenient to interpret M as the magnetization per mol (rather than per unit volume); thus eq. (1) represents the work performed by the external field when changing the magnetization by dM per mol, i. e. a quantity of energy added to the system.

We now have to consider two magnetic variables, H, M [1] in addition to the thermal variables T, s. We can also disregard the two mechanical variables p, V because they are unimportant here (they would come into play only if the phenomena of magnetostriction were considered). Combining the First and the Second Laws we have

(2) $$du = T \, ds + \mu_0 \, H \, dM$$

or

(2 a) $$ds = \frac{du}{T} - \frac{\mu_0 \, H}{T} dM.$$

In his first paper on this subject (1905) Paul Langevin made the tentative assumption that both terms in ds are perfect differentials. Hence $\mu_0 \, H/T$ must be independent of T, and so it must be a function of M alone. This is equivalent to saying that M is a function of H/T only:

(3) $$M = f\left(C \cdot \frac{H}{T}\right).$$

At the same time u in eq. (2 a) must be independent of M and a function of T alone:

(3 a) $$\left(\frac{\partial u}{\partial M}\right)_T = 0.$$

The argument of the function f in eq. (3) contains a constant C which is characteristic of the material and which can be assumed to include the coefficient μ_0 from eq. (2 a); furthermore, this constant must be such as to

---

[1] It would be more consistent to choose $\mu_0 H$ instead of H itself as the first magnetic variable. Being an intensive quantity it corresponds to the variables T and p in the two other pairs of variables. However, in order not to obscure the text with the unimportant coefficient μ_0 we shall use H alone.

render the argument of the function f dimensionless. In connection with eq. (3 a) it is necessary to remark that u is here considered to be a function of T and M and not of s and M as in eq. (2) (disregarding its dependence on p, v which we continue to neglect).

The preceding equations are reminiscent of those for a perfect gas if the magnetic properties H, M, C are imagined replaced by the mechanical quantities $1/v$, p, $1/R$ respectively. Equation (3 a) is seen to transform into the fundamental caloric equation for perfect gases $(\partial u/\partial v)_T = 0$ and eq. (3) becomes the equation of state if $f(x)$ is replaced by x. Thus .

$$\frac{1}{v} = \frac{p}{RT}.$$

This analogy between the perfect gas and the type of magnetic substance which we are now considering suggests that eq. (3) may be termed "the equation of state of a perfect magnetic substance." In any case it should be borne in mind that this analogy applies to *paramagnetic* substances only. Putting again $f(x) = x$, we obtain

(4) $$M = C \cdot \frac{H}{T}; \qquad \frac{M}{H} = \chi = \frac{C}{T}.$$

This equation is known as *Curie's law for paramagnetic solids*, already mentioned in Vol. III, eq. (13.10); C is *Curie's constant* and χ is the *magnetic susceptibility per mol*. It is seen from (4) that C has the dimension of temperature.

The equation of state for diamagnetic substances does not conform to the scheme of eq. (3), because M is proportional to H and independent of temperature. However (2) is valid for diamagnetic substances. The equation for *ferromagnetic substances* which is of great interest to us here also differs from the scheme in (3). Consequently, eq. (3 a) does not apply to ferromagnetic substances as will be explained in Section D.

B. Langevin's Equation for Paramagnetic Substances

A pivoted elementary magnet of moment m placed in an external magnetic field will point in the latter's direction. If all n elementary magnets contained in one mol were so directed, the magnetization would assume its saturation value

(5) $$M_\infty = m\,n.$$

This state is resisted by thermal agitation. It is now our task to determine that state of compromise between saturation $M = M_\infty$ and complete disorder

$M = 0$ which corresponds to a given temperature T. Langevin deduced the result under consideration by a comparatively simple application of Boltzmann's statistics, *cf.* Sec. 25. It will be shown there that his result has the form

(5 a) $$\frac{M}{M_\infty} = \frac{\cosh \alpha}{\sinh \alpha} - \frac{1}{\alpha}; \quad \alpha = \frac{\mu_0 M_\infty H}{RT},$$

where α is dimensionless, because the numerator as well as the denominator in eq. (5 a) defining α have the dimensions of energy.

The expression

(5 b) $$L(\alpha) = \frac{\cosh \alpha}{\sinh \alpha} - \frac{1}{\alpha}$$

is known as *Langevin's function*. It is represented graphically by the monotonically increasing curve OBA in Fig. 18 and corresponding to the following approximations:

(6) $$\alpha \to 0, \quad L(\alpha) = \frac{1 + \frac{1}{2}\alpha^2 + \cdots}{\alpha + \frac{1}{6}\alpha^3 + \cdots} - \frac{1}{\alpha} = \frac{1}{\alpha} \frac{\frac{1}{3}\alpha^2}{1 + \frac{1}{6}\alpha^2} \to \frac{1}{3}\alpha$$

(6 a) $$\alpha \to \infty; \quad L(\alpha) = \frac{e^\alpha + e^{-\alpha}}{e^\alpha - e^{-\alpha}} - \frac{1}{\alpha} \approx 1 - \frac{1}{\alpha} \to 1.$$

From (6) and (5 a) we obtain the following expression which is valid for almost all values of α which can be attained in practice:

(7) $$\chi = \frac{M}{H} = \frac{M_\infty \alpha}{H \cdot 3} = \frac{\mu_0 M_\infty^2}{3RT}.$$

This is Curie's law (4) where the Curie constant C has the value

(7 a) $$C = \frac{\mu_0 M_\infty^2}{3R}.$$

It breaks down only for $T \to 0$, when eq. (4) leads to $\chi \to \infty$ instead of the correct finite value

$$\chi = M_\infty / H$$

which follows from (7) and (6 a).

The approximations (6) and (6 a) are seen plotted in Fig. 18. They are represented, respectively, by the dash-dot lines near the origin 0 and by the asymptote. Generally speaking (*cf. infra*), all paramagnetic states which are attainable in practice lie near the lowest end of the tangent curve.

The statistical theory due to Langevin neglects mutual interactions between the elementary magnets. It assumes that they are only subjected to the influence of the external field, which, obviously, represents a far-reaching idealization. This idealization is equivalent to the assumption in (3 a) that the energy u is independent of M; this would not be true if interactions between elementary magnets were taken into account. It is now clear that Langevin's equation of state (5) is compatible with the scheme (3) because the latter was thermodynamically linked with the condition (3 a). In order to justify this drastic simplification we may mention that saturation effects are, generally speaking, unobservable in paramagnetic substances, and can only be expected to occur at the lowest temperatures. (This is confirmed by the observations on gadolinium sulphate made at temperatures down to 1.3 K by Woltjer and Kammerlingh Onnes at the Cryogenic Laboratory in Leiden, cf. Section E.) Debye[1] demonstrated that Langevin's function ceases to be applicable at such low temperatures because it contradicts Nernst's Third Law as T is made to tend to zero.

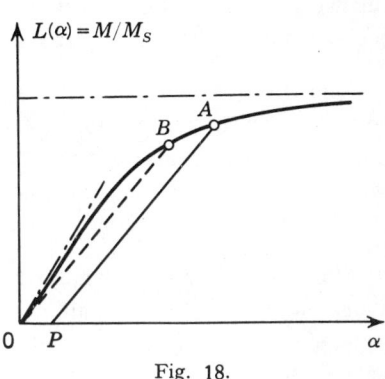

Fig. 18.

Langevin's curve for paramagnetic substances and its application in Weiss' theory of ferromagnetic phenomena.

C The Theory of ferromagnetic phenomena due to Weiss

Pierre Weiss stated the extremely fruitful supposition that in ferromagnetic bodies it is possible to discern small regions or domains in which the elementary magnets cause each other to become parallel, thus giving rise to an internal, molecular field H_m whose strength exceeds observed, external fields by many orders of magnitude, cf. Vol. II, Sec. 14 A. Weiss assumes that this field is proportional to the magnetization, M, present in it, the coefficient of proportionality, N, being very large and depending on the material under consideration:

(8) $$H_m = N \cdot M.$$

---

[1] Ann. d. Phys. **81**, 1154 (1926). The required simplification of Langevin's function is furnished by quantum mechanics, see *infra*, Section D.

These "Weiss domains" are aligned individually but the direction of a molecular field H_m varies from domain to domain. Consequently the body appears to be non-magnetic in the absence of an external field H. The moment exerted by such an external field H on the domains is quite different from that exerted on individual elementary magnets; the external magnetic influence of a ferromagnetic order of magnitude is due to the alignment of these domains with the external field.

The change in direction, particularly in the case of weak fields, is not due to a rotation in the directions of the Weiss domains; the principal effect is due to irreversible, abrupt turns performed by the elementary magnets at the boundaries of Weiss' domains and to the wall displacements connected with them (Vol. III Sec. 14 C).

We shall refrain from analyzing the interactions which occur between the Weiss domains and we shall restrict ourselves to the consideration of the influence of the external magnet on the magnetization in a single domain.

Quantitatively Weiss superimposes the inner field H_m on the external field H in Langevin's expression for α:

$$(9) \qquad \alpha = \frac{\mu_0 M_\infty}{RT}(H + H_m) = \frac{\mu_0 M_\infty}{RT}(H + NM).$$

It is seen that the value of α becomes much larger than in the paramagnetic case because $H_m \approx H$. Furthermore, on substituting (9), Langevin's assumption (5 a) becomes an implicit equation in M, because M appears not only explicitly on the left-hand side of (5 a) but also on the right-hand side as given in (9).

Figure 18 shows a graphical method of solving this equation. On the one hand, the point which is determined by the two unknowns M/M_∞ and α must lie on Langevin's curve and on the straight line defined by eq. (9), on the other. It must, therefore, satisfy the two equations

$$(9\text{ a}) \qquad \frac{M}{M_\infty} = L(\alpha),$$

$$(9\text{ b}) \qquad \alpha = \alpha_0 + \beta \frac{M}{M_\infty}; \quad \alpha_0 = \frac{\mu_0 M_\infty H}{RT}; \quad \beta = N\frac{\mu_0 M_\infty^2}{RT}.$$

The straight line (9 b) intersects the axis of abscissae at the point $\alpha = \alpha_0$ (denoted by P in Fig. 18); according to the definition of α in eq. (5 a) this is also the abscissa in the paramagnetic case and lies very near the origin.

According to (9 b) the tangent of the angle which it includes with the axis of abscissae is given by

$$\text{(9 c)} \qquad \frac{1}{\beta} = \frac{1}{N} \frac{RT}{\mu_0 M_\infty^2}.$$

This slope depends on temperature and decreases with it. The point of intersection A of this straight line with Langevin's curve moves to the right as T is decreased; in this way M approaches to the saturation value M_∞ which corresponds to perfectly aligned elementary magnets.

The position of the point of intersection changes very little when the external field is removed, i. e. when we make $H = 0$. The straight line PA is then translated parallel to itself until it passes through the origin 0 (since $\alpha_0 = 0$) and the new point of intersection is at B. The field of a single Weiss domain remains almost unchanged. Thus there exists the possibility of a residual *spontaneous magnetization* of Weiss' domains leading to *permanent magnetization*. The preceding argument has shown that eq. (9) reproduces the essential features of ferromagnetic behavior at *low temperatures*, such as *the existence of spontaneous magnetization which increases as T is decreased*. The fact that this spontaneous magnetization cannot always be observed follows from the interaction between the individual fields whose directions may differ from domain to domain causing them to cancel each other.

Until now we have implied that the straight line PA is less steep than e. g. the tangent to the Langevin's curve at 0. When the opposite is true the point of intersection will lie near the origin 0 provided that the external field is sufficiently weak. In this case the approximation (6) for $L(\alpha)$ may be used and eqs. (9 a, b) yield

$$\frac{M}{M_\infty} = \frac{1}{3} \cdot \frac{\mu_0 M_\infty}{RT} (H + NM).$$

It follows that

$$\text{(10)} \qquad M \left(T - \frac{\mu_0}{3} \frac{M_\infty^2 N}{R} \right) = \frac{\mu_0}{3} \frac{M_\infty^2}{R} H.$$

The coefficient of H on the right-hand side is equal to *Curie's constant* C from eq. (7 a). The left-hand side contains its multiple NC which we shall denote by Θ, or

$$\text{(11)} \qquad \Theta = \frac{\mu_0 M_\infty^2 N}{3R}.$$

Θ is known as the *Curie point*.

Since the factor 1/3 in the preceding equations derives from the first term in the series expansion for $L(\alpha)$ and is identical with $L'(0)$, we can replace (11) by

$$(11\text{ a}) \qquad \Theta = \mu_0 L'(0) \frac{M_\infty{}^2 N}{R}.$$

It is convenient to retain this form rather than that in (11) because some of the succeeding calculations can thus be made independent of the particular choice of Langevin's function $L(\alpha)$. This will make the results more suitable for the introduction of generalizations which are suggested by quantum mechanics.

Substituting the abbreviation Θ from (11) or (11 a) we can transform eq. (10) into

$$(12) \qquad M = \frac{CH}{T-\Theta}.$$

It is seen that above the Curie point the body behaves like a paramagnetic substance and obeys the Curie-Weiss law (12). Its graphical representation is given by a straight line on plotting H/M in terms of T. Strictly speaking experiments suggest the existence of two slightly different Curie points depending on whether Θ is defined with the aid of this straight line or on the basis of the disappearance of spontaneous magnetization. A more thorough consideration of these and other details of the extensive field of ferromagnetic experimental data would exceed the scope of these lectures.

We must, however, examine a little more closely the neighborhood of the Curie point. It is evident that eq. (12) remains valid for $T \to \Theta$ only on condition that H tends to zero sufficiently strongly at the same time. Thus we may apply eq. (12) above the Curie point when $H = 0$ so that we obtain $M = 0$ denoting *no* spontaneous magnetization.

We shall now consider the magnetization when $H = 0$, i. e. we shall investigate the spontaneous magnetization close to but below the Curie point; the corresponding value of α will be denoted by α_{sp}. It is no longer permissible to approximate the Langevin function by a straight line even though $H = 0$ if the point of intersection B from Fig. 18 is not to be lost. Moreover, it is necessary to take into account the higher derivatives of $L(\alpha)$ at the origin. All even derivatives vanish at the origin since $L(\alpha)$ is an odd function of α. Neglecting the derivatives of the 5-th, 7-th, etc. order, we obtain from (9 a) that

$$(13) \qquad \frac{M_{sp}}{M_\infty} = \alpha_{sp} L'(0) + \frac{\alpha_{sp}{}^3}{6} L'''(0).$$

On the other hand it follows from (9) with $H = 0$ and in view of (11 a) that:

$$(14) \qquad \frac{M_{sp}}{M_\infty} = \alpha_{sp} \frac{RT}{\mu_0 N M_\infty^2} = \alpha_{sp} L'(0) \frac{T}{\Theta}.$$

On comparing (13) with (14) we obtain an equation whose non-vanishing solution can be written

$$(15) \qquad \alpha_{sp} = \sqrt{\frac{6 L'(0)}{-L'''(0)}\left(1 - \frac{T}{\Theta}\right)}.$$

The value of $L'''(0)$ calculated from (5 a) is equal to $-2/15$. Substituting (15) into (13) we obtain, after a short calculation, that the spontaneous magnetization just below the Curie point is given by

$$(16) \qquad M_{sp} = M_\infty \sqrt{\frac{6[L'(0)]^3}{-L'''(0)} \cdot \frac{T}{\Theta}} \cdot \sqrt{\left(1 - \frac{T}{\Theta}\right)}.$$

Figure 19 shows that the plot of M_{sp} in terms of decreasing T has a vertical tangent at the Curie point, in agreement with eq. (16). The curve increases up to $T = 0$ where $M_{sp} = M_\infty$. It must, however, be noted that the diagram is only qualitatively correct. In actual fact its shape must be changed due to quantum effects (directional quantization of spin moments; the curve obtains a horizontal tangent at $M_{sp} = M_\infty$ and its slope is not like that shown in Fig. 19).

It is clear that the diagram is true for a single Weiss domain. The degree of permanent magnetization which will be discernible in a macroscopic aggregate depends on the structure of the material and cannot be described with the aid of the present theory.

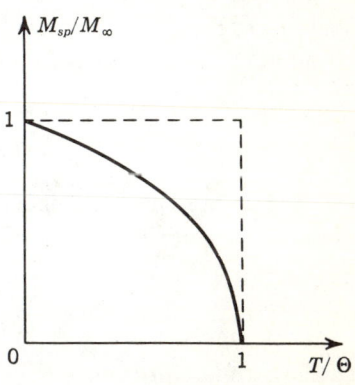

Fig. 19.

Spontaneous magnetization below the Curie point according to Weiss' theory (subject to quantum corrections).

D. The specific heats c_H and c_M

The magnetic equation of state is no longer of the simple type (3), owing to the presence of the term NM in (9). Consequently the caloric equation (3 a) ceases to apply. The equation which takes its place can be deduced from the expression for entropy (2 a) by writing it in terms of T and M as the independent variables:

(17) $$ds = \frac{1}{T}\left(\frac{\partial u}{\partial T}\right)_M dT + \frac{1}{T}\left[\left(\frac{\partial u}{\partial M}\right)_T - \mu_0 H\right] dM.$$

Taking the partial derivative of the factor before dT with respect to M, and of that before dM with respect to T we have

$$\frac{1}{T}\frac{\partial^2 u}{\partial M\, \partial T}$$

and

$$\frac{1}{T}\left[\frac{\partial^2 u}{\partial T\, \partial M} - \mu_0 \left(\frac{\partial H}{\partial T}\right)_M\right] - \frac{1}{T^2}\left[\left(\frac{\partial u}{\partial M}\right)_T - \mu_0 H\right].$$

Since both derivatives must be equal to each other, we obtain

(18) $$\left(\frac{\partial u}{\partial M}\right)_T = \mu_0 \left\{H - T\left(\frac{\partial H}{\partial T}\right)_M\right\}.$$

The addition of the subscripts T and M in the last formula is required for the sake of clarity. First we note that the preceding derivation is completely analogous to that of $\partial u/\partial v$ in (9.6) from van der Waals' equation. Substituting (18) into (17) we find that the heat added reversibly is given by

(19) $$T\, ds = \left(\frac{\partial u}{\partial T}\right)_M dT - T\mu_0 \left(\frac{\partial H}{\partial T}\right)_M dM.$$

If T and H are chosen as independent variables instead of T and M it is only necessary to substitute

$$dM = \left(\frac{\partial M}{\partial T}\right)_H dT + \left(\frac{\partial M}{\partial H}\right)_T dH.$$

It follows from (15) that

(20) $$T\, ds = \left\{\left(\frac{\partial u}{\partial T}\right)_M - T\mu_0 \left(\frac{\partial H}{\partial T}\right)_M \left(\frac{\partial M}{\partial T}\right)_H\right\} dT - T\mu_0 \left(\frac{\partial H}{\partial T}\right)_M \left(\frac{\partial M}{\partial H}\right)_T dH.$$

In strict analogy with the molar specific heats, c_v, and, c_p, we now define the molar specific heats at constant magnetization, c_M, and that at constant field intensity, c_H. According to (19) and (20), we obtain

(21) $$c_M = T\left(\frac{\partial s}{\partial T}\right)_M = \left(\frac{\partial u}{\partial T}\right)_M,$$

(21 a) $$c_H = T\left(\frac{\partial s}{\partial T}\right)_H = \left(\frac{\partial u}{\partial T}\right)_M - T\mu_0 \left(\frac{\partial H}{\partial T}\right)_M \left(\frac{\partial M}{\partial T}\right)_H,$$

whence, by subtraction

$$(22) \quad c_H - c_M = -T\mu_0 \left(\frac{\partial H}{\partial T}\right)_M \left(\frac{\partial M}{\partial T}\right)_H.$$

This, again, is an exact analogue of the already familiar general expression for $c_p - c_v$ (eq. (7.9) with $-p, v$ replaced by $\mu_0 H, M$).

The two derivatives on the right-hand side of (22) can be found from the parametric representation (9 a), (9 b); on differentiating with respect to T at constant M, we obtain

$$(23) \quad 0 = L'(\alpha) \left\{ \frac{\mu_0 M_\infty}{RT} \left(\frac{\partial H}{\partial T}\right)_M - \frac{\alpha}{T} \right\}$$

and at constant H we have

$$(24) \quad \left(\frac{\partial M}{\partial T}\right)_H = M_\infty L'(\alpha) \left\{ \frac{\mu_0 N M_\infty}{RT} \left(\frac{\partial M}{\partial T}\right)_H - \frac{\alpha}{T} \right\}.$$

It follows at once from (23) that

$$(25) \quad \left(\frac{\partial H}{\partial T}\right)_M = \frac{R}{\mu_0 M_\infty} \alpha.$$

Taking into account the definition (11 a) of Θ, we calculate from (24) that

$$(26) \quad \left(\frac{\partial M}{\partial T}\right)_H = \frac{M_\infty L'(0) L'(\alpha) \alpha}{\Theta L'(\alpha) - T L'(0)}.$$

Hence, according to (22) the difference of the specific heats becomes:

$$(27) \quad c_H - c_M = \frac{R L'(0) L'(\alpha) \times \alpha^2}{L'(0) - (\Theta/T) L'(\alpha)}.$$

We shall now discuss this result for the special case when $H = 0$ (removal of external field) putting $M = M_{sp}$ accordingly (spontaneous magnetization), see Fig. 20. As we already know at $T > \Theta$ we have $M_{sp} = 0$ (paramagnetic behavior). Consequently it follows from (9) that $H = 0$ implies $\alpha = 0$. Hence eq. (27) yields

$$(28) \quad c_H = c_M$$

for $T > \Theta$ and $H = 0$.

Of the region $T \leqslant \Theta$ we shall first consider case a) $T \ll \Theta$ which, according to (9), implies $\alpha \gg 1$. It then follows from (6 a) that

$$L'(\alpha) \approx \frac{1}{\alpha^2}$$

and from (27) that

(29) $$c_H - c_M = \frac{R\,L'(0)}{L'(0) - \dfrac{\Theta}{T} \cdot \dfrac{1}{\alpha^2}}.$$

From (9) with $H = 0$ and $M = M_{sp} = M_\infty$ (cf. Fig. 17), we have:

$$\alpha = \frac{\mu_0\,M_\infty{}^2\,N}{R\,T}$$

which can be replaced by

$$\alpha = \frac{\Theta}{T} \bigg/ L'(0)$$

in view of (11 a). Substituting this into (29), we have

(29 a) $$c_H - c_M \approx \frac{R}{1 - \dfrac{T}{\Theta} L'(0)} \approx R.$$

Considering the neighborhood of the Curie point we assume

(29 b) $$\Theta - T \ll \Theta.$$

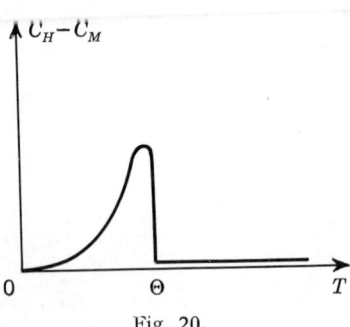

Fig. 20.

Qualitative representation of the maximum in $c_H - c_M$ at the Curie point. The behavior near $T = 0$ is sketched in accordance with the results of quantum mechanics and unlike eq. (29). For $T > \Theta$ we have $c_H - c_M = 0$.

It is now necessary to proceed in the same way as in connection with eq. (13) and to take into account the third derivative of the Langevin function in the denominator of (27), replacing $L'(\alpha)$ in the numerator by $L'(0)$ which is permissible because $\alpha = \alpha_{sp} \ll 1$: Thus we obtain

(30) $$c_H - c_M = \frac{R\,\alpha^2 [L'(0)]^2}{L'(0) - \dfrac{\Theta}{T}\left[L'(0) + \dfrac{1}{2}\alpha^2\,L'''(0)\right]}.$$

Substituting α from (15) and cancelling the common factor $\Theta - T$ in the numerator and in the denominator, we find that

(31) $$c_H - c_M = 3\,R\,\frac{[L'(0)]^2}{-L'''(0)} \cdot \frac{T}{\Theta}.$$

Thus there is a jump in the specific heats at $T = \Theta$, because, as already mentioned, $c_H = c_M$ for $T > \Theta$.

Making use of the preceding numerical values $L'(0) = 1/3$, $L'''(0) = -2/15$ we find that it is

(31 a) $$c_H - c_M = \frac{5}{2} R.$$

In actual fact this sudden jump is smoothed out into a maximum (owing to the small latitude in the value of the Curie point, *cf. supra*) which decreases steeply on the side $T > \Theta$ and which is much more gradual on the side $T < \Theta$. Such a maximum persists also in the case when $H \neq 0$, i. e. in the case when the magnetization is influenced by the external field and is not spontaneous.

It will be recalled that in Section C it was found necessary to stress the fact that the results contained in it were restricted in their application to a single Weiss domain and that they were less pronounced in the case of a complete macroscopic system and depended on the particular material. This restriction is unnecessary as far as the specific heats are concerned. The specific heats superimpose themselves one on the other like *scalars* and not like the fields whose summation obeys the laws of *vectors*. Consequently our present formulae remain valid for the macroscopic system.

However, it is necessary to remember that our present results must be corrected in the light of *quantum mechanics*. This may be inferred at once from the fact that eq. (29 a) implied $c_M - c_H \to R$ for $T \to 0$ whereas Nernst's Third Law requires that $c_M - c_H \to 0$; *cf.* clause 3 in Sec. 12.

The quantum theory leads to much lower values than the value $5\,R/2$ in eq. (31 a); for example, the value $3\,R/2$ may be obtained depending on the kind of quantization of direction which must be assumed on atomistic grounds for a given choice of the Langevin function L.

A comprehensive and critical presentation of ferromagnetic phenomena is given in a book by Becker and Doering[1] which has already been quoted in Vol. III Sec. 14 D. We have assumed in the preceding argument that Weiss' model provides a sufficiently accurate approximation to reality; the book by Becker and Doering does, on the other hand, contain a detailed comparison of this model with the existing pertinent experimental material.

---

[1] R. Becker and W. Doering, Ferromagnetismus, Berlin 1939. The book also discusses the atomistic aspects of the problem which had to be omitted from this course of lectures.

E. The Magneto-Caloric Effect

Isentropic demagnetization causes a drop in temperature in the case of ferro- and paramagnetic substances. According to (20) and (21) we can calculate it from

(32) $$c_H \left(\frac{\partial T}{\partial H}\right)_s = T\mu_0 \left(\frac{\partial H}{\partial T}\right)_M \left(\frac{\partial M}{\partial H}\right)_T = -T\mu_0 \left(\frac{\partial M}{\partial T}\right)_H.$$

It is known as the *magneto-caloric effect*. (Conversely, a sudden, and therefore adiabatic, magnetization involves a corresponding increase in temperature.) We now proceed to calculate this effect, having described it qualitatively in terms of the disorder associated with demagnetization at the end of Sec. 11 and Sec. 12.

We shall restrict our considerations to the particularly interesting case of a paramagnetic salt (e. g. gadolinium sulphate) and assume that it obeys Curie's law down to the lowest temperatures. We thus have

$$M = \frac{C}{T} H, \quad \left(\frac{\partial H}{\partial T}\right)_M = \frac{M}{C}, \quad \left(\frac{\partial M}{\partial H}\right)_T = \frac{C}{T},$$

and from (32), we find that

$$c_H \left(\frac{\partial T}{\partial H}\right)_s = T\mu_0 \frac{M}{C} \cdot \frac{C}{T} = \mu_0 \frac{C}{T} H.$$

It follows that the isentropic process under consideration is governed by the differential equation

(33) $$T\, dT = \frac{\mu_0 C}{c_H} H \cdot dH.$$

The process is described by

$$H \to 0, \quad T \to T_0.$$

Assuming that the coefficients C and c_H are constant we can integrate (33) to obtain

(34) $$T_0^2 - T^2 = -\frac{\mu_0 C}{c_H} H^2, \quad T_0 = T\sqrt{1 - \frac{\mu_0 C}{c_H} \cdot \frac{H^2}{T^2}}.$$

It is seen that the temperature does, in fact, decrease, the drop in temperature being larger for a stronger original field H and for a lower initial temperature T.

The preceding calculation is also superficial to a certain extent because of the extrapolation of Curie's law to the lowest temperatures. This implies that interactions between elementary magnets have been neglected and this is no longer permissible. Nevertheless, eq. (34) does give an idea of the very effective step taken by Debye, Giauque and Kammerlingh Onnes in order to come nearer to the absolute zero of temperature.

20. Black body radiation

All hot bodies emit electromagnetic radiation. As the temperature increases the body changes from a red through a yellow to a white glow. It must be realized, however, that bodies emit radiation even at ordinary or low temperatures, except that the wavelength then lies in the infrared region. All thermal radiations are wave-like in their character but within the field it is possible to analyze them exactly according to the laws of geometrical optics which means that they may be resolved into pencils of rays.

Let us now imagine a hollow box whose walls are maintained at a constant temperature. The radiation present inside it is in thermal equilibrium with its walls. Consequently we must ascribe to it the same temperature T as that possessed by the walls. This is true for every element of volume in the cavity and specifies *homogeneous* radiation throughout, i. e. one which is independent of the space coordinates. The cavity constitutes a thermodynamic system (proof in Section A) which is independent of the particular physical and chemical processes of emission and absorption taking place in the walls.

It is found that the internal equilibrium is not disturbed appreciably if a small hole is made in the box so that the radiation can leave the cavity and can thus be made accessible to observation. Radiation from outside which may fall on the opening will not be reflected; it is completely absorbed by the walls after having reflected from them a large number of times and after having been partly absorbed on each reflection. Since a surface which absorbs radiation completely is usually called "black" it is natural to call the radiation emitted through an opening in the box "black body radiation."

The introduction of a "speck of soot," i. e. of a perfectly absorbing body of very small heat capacity into the cavity does not disturb the state of equilibrium. On the other hand when the inner walls of the cavity are made of a perfectly reflecting material and cannot, therefore, influence the rays falling on them, the radiation filling the cavity may become one which is not in equilibrium. The introduction of a speck of soot into the cavity will turn the radiation into black body radiation. (The speck of dust performs the role of a catalyzer.)

The view which presents itself to an observer inside the box is not very interesting: he perceives the same luminosity at every point and in all directions. He cannot see the shape of the cavity and is not aware of the differences in the distances from the walls in varying directions. Using a Nicol prism he could verify that the radiation is not polarized. On changing the temperature he will only notice a change in the intensity and in the color of the radiation.

A. Kirchhoff's law

We recall from Electrodynamics that electromagnetic radiation carries energy and momentum, *cf*. Vol. III. Sec. 31. The energy density was denoted there by W. In a monochromatic field of radiation its average with respect to time depends on the space coordinate, on the frequency, v, and on the amplitude, or, more precisely, on its square, the intensity. We now consider all radiation within a small spectrum interval, dv, as distinct from monochromatic radiation. The energy density contained within this interval will be denoted by $\mathsf{u}\, dv$, and that of the whole spectrum will be denoted by u. We then have

(1) $$u(T) = \int_0^\infty \mathsf{u}(v, T)\, dv.$$

The argument, T, has been added here to emphasize the fact that the amplitude (or intensity) of black body radiation depends only on temperature if equilibrium prevails and is the same at every point in the cavity. The symbol u is here used in a slightly different sense, because u denotes now energy per unit volume and not per unit mass (or per mol):

(1 a) $$[u] = \frac{\text{erg}}{\text{cm}^3}.$$

It follows from (1) that the dimension of u is

(1 b) $$[\mathsf{u}] = \frac{\text{erg sec}}{\text{cm}^3}.$$

Kirchhoff (1859) proved that u is a function of the arguments v and T and that it is *independent of the nature of the walls of the cavity*. This proposition is known as Kirchhoff's law. In order to indicate the method of proving it let us consider two hollow boxes A and B whose walls are different. Let us assume that u in A is larger than in B, in a certain spectral region (v, dv). We now connect A to B through a small tube which is opaque to all wavelengths except v (color filter). In such an arrangement more heat would flow from A to B than in the reverse direction, thus upsetting the state of equilibrium; the temperature of B would increase and that of A would decrease until the two values of u would have become equal. In this way a temperature difference would be created "spontaneously" (without work being done on the system) and such a result is inconsistent with the Second Law. *We conclude that* u *must be a universal function of v and T*; it follows from (1) that *u is a universal function of T*.

We now consider the *flux* of energy (denoted by S in electrodynamics) as distinct from its density. It is defined as the amount of energy radiated per unit area and time. The vectorial character of S corresponds to the direction of the normal to the unit of area under consideration. Since black body radiation is *isotropic* (uniform in all directions) it loses its vectorial character and we are justified in speaking of a scalar radiation intensity. We shall not associate it with a discrete direction (the flux of energy in any given single direction is zero) but with a small cone of radiation $d\Omega$. We now imagine that the direction is enclosed in such a cone and denote the energy radiated through $d\Omega$ by $K\,d\Omega$. Consequently, the energy radiated through an elementary cone which forms an angle θ with the normal is given by

(2) $\qquad K\cos\theta\,d\Omega \qquad$ where $\qquad d\Omega = \sin\theta \cdot d\theta \cdot d\phi.$

The amount of radiation passing through an elementary area da during a time dt in a "forward" (or "rearward") direction is then

(2 a) $\qquad K\,da\,dt \int_0^{\pi/2} \cos\theta \sin\theta\,d\theta \int_0^{2\pi} d\phi = \pi K\,da\,dt.$

If K is analyzed spectrally and if the two directions of polarization are distinguished by a dash we may write

(3) $\qquad K(T) = \int_0^\infty [\mathsf{K}(v, T) + \mathsf{K}'(v, t)]\,dv = 2\int_0^\infty \mathsf{K}(v, T)\,dv,$

the last equality being a consequence of the absence of polarization in black body radiation. The dimension of K is the same as that of S; that of K follows from (3) and is

(3 a) $\qquad\qquad [K] = \dfrac{\text{erg}}{\text{cm}^2\,\text{sec}} \ ; \qquad [\mathsf{K}] = \dfrac{\text{erg}}{\text{cm}^2}.$

The quantities u and K satisfy the relation

(4) $$u = 4\pi K/c.$$

We shall refrain here from giving the proof because it can be deduced from a simple premiss of geometrical optics, assuming the cavity to be evacuated. If this were not so it would be necessary to replace c by c/n. In view of (3) and (1) we have from (4) that

(4 a) $$u = 8\pi K/c.$$

We now proceed to obtain an *extension of Kirchhoff's law* by applying the equilibrium principle to the walls of the cavity.

The *absorptive power* of a wall element da will be denoted by A; in other words A denotes that fraction of the impinging radiation $K(\nu, T)$ (assumed spectrally decomposed) which is converted into heat as it penetrates into the wall. Thus the amount of energy (per unit area and time and per solid angle $d\Omega$) which is deducted from the system at equilibrium is

(5) $$A\,K(\nu, T).$$

This energy must be replaced in the cavity by the *emissivity* E of the same element of wall. In the case of a blackened surface ($A = 1$) we have

(5 a) $$E = K(\nu, T).$$

The emissivity of a perfectly white perfectly reflecting surface ($A = 0$) must be $E = 0$. In such a case, as already stated previously, the wall cannot contribute to the establishment of thermodynamic equilibrium. In an average surface E must replace the amount (5) withdrawn from the cavity. Thus we must have

(6) $$\frac{E}{A} = K(\nu, T).$$

For pure thermal radiation the ratio of the emissivity to the absorptive power is a universal function of the frequency and temperature.

Kirchhoff's law and the present extension thereof have now become very important not only in problems of black body radiation but also in illumination engineering. It contributed to the discovery of spectral analysis which was made by Kirchhoff and Bunsen at about that time.

B. The Stefan–Boltzmann law

It has already been stated at the beginning of Section A that radiation carries with it momentum in addition to energy. This is the cause of the pressure of light discovered by Maxwell. According to the last equations in Sec. 31 of Vol. III, the pressure exerted by a wave forming an angle θ with the normal to an element of area da is $u \cos^2 \theta$; it follows that for radiation coming from all sides the pressure is

$$(7) \qquad p = u \int_0^{\pi/2} \cos^2 \theta \sin \theta \, d\theta = u/3.$$

The preceding equation is valid for a partly reflecting surface as well as for a black one because the thrust due to the reflected radiation is added to that due to emitted radiation.

Let us now imagine an evacuated cylindrical vessel fitted with a sliding piston and filled with black body radiation at a temperature T. The volume V can be changed at will by moving the piston (infinitely slowly). The preceding constitutes a thermodynamic system with two variables and its energy is

$$U = V u(T),$$

whereas, according to (7), the work on the piston is given by

$$dW = p \, dV = \frac{1}{3} u(T) \, dV.$$

The change in entropy is

$$(8) \qquad dS = \frac{dU + dW}{T} = \frac{V}{T} \frac{du}{dT} dT + \frac{4}{3} \frac{u}{T} dV.$$

Since dS is a perfect differential we must have

$$\frac{1}{T} \frac{du}{dT} = \frac{4}{3} \frac{d}{dT} \left(\frac{u}{T} \right);$$

after a short calculation we find that

$$\frac{du}{u} = 4 \frac{dT}{T}; \qquad \log u = 4 \log T + \text{const, or}$$

$$(9) \qquad u = a T^4.$$

In order to determine the constant of integration a we substitute K for u from eq. (4) and we obtain

$$K = \frac{c\,a}{4\pi} T^4. \tag{10}$$

According to (2 a) the left-hand side represents the total radiation of a black surface (e. g. the hole in the wall of any black body cavity) per unit area and time. The constant $c\,a/4\pi$ which appears on the right-hand side is usually denoted by σ; it can be determined from observation. Equation (10) contains a statement of the law of radiation discovered empirically by Stefan. The preceding thermodynamic derivation was first given by Boltzmann in 1884. In his memorial address devoted to Boltzmann, H. A. Lorentz called it "a veritable pearl of theoretical physics."[1]

Substituting (9) into (8) we obtain

$$dS = 4\,a\left(V\,T^2\,dT + \frac{1}{3} T^3\,dV\right) = \frac{4}{3}\,a\,d(T^3\,V). \tag{11}$$

The integration of (11) does not lead to a new constant because according to the Third Law we must have $S = 0$ for $T = 0$. Thus we obtain

$$S = \frac{4}{3}\,a\,T^3\,V. \tag{12}$$

The equation of an isentrope in the T, V plane is represented by

$$T^3\,V = \text{const.} \tag{12 a}$$

It describes the change in temperature which accompanies an adiabatic and reversible change in volume (and hence, according to (9), also the change in the energy density u). Equation (12 a) is seen to be identical with the isentropic equation for a perfect gas whose ratio of specific heats $K = 4/3$.

C. Wien's law

The most significant idea which W. Wien used to determine the relation between frequency and temperature for black-body radiation consisted in his inquiring into the change in the spectrum of radiation on reflection from a moving mirror. It will be recalled from Vol. IV, Sec. 13, that the frequency of reflected light differs from that of incident light when the mirror moves

---

[1] Verh. d. Deutsch. Physik. Gesellschaft, 1907.

in the direction of its normal. The same is true of the intensity of incident and reflected radiation. Making use of the premiss that the modified spectrum must retain the properties of equilibrium radiation if the process is conducted in a suitable way it is possible to deduce the *shift* in the maximum of intensity and hence the color of the radiation which accompanies a change in temperature.

We shall refrain from proving Wien's law on the basis of a suitable model[1] and will concentrate on the widely discussed problem of whether it can be made plausible with the aid of dimensional analysis, that is on the basis of considerations of similarity.[2] As always we shall assume four fundamental units, one of them being temperature (symbol θ). The remaining three are the mechanical units, it being convenient to replace the unit of mass by the unit of energy (erg, symbol e).[3] Time and length will be denoted by t and l respectively.

According to (1 b) the dimension of u is $e\,t/l^3$. It is now necessary to express u in terms of ν and T (dimension t^{-1} or θ, respectively) and of certain universal constants. The latter include the speed of light c (dimension $l\,t^{-1}$) and the universal constant R, whose dimension is $e\,\theta^{-1}$ because $R\,T$ denotes an energy, as seen from the equation of state of a perfect gas. R is usually referred to one mol of some substance. In what follows, however, it is more convenient to refer it to a single molecule which can be effected by dividing it by the number of molecules per mol. It is known as Boltzmann's constant, k, and its dimension $e\,\theta^{-1}$ is the same as that of R.

The five quantities in question are shown listed together with their dimensions in the following (we shall refer to the last column presently):

(13)

u	ν	T	c	k	$h = k\alpha$
$e\,l^{-3}\,t$	t^{-1}	θ	$l\,t^{-1}$	$e\,\theta^{-1}$	$e\,t$

We now try to form a product of these five quantities, each raised to a certain (positive or negative) power satisfying the condition that it has the dimension zero in all four units

(14) $$e, l, t, \theta.$$

---

[1] The simplest proof of this kind was given by von Laue, Ann. d. Phys. (5) **43**, 220, (1943). The model consists of a single pencil of monochromatic rays and the proof is based on its invariance with respect to Lorentz transformations. Our argument only assumes invariance with respect to change of scale.

[2] *Cf.* a note by Glaser, Sitzungsber. d. Akad., Wien, Vol. 156, p. 87; our considerations are partly based on this note.

[3] It is assumed that the fourth unit of our electrodynamical system, the unit of electricity Q, does not occur in the argument.

We can assume that one of the exponents has a prescribed value, say unity, without loss of generality. In this manner the four remaining exponents are seen to be uniquely determined by the four equations which result from equating to zero the sum of the exponents for each of the units (14). There exists only *one* such product. Assuming that the exponent of u is equal to 1 we can deduce from Table (13) that $u\, c^3$ and $v^2\, k\, T$ have the dimension $e\, t^{-2}$, so that the product in question becomes

(15) $$\Pi = \frac{u\, c^3}{v^2\, k\, T},$$

and Π denotes an unknown universal number. The spectral distribution function becomes

(16) $$u = \frac{v^2\, k\, T}{c^3}\, \Pi.$$

This is the *unique* (except for an undetermined factor) answer which is supplied by *classical* physics to the problem of the spectrum of black body radiation. The adjective "classical" means here that the argument is confined to the application of the two universal constants c and k which have been in use in physics for a very long time.

Equation (16) was first deduced by Lord Rayleigh in 1900 who obtained it from classical statistics, finding at the same time that the numerical constant Π was equal to 8π. The equation was further developed by J. H. Jeans (the Rayleigh-Jeans radiation formula). It is, however, clear that the equation gives absurd results for large values of v, because it leads to an infinite value of u for $v \to \infty$, and because the integral for total radiation, $u = \int u\, dv$ is divergent.

In order to reach agreement with experiment we are forced to give up the limitation of using only two universal constants. There must be a third such constant, because it follows from Kirchhoff's law that apart from u, v, and T no other *variables* enter into the problem.

The third constant will lead to an additional dimensionless group Π' which is independent of eq. (15) and which may be assumed independent of u and depending on the first power of v without any loss in generality[1]. Thus we find that

(17) $$\Pi' = \alpha\, v\, T^n.$$

---

[1] If this were not the case it would suffice to multiply the number Π' by a suitable power of Π in order to eliminate u and to raise the result to such a power as to render the exponent of v equal to unity. The last operation is always possible because experiments show that Π' cannot be independent of v.

The constant α in this equation is a combination of c, k together with the new universal constant. Consequently

(17 a) $$\Pi = f(\Pi')$$

or

(17 b) $$\mathsf{u}(\nu, T) = \frac{\nu^2}{c^3} \bar{k}\, T \cdot f(\alpha\, \nu\, T^n).$$

The exponent n must be so selected as to yield eq. (9) on integration over all frequencies. From

$$u = \frac{kT}{c^3} \int_0^\infty f(\alpha\, \nu\, T^n)\, \nu^2\, d\nu$$

with the abbreviation $\alpha\, \nu\, T^n = x$, we have

$$u = \frac{k\, T^{1-3n}}{\alpha^3\, c^3} \int_0^\infty f(x)\, x^2\, dx.$$

The result will be proportional to T^4 only if we put $n = -1$. In this way eq. (17 b) leads to Wien's law:

(18) $$\mathsf{u}(\nu, T) = \frac{\nu^2\, k\, T}{c^3} f\left(\frac{\alpha\, \nu}{T}\right).$$

The unknown function of two variables, $\mathsf{u}(\nu, T)$, has thus been reduced to the unknown function, f, of a single variable, $\alpha\, \nu/T$. This is the great achievement of Wien's law.

It is convenient to include Boltzmann's constant, k, in the argument of f and to put $k\alpha = h$. This gives the more familiar form

(18 a) $$\mathsf{u}(\nu, T) = \frac{\nu^2\, k\, T}{c^3} f\left(\frac{h\, \nu}{k\, T}\right).$$

The quantity h represents a new constant and has the dimension of "action" i. e. $e\, t$. It completes our Table (13). We add here parenthetically that h is Planck's quantum of action which has now become a familiar fundamental constant and which has been anticipated, at least as far as its dimension is concerned, by Wien's law. Multiplying and dividing the coefficient of f in eq. (18 a) by $h\, \nu$, we obtain

(18 b) $$\mathsf{u}(\nu, T) = \frac{h\, \nu^3}{c^3} \cdot \frac{f(x)}{x} = \frac{h\, \nu^3}{c^3} f_1(x); \qquad x = \frac{h\, \nu}{k\, T}.$$

Consequently, the Stefan-Boltzmann constant a from eq. (9) becomes

$$(19) \qquad a = k \left(\frac{k}{hc}\right)^3 \times F, \qquad F = \int_0^\infty x^3 f_1(x)\,dx.$$

To conclude we shall give reasons for describing the preceding law as "Wien's displacement law." We now ask for that value of ν which corresponds to the maximum in the intensity u for a given temperature, that is that value of ν for which $\partial u/\partial \nu = 0$. From eq. (18 a), we find that it is given by

$$(20) \qquad 2f(x) + xf'(x) = 0, \qquad x = \alpha \nu/T.$$

We shall denote the real positive root of this equation by $x = x_m$ corresponding to $\nu = \nu_m$. Thus

$$(20\text{ a}) \qquad \nu_m = x_m T/\alpha.$$

As T increases the point of maximum intensity is "displaced" towards larger values of ν. Since the value of ν_m determines the general coloring perceived on observing the whole spectrum, eq. (20 a) is seen to supply an explanation for the transition from a red to a white glow at increasing temperature.

It has become customary to associate the values of λ rather than those of ν with our color perception. Since

$$\nu = \frac{c}{\lambda}; \qquad |d\nu| = \frac{c}{\lambda^2}|d\lambda|; \qquad \mathsf{u}\,|d\nu| = \mathsf{u}_\lambda\,|d\lambda|$$

the variation of the intensity u_λ on the scale of λ becomes

$$(21) \qquad \mathsf{u}_\lambda = \frac{kT}{c\lambda^2} f\left(\frac{\alpha c}{\lambda T}\right) \frac{|d\nu|}{|d\lambda|} = \frac{kT}{\lambda^4} f\left(\frac{\alpha c}{\lambda T}\right),$$

as seen from eq. (18). Introducing a new variable y and a new function $g(y)$ by

$$(21\text{ a}) \qquad y = \frac{\lambda T}{\alpha c}; \qquad g(y) = y f\left(\frac{1}{y}\right)$$

we obtain

$$(22) \qquad \mathsf{u}_\lambda = \frac{\alpha k c}{\lambda^5} g(y)$$

as seen from (21). Hence

$$(22\text{ a}) \qquad \frac{\partial \mathsf{u}_\lambda}{\partial \lambda} = -\frac{\alpha k c}{\lambda^6} [5\,g(y) - y\,g'(y)],$$

and the position of maximum intensity is given by the equation

(23) $$5\,g(y) - y\,g'(y) = 0.$$

Making use of the real positive value y_m of the root of this equation we find from eq. (21) that

(23 a) $$\lambda_m T = \alpha\,c\,y_m.$$

The root $y = y_m$ differs from the root x_m in eq. (20) because y and x have different meanings. Qualitatively the conclusion regarding the displacement in color is, evidently, the same as before: As the temperature is increased the value of λ_m is shifted towards shorter wavelengths (higher frequencies ν).

D. Planck's law of radiation

Planck inserts into the field of radiation a linear oscillator which reacts with it to a certain extent: it is a Hertz dipole of a definite natural frequency ω_0 whose dimensions are small compared with the relevant wavelengths. If the oscillator were free it would perform damped oscillations because of the electromagnetic radiation and with small damping it would react sharply on the frequencies ω of incident radiation which lie in the neighborhood of ω_0. Assuming that the incident and the excited oscillation are given by

(24) $$C \sin \omega t \quad \text{and} \quad D \sin(\omega t + \delta),$$

respectively, and applying the result in Vol. I, eq. (19.10) we find that

(24 a) $$D = \frac{C}{M} \{(\omega^2 - \omega_0^2)^2 + 4\rho^2 \omega^2\}^{-\frac{1}{2}}.$$

The oscillation equation must be assumed to be of the form

(25) $$m(\ddot{x} + 2\rho\,\dot{x} + \omega_0^2\,x) = e\,\mathsf{E}_x$$

in accordance with eq. (19.9) of Vol. I. E_x denotes here the component of the electrical field of radiation, E, which coincides with the direction of motion x; e and m denote the charge and mass of the oscillating electron. According to (25) the opposing damping force is equal to

(26) $$\mathsf{R} = -2\rho\,m\,\dot{x}.$$

Comparing it with the damping force ("reaction force") of radiation from Vol. III, eq. (36.4), we have

$$\mathsf{R} = \frac{e^2}{6\pi\,\varepsilon_0\,c^3}\,\dddot{x},$$

which can also be written

(26 a) $$R = -\frac{e^2}{6\pi\varepsilon_0 c^3}\omega^2 \dot{x}$$

because of the dependence of $x = D\sin(\omega t + \delta)$ on time. It follows from eqs. (26) and (26 a) that

(26 b) $$\rho = \frac{1}{12\pi m}\frac{e^2}{\varepsilon_0 c^3}\omega^2.$$

We now proceed to calculate the energy of the oscillator. Its kinetic energy is

$$\frac{m}{2}\dot{x}^2 = \frac{m}{2}D^2\omega^2\cos^2(\omega t + \delta),$$

and its potential energy can be found from (25):

$$\frac{m}{2}\omega_0^2 x^2 = \frac{m}{2}D^2\omega_0^2\sin^2(\omega t + \delta).$$

Taking into account eq. (24 a) we find that their sum averaged over time is equal to

(27) $$U_\omega = \frac{m}{4}D_\omega^2(\omega^2 + \omega_0^2) = \frac{C_\omega^2}{4m}\frac{\omega^2 + \omega_0^2}{(\omega^2 - \omega_0^2)^2 + 4\rho^2\omega^2}.$$

We have added the subscript ω to the energy U and to the amplitudes C, D in order to emphasize that, so far, we have been considering a single monochromatic oscillation. However, an oscillator placed in a field of radiation is excited by a continuous spectrum of mutually incoherent oscillations C_ω. The requirement of incoherence is as essential for our black body radiation as it was for natural "white" light in Vol. IV, Sec. 49.

It follows that in this case the squares of the amplitudes (intensities) are added, and not the amplitudes themselves as for coherent light. Hence (27) yields

(28) $$U = \int U_\omega d\omega = \frac{1}{4m}\int C_\omega^2 \frac{\omega^2 + \omega_0^2}{(\omega_0^2 - \omega^2)^2 + 4\rho^2\omega^2}d\omega.$$

The fraction in the integrand on the right-hand side varies strongly with ω and possesses a sharp maximum in the neighborhood of $\omega = \omega_0$ (the maximum is sharp owing to the smallness of the term $\rho^2\omega^2$). On the other hand C^2

varies slowly and may be replaced by its value C^2 at $\omega = \omega_0$. Thus instead of (28) we may write

(28 a) $$U = \frac{C^2}{4\,m} \int_0^\infty \frac{\omega^2 + \omega_0^2}{(\omega_0^2 - \omega^2)^2 + 4\,\rho^2\,\omega^2}\,d\omega.$$

This integral can be further simplified because the numerator and the term $4\,\rho^2\,\omega^2$ vary slowly. They may be replaced by

(28 b) $\qquad 2\,\omega_0^2$ and $4(\sigma\,\omega_0^2)^2\,\omega_0^2$ respectively with

(28 c) $$\sigma = \frac{1}{12\pi}\,\frac{e^2}{m\,\varepsilon_0\,c^3} \qquad \text{from eq. (26 b).}$$

We may further write

$$(\omega_0^2 - \omega^2)^2 = (\omega - \omega_0)^2\,4\,\omega_0^2.$$

In this way the integral under consideration becomes

$$\frac{1}{2}\int_0^\infty \frac{d\omega}{(\omega - \omega_0)^2 + (\sigma\,\omega_0^2)^2} = \frac{1}{2\,\sigma\,\omega_0^2}\int_{-1/\sigma\omega_0}^\infty \frac{d\xi}{\xi^2 + 1}\,;\qquad \xi = \frac{\omega - \omega_0}{\sigma\,\omega_0^2}.$$

Since $\sigma\,\omega_0 \ll 1$ we obtain

$$\frac{1}{2\,\sigma\,\omega_0^2}\,\tan^{-1}\xi\,\Big|_{-\infty}^{+\infty} = \frac{\pi}{2\,\sigma\,\omega_0^2},$$

and (28 a) transforms into

(29) $$U = \frac{\pi}{8\,m\,\sigma\,\omega_0^2}\,C_0^2.$$

It remains now to express C_0 in terms of the energy density, u, of black body radiation. The energy density is equal to twice its electrical contribution, i. e. to

$$(\mathsf{E},\,\mathsf{D}) = \varepsilon_0\,\mathsf{E}^2.$$

Taking a time average for black, isotropic radiation, we have

(30) $$\varepsilon_0\,\overline{\mathsf{E}^2} = 3\,\varepsilon_0\,\overline{\mathsf{E}_x^2} = u_\omega.$$

The last term on the right-hand side denotes the energy density on the ω-scale as distinct from that on the ν-scale used before. It will be recalled from (24) and (25) that C is equal to the amplitude of eE_x; averaging over time, we find that at $\omega = \omega_0$ we have

$$\frac{1}{2} C_0^2 = e^2 \, \overline{E_x^2}.$$

Substituting $\overline{E_x^2}$ from (30), we obtain

(30 a) $$C_0^2 = \frac{2}{3} \frac{e^2}{\varepsilon_0} \mathsf{u}_\omega$$

where u_ω denotes the energy density over the interval $d\omega$ at $\omega = \omega_0$. Consequently

$$\mathsf{u}_\omega \, d\omega = \mathsf{u}_\nu \, d\nu; \qquad \mathsf{u}_\omega = \frac{1}{2\pi} \mathsf{u}_\nu.$$

Hence (30 a) can be replaced by

(30 b) $$C_0^2 = \frac{1}{3\pi} \frac{e^2}{\varepsilon_0} \mathsf{u}_\nu.$$

Substituting this expression into (29) and taking into account (28 c) we find that the quantities e and m which refer to the specific model of the oscillator are cancelled (it is seen that the constant ε_0 also vanishes as might have been expected on dimensional grounds), and we obtain simply:

(31) $$U = \frac{\pi c^3}{2 \omega_0^2} \mathsf{u}_\nu = \frac{c^3}{8\pi \nu^2} \mathsf{u}_\nu.$$

The preceding argument shows that the energy of the *oscillator* is just as universal as the energy density of *black body radiation* so that in the succeeding reasoning Planck could use the former instead of the latter. He associated with the oscillator the entropy S, in addition to the temperature T, the former being given by

(32) $$dS = \frac{dU}{T},$$

at constant radiation volume ($dV = 0$).

In his Nobel Prize inaugural address delivered in 1920 Planck gave a fine example of objectivity; his law of radiation is described at first as "an interpolation formula which resulted from a lucky guess."

The experimental results obtained before 1900 (Paschen, Lummer, and Pringsheim) for the case of short wavelengths appeared to confirm an empirical hypothesis advanced by W. Wien. It follows from (18 b) by putting

$$f_1(x) = A\, e^{-x}, \qquad x = \alpha\, \nu/T.$$

Thus

(33) $$u(\nu, T) = \frac{\alpha\, k\, A}{c^3}\, \nu^3\, e^{-\alpha \nu/T}.$$

Correspondingly we obtain from (31)

(33 a) $$U = A_1\, e^{-\alpha \nu/T}; \qquad A_1 = \frac{\alpha\, k}{8}\, \nu\, A.$$

Evaluating $1/T$ and taking into account (32), we have

(33 b) $$\frac{dS}{dU} = -\frac{1}{\alpha\, \nu} \log\left(\frac{U}{A_1}\right)$$

and

(33 c) $$\frac{d^2 S}{dU^2} = -\frac{1}{\alpha\, \nu\, U}.$$

Later measurements performed at long wavelengths (infrared region) by Rubens and Kurlbaum revealed a completely different pattern of behavior which appeared to confirm the Rayleigh-Jeans equation (16). According to (31) the corresponding oscillator energy becomes

(34) $$U = k\, T,$$

if the numerical factor Π in (16) is given its value of 8π calculated by Rayleigh. Hence, according to (32), we find

(34 a) $$\frac{dS}{dU} = \frac{1}{T} = \frac{k}{U},$$

(34 b) $$\frac{d^2 S}{dU^2} = -\frac{k}{U^2}.$$

Planck now uses the following formula

(35) $$\frac{d^2 S}{dU^2} = -\frac{1}{\alpha\, \nu\, U + U^2/k}$$

to interpolate between (33 c) and (34 b). The right-hand side can be written

$$(35\text{ a}) \qquad -\frac{1}{\alpha\nu}\cdot\frac{1}{U+\beta U^2} = -\frac{1}{\alpha\nu}\left(\frac{1}{U}-\frac{\beta}{1+\beta U}\right); \qquad \beta = \frac{1}{\alpha\nu k},$$

and it is seen that (35) can be integrated. The constant of integration will be determined from the condition that for $U = \infty$ we must have $T = \infty$ so that $dS/dU = 0$. Hence

$$(36) \qquad \frac{dS}{dU} = -\frac{1}{\alpha\nu}\log\frac{\beta U}{1+\beta U}.$$

As seen from (32), the derivative dS/dU can be replaced by $1/T$ so that

$$(36\text{ a}) \qquad \log\frac{\beta U}{1+\beta U} = -\frac{\alpha\nu}{T}; \qquad \beta U = \frac{1}{e^{\alpha\nu/T}-1}.$$

Substituting from (35 a) we obtain

$$(36\text{ b}) \qquad U = \frac{\alpha\nu k}{e^{\alpha\nu/T}-1}.$$

According to Table (13) αk has the dimension of energy $\times$ time = action. It is seen that the new universal constant, the *quantum of action*

$$(37) \qquad h = \alpha k$$

which has already been mentioned previously, now makes its appearance. The energy of the oscillator becomes

$$(38) \qquad U = \frac{h\nu}{e^{h\nu/kT}-1}$$

and (31) leads to Planck's law of radiation

$$(39) \qquad u_\nu = \frac{8\pi\nu^2}{c^3}\frac{h\nu}{e^{h\nu/kT}-1}.$$

The statistical derivation of the same law, *cf.* Sec. 33, goes much deeper than this somewhat cumbersome argument and places the revolutionary character of the constant h in its proper light. The preceding argument outlined Plancks original train of thought and the reason for describing it here lies not only in its very great historical importance; it has been quoted also in order to demonstrate that the application of the concept of entropy to the oscillator plays a very important part in it.

Figure 21 shows a three-dimensional model of Planck's law of radiation in which u_ν has been plotted vertically upwards, ν to the right in the horizontal plane, and T has been plotted rearwards. The model consists of six plane profiles placed one behind the other. The profiles represent the dependence of u_ν on ν for $T = 100, 200, \ldots, 600$ K. The vertical profile which passes through the maxima ν_m as given by eq. (20 a) is developable owing to the linearity of the equation linking ν_m with T.

We now proceed to show how the limiting cases represented by the equations of Rayleigh-Jeans and Wien, respectively, can be deduced from (39):

For small values of ν and a fixed value of T we can expand the denominator of (39) into a series and obtain

(40) $$u = \frac{8\pi k}{c^3} \nu^2 T.$$

For large values of ν and a fixed value of T we can neglect 1 in the denominator of (39), and we have

(40 a) $$u = \frac{8\pi h}{c^3} \nu^3 e^{-h\nu/kT}.$$

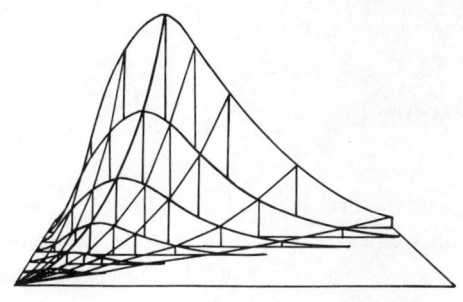

Fig. 21.

Cardboard model of Planck's law of radiation $u_\nu = f(\nu, T)$; ν is measured to the right, T is measured rearwards. The graded shading in the photograph is due to the light falling on the model. At $T = 600$ K the maximum is at $\nu_m = 4 \times 10^{13}$ sec^{-1}, at $T = 200$ K the much lower maximum is at $\nu_m = 12 \times 10^{12}$ sec^{-1}. The profile corresponding to $T = 100$ K protrudes so little that is is hardly visible.

Equation (40) is identical with (16) if we put $\Pi = 8\pi$, as already stated; eq. (40 a) transforms into eq. (33) if the previous constant A is also replaced by 8π. Finally the constant a in the Stefan-Boltzmann law obtains a definite theoretical justification. On comparing eq. (18.6) with (39) we obtain the following expression for the function f_1:

$$f_1(x) = \frac{8\pi}{e^x - 1}.$$

Hence the integral F in eq. (19) becomes:

(41) $$F = 8\pi \int_0^\infty \frac{x^3}{e^x - 1} dx.$$

Since for all values of $x > 0$ e^{-x} is less than unity, we can rewrite eq. (41) to give

$$(41\,a) \qquad \frac{F}{8\pi} = \int_0^\infty \frac{e^{-x}}{1-e^{-x}} x^3\, dx = \int_0^\infty (e^{-x} + e^{-2x} + e^{-3x} + \ldots)\, x^3\, dx.$$

Denoting $2x, 3x, \ldots$ in the 2nd, 3rd, $\ldots$ term of the series respectively by ξ, we have

$$(41\,b) \qquad \left(1 + \frac{1}{2^4} + \frac{1}{3^4} + \ldots\right) \int_0^\infty \xi^3 e^{-\xi}\, d\xi.$$

The integral is equal to $\Gamma(4) = 3!$, and the value of the series in the brackets in front of the integral can be taken from Vol. VI, eq. (2.18), where it has been shown to be equal to $\pi^4/90$. Hence eq. (41 b) gives $\pi^4/15$ and eq. (41 a) yields

$$(41\,c) \qquad F = 8\pi^5/15.$$

Substituting this value into (19) we obtain the following theoretical value of the Stefan-Boltzmann constant a:

$$(42) \qquad a = \frac{8\pi^5}{15} \frac{k^4}{(hc)^3}.$$

Since a and the constant $\alpha = h/k$ from Wien's displacement law are known from measurements, eqs. (42) and (23 a) with (37) can be, in turn, used to evaluate h and k. At the present time the following are regarded as their most accurate values

$$(43) \qquad h = 6.624 \times 10^{-27}\ \text{erg sec}; \qquad k = 1.380 \times 10^{-16}\ \text{erg/deg}.$$

21. Irreversible processes. Thermodynamics of near-equilibrium processes

A. Conduction of heat and local entropy generation

So far we have considered, essentially, only states of thermodynamic equilibrium. Concerning irreversible processes we were able only to establish that they are associated with an increase in entropy, provided that they take place in a closed system within an adiabatic boundary. We now propose to determine in greater detail where that entropy increase is located and how it depends on the parameters of the system.

We shall begin by considering a particularly simple example, namely the conduction of heat through a homogeneous and isotropic solid body disregarding its thermal expansion. If the temperature varies from point to point then, generally speaking, the internal energy per unit mass $u(x, y, z, t)$ will depend on the space coordinates and on time. The same is true of the flux of heat $\mathbf{W}$. The principle of the conservation of energy (see Vol. VI, eq. (7.11); it should be noted that the symbol u in that equation denoted temperature) can be written as

$$\rho \frac{\partial u}{\partial t} + \operatorname{div} \mathbf{W} = 0 \tag{1}$$

where ρ denotes density. To complete the description of the process of heat conduction it is necessary to write down the relation between internal energy and temperature, e. g. in differential form

$$du = c\, dT, \tag{2}$$

where c denotes the specific heat, and Fourier's hypothesis for the relation between the heat flux and temperature gradient (Secs. 44, 45 and Vol. VI, eq. (7.12)):

$$\mathbf{W} = -\varkappa \operatorname{grad} T, \tag{3}$$

where $\varkappa$ is the thermal conductivity.

For the time being we shall disregard eqs. (2) and (3) and we shall concentrate our attention on eq. (1). The internal energy and entropy are connected through the relation

$$du = T\, ds \tag{4}$$

because changes in volume have been neglected. From eqs. (1) and (4) we obtain

$$\rho \frac{\partial s}{\partial t} = -\frac{1}{T} \operatorname{div} \mathbf{W}, \tag{5}$$

or rearranged:

$$\rho \frac{\partial s}{\partial t} + \operatorname{div} \frac{\mathbf{W}}{T} = -\frac{1}{T^2} (\mathbf{W} \cdot \operatorname{grad} T). \tag{6}$$

Equation (1) is an equation of continuity which means that it expresses a conservation principle; in this case, that of energy. Equation (6) would also express a principle of conservation if its right-hand side vanished. Now it is known that entropy satisfies no conservation principle, moreover, it increases in an isolated system during an irreversible process, such as the conduction of heat, within it. This increase in entropy must now be related to the right-hand side of eq. (6), and to achieve it we shall integrate eq. (6) over the heat conducting body. Using Gauss' theorem (see Vol. II, eq. (3.1)) we obtain

$$\rho \frac{\partial}{\partial t} \int s \, dV + \oint \frac{W_n}{T} dA = - \int \frac{1}{T^2} (\mathbf{W} \cdot \text{grad } T) \, dV. \tag{7}$$

Assuming, at first, that the surface of the heat conducting body is adiabatically insulated we shall find that W_n vanishes. Then the term on the left-hand side of eq. (7) is the change in the entropy of the body per unit time. It is expressed on the right-hand side in terms of temperature, temperature gradient and heat flux. Since by Clausius' principle heat cannot flow spontaneously from a lower to a higher temperature ($\mathbf{W} \cdot \text{grad } T$) must be negative if $\mathbf{W} \neq 0$ and grad $T \neq 0$. Hence the right-hand side of eq. (6) is positive as required by the Second Law.

The change in entropy per unit time is thus given by a volume integral; it is natural to define the integrand as the change in entropy per unit time and per unit volume. We shall regard this quantity as the *entropy generated locally*. Hence the local entropy increase is defined as

$$\theta = -\frac{1}{T^2} (\mathbf{W} \cdot \text{grad } T) \tag{8}$$

and depends only on the state which prevails at the given instant and at the given point. In this context it is necessary to interpret the concept of state more widely than hitherto. We have, namely, stipulated constant temperature in eq. (8) and it is noted that in order to specify the state it is necessary to specify in addition the temperature gradient and hence (see eq. (3)) the heat flux.

If we now drop the restriction regarding the adiabatic nature of the boundary and imagine that it is in contact with heat sources, then W_n denotes the heat transferred to the respective source of heat per unit area of the boundary and per unit time (= energy, since work = 0), and W_n/T represents the entropy transferred from the body to the source. It is, therefore, natural to define W_n/T as the *entropy flux*

$$S = \frac{W}{T}. \tag{9}$$

With these conventions eq. (6) becomes

(10) $$\rho \frac{\partial s}{\partial t} + \operatorname{div} \mathbf{S} = \theta$$

and integration over any portion of the body gives

(11) $$\rho \frac{\partial}{\partial t} \int s\, dV + \oint S_n\, dA = \int \theta\, dV.$$

Reading from right to left we can now interpret the physical meaning of this equation: The quantity of entropy generated inside the volume of integration is partly conducted away from it and partly contributes to the change in the entropy of the volume. Evidently the latter contribution can also be negative.

The preceding argument allowed us to determine the sources of entropy and their output for the case of the conduction of heat. In accordance with the Second Law we can, further, establish that this output can never be negative. Equation (10) together with the inequality $\theta \geqslant 0$ can be regarded as the *differential formulation of the Second Law of Thermodynamics*. The statement in integral form, namely that the entropy in an isolated system cannot decrease, can be replaced by its corollary in differential form which asserts that the quantity of entropy generated locally cannot be negative irrespective of whether the system is isolated or not, and irrespective of whether the process under consideration is irreversible or not.

We shall now compare Fourier's hypothesis, eq. (3), with the expression for the quantity of entropy generated locally in eq. (8). We find that the latter contains as factors precisely the quantities $\mathbf{W}$ and $\operatorname{grad} T$ which enter into Fourier's equation. This fact, as will be seen later, has a more general significance. Moreover, from

$$\theta = \frac{\varkappa}{T^2} (\operatorname{grad} T)^2 \geqslant 0$$

we deduce that

$$\varkappa \geqslant 0.$$

B. The conduction of heat in an anisotropic body and Onsager's reciprocal relations

We now proceed to consider the more general case of the conduction of heat in an anisotropic body such as a crystal of arbitrary constitution. The preceding argument remains unchanged except for eq. (3) which must now be replaced by a tensor relation between the components of the temperature gradient and of the heat flux (we now denote the coordinates by x_1, x_2, and x_3).

(12)
$$\begin{cases} W_1 = -\varkappa_{11}\dfrac{\partial T}{\partial x_1} - \varkappa_{12}\dfrac{\partial T}{\partial x_2} - \varkappa_{13}\dfrac{\partial T}{\partial x_3}\,, \\ W_2 = -\varkappa_{21}\dfrac{\partial T}{\partial x_1} - \varkappa_{22}\dfrac{\partial T}{\partial x_2} - \varkappa_{23}\dfrac{\partial T}{\partial x_3}\,, \\ W_3 = -\varkappa_{31}\dfrac{\partial T}{\partial x_1} - \varkappa_{32}\dfrac{\partial T}{\partial x_2} - \varkappa_{33}\dfrac{\partial T}{\partial x_3}\,. \end{cases}$$

It expresses the fact that in a crystal the temperature gradient and heat flux are not, generally speaking, parallel (more precisely anti-parallel).

If, as before, we now compare the assumption in eq. (12) (such an assumption, as well as similar ones made in connection with the other irreversible processes, is known as a phenomenological hypothesis) with the expression for local entropy generation in eq. (8), rewriting the latter as

(13)
$$\theta = -\frac{1}{T^2}\left(W_1\frac{\partial T}{\partial x_1} + W_2\frac{\partial T}{\partial x_2} + W_3\frac{\partial T}{\partial x_3}\right),$$

we notice that the phenomenological hypothesis, eq. (12), expresses the first factors W_1, W_2, W_3 in eq. (13) in terms of the second factors, $\partial T/\partial x_1$, $\partial T/\partial x_2$ and $\partial T/\partial x_3$, as linear homogeneous functions. In any case the phenomenological hypothesis in eq. (12) is not arbitrary. It must, first, satisfy the condition that $\theta \geqslant 0$ for any temperature gradient, i. e.

(14)
$$\sum_i \sum_k \varkappa_{ik}\frac{\partial T}{\partial x_i}\frac{\partial T}{\partial x_k} \geqslant 0$$

which shows that the tensor $\varkappa_{ik}$ turns out to be non-negative definite. Secondly, we must have

(15)
$$\varkappa_{ik} = \varkappa_{ki} \qquad (i, k = 1, 2, 3)$$

which shows that the tensor is symmetrical.

The last relation is confirmed by experiment[1] and follows from kinetic theories of heat conduction. It is, finally, a particular case of quite general symmetrical relations which were postulated by Onsager[2]. We shall revert to a more general formulation of these reciprocal relations later.

---

[1] M. Voigt, Nachr. Ges. Wiss. Göttingen, Math. Phys. Class, p. 87 (1903); Ch. Soret, Arch. de Genève, Vol. 29, p. 355 (1893), Vol. 32, p. 611 (1894).
[2] L. Onsager, Phys. Rev. Vol. 37, p. 405, Vol. 38, p. 2265 (1931).

In accordance with the language of the general thermodynamic theory of irreversible processes we consider that in the preceding example there are three elementary irreversible processes which are superimposed on one another. Each of them corresponds to an elementary irreversible process, associated with one coordinate direction. Furthermore, in accordance with the hypothesis in eq. (12) and from eq. (13) it is seen that when several elementary irreversible processes interact with each other the quantity of entropy generated locally can be split into three terms each of which is due to one irreversible process only. On the other hand the phenomenological hypothesis shows that such elementary processes may be coupled, meaning that one temperature gradient, e. g. that in the direction x_1, can give rise to a heat flux in another direction, such as x_2 and x_3.

This feature is quite general and may appear during an interaction of completely dissimilar irreversible processes such as heat conduction and diffusion, the conduction of heat and electricity, etc. In these cases the coupling of irreversible processes through the respective phenomenological hypotheses leads to thermal diffusion (known as the Soret effect when condensed phases are concerned) and the Dufour effect or thermal effusion (temperature gradient evoked during diffusion), or to thermoelectric phenomena.

C. Thermoelectric phenomena

In the case of thermoelectric effects we are dealing, on the one hand, with the flux of energy and electricity and, on the other, with temperature gradients and electric field intensities as their causes. We consider a metal which carries an electric current and throughout which there exists a temperature gradient. We can write down the principle of the conservation of energy from which, in turn, we can deduce the entropy equation which corresponds to eq. (6). We assume here that the specific internal energy, u, and the specific entropy, s, are independent of the current density **I**. This hypothesis is of the same nature as the implicit assumption in Section A that the internal energy depends on temperature but not on the heat flux or temperature gradient. The electron theory of metals furnishes further justification for such an assumption (see also Sec. 45).

We shall now postulate the existence of only one kind of mobile carriers of electricity. These can be regarded as being endowed with a negative charge $-e(e > 0)$ without introducing any essential limitation into our argument. The same idea forms also the basis of the electron theory of metals.

In formulating the energy equation it is necessary to take into account that the metal receives a quantity (**I · E**) of electrical energy per unit time and volume and that due to charging with $-\operatorname{div} \mathbf{I}$ per unit time there is an increase in potential energy by an amount $-\Phi \operatorname{div} \mathbf{I}$, where Φ denotes the electrical potential and $\mathbf{E} = -\operatorname{grad} \Phi$ is the electric field strength. It is hereby implied that the electric currents vary at a slow rate. The principle of the conservation of energy, in analogy with eq. (1), now becomes

$$(16) \qquad \rho \frac{\partial u}{\partial t} = -\operatorname{div} \mathbf{W} + (\mathbf{I} \cdot \mathbf{E}) - \Phi \operatorname{div} \mathbf{I}.$$

The differential of specific entropy from eqs. (18.2) and (18.3) assumes the form

$$(17) \qquad T\,ds = du - (\mu - F\Phi)\,dn$$

on the assumption that changes in volume are negligible and taking into account that $z = -1$. Here Ln is the number of carriers of electricity per gram (L = Avogadro's number), μ denotes the chemical and $\mu - F\Phi$ the electrochemical potential of the carriers. Since $F = Le$ and since $-\rho L n e$ represents the charge per unit volume, we have

$$\rho F \frac{\partial n}{\partial t} = -\rho \frac{\partial(-L n e)}{\partial t} = \operatorname{div} \mathbf{I}.$$

Hence

$$(17\text{ a}) \qquad T \rho \frac{\partial s}{\partial t} = \rho \frac{\partial u}{\partial t} + \left(\Phi - \frac{\zeta}{e}\right) \operatorname{div} \mathbf{I},$$

where we have put $\mu/F = \zeta/e$ i. e. $\zeta = \mu/L$. In the electron theory of metals the quantity ζ is referred to as the chemical potential per electron. Eliminating $\partial u/\partial t$ from eqs. (16) and (17 a) and rearranging slightly, we obtain

$$(17\text{ b}) \qquad \rho \frac{\partial s}{\partial t} + \operatorname{div} \frac{1}{T}\left(\mathbf{W} + \frac{\zeta}{e}\mathbf{I}\right) =$$

$$= \frac{1}{T}\left(\mathbf{W} \cdot -\frac{1}{T}\operatorname{grad} T\right) + \frac{1}{T}\left(\mathbf{I} \cdot \mathbf{E} + T \operatorname{grad} \frac{\zeta}{eT}\right).$$

This equation is the counterpart of eq. (6). Here again we shall regard the quantity $\dfrac{1}{T}\left(\mathbf{W} + \dfrac{\zeta}{e}\mathbf{I}\right)$ as the flux of entropy, whereas the right-hand side of the equation represents the quantity of entropy generated locally, θ. The results derived in Sections A and B show how to obtain the assumptions for the fluxes of heat and current of electricity from the local entropy increase.

First we find from eq. (17 b) that θ is a linear function of the fluxes **W** and **I** and express them again as linear functions of their coefficients $-1/T \operatorname{grad} T$ and $\mathbf{E} + T \operatorname{grad}(\zeta/e\, T)$, namely

$$(18) \quad \begin{cases} \mathbf{W} = -\dfrac{\alpha}{T} \operatorname{grad} T + \beta \left(\mathbf{E} + T \operatorname{grad} \dfrac{\zeta}{eT} \right) \\ \mathbf{I} = -\dfrac{\gamma}{T} \operatorname{grad} T + \delta \left(\mathbf{E} + T \operatorname{grad} \dfrac{\zeta}{eT} \right). \end{cases}$$

Solving for **E** and **W**, we have, with the usual notation

$$(18\text{ a}) \quad \mathbf{W} = -\varkappa \operatorname{grad} T - \left(\Pi + \dfrac{\zeta}{e} \right) \mathbf{I}$$

$$(18\text{ b}) \quad \mathbf{E} = \dfrac{1}{\sigma} \mathbf{I} - \varepsilon \operatorname{grad} T - \operatorname{grad} \dfrac{\zeta}{e}$$

with the coefficients $\varkappa$, Π, ε, $1/\sigma$ whose significance will be further investigated later. Their relation to β and γ is noted for further reference:

$$(18\text{ c}) \quad \beta = -\sigma \left(\Pi + \dfrac{\zeta}{e} \right); \qquad \gamma = -\sigma \left(\varepsilon\, T + \dfrac{\zeta}{e} \right).$$

Equation (18 b) connects the jump in potential across a boundary between two metals with the jump in ζ. Namely, if eq. (18 b) is integrated along a very small path crossing the boundary between metals I and II, we obtain

$$(18\text{ d}) \quad \Phi_{\mathrm{II}} - \Phi_{\mathrm{I}} = \dfrac{1}{e} (\zeta_{\mathrm{II}} - \zeta_{\mathrm{I}})$$

because the contributions of the first two terms on the right-hand side of the equation can be made as small as we please. The difference $\Phi_{\mathrm{II}} - \Phi_{\mathrm{I}}$ is the contact potential between the two metals. At equilibrium $\mathbf{I} = 0$, $\mathbf{W} = 0$, $\operatorname{grad} T = 0$ and hence $\mathbf{E} = -\operatorname{grad} \zeta/e$. It is found from eq. (18 b) that at equilibrium electrical field strengths are only present in regions where ζ varies at constant temperature. In other words they exist only in regions where the material is non-homogeneous, i. e. in particular across the boundary between two different homogeneous materials.

If no electric current is present eq. (18 a) reduces to Fourier's law of heat conduction and $\varkappa$ is the thermal conductivity as seen upon comparing with eq. (3). Equation (18 b) shows that in such a case there exists everywhere a field of strength $\mathbf{E} = -\varepsilon \operatorname{grad} T - \operatorname{grad} \zeta/e$. The coefficient ε is known as the absolute *thermal force* (see also eq. (25) in Sec. 45, which contains an explicit expression for ε).

If the temperature is uniform everywhere and if the material is homogeneous so that grad $\zeta = 0$, eq. (18 b) will reduce to Ohm's law with σ denoting the electrical conductivity. In general, i. e. when the temperature is not constant throughout the space, eq. (18 b) can be rewritten as

$$\text{(19)} \qquad \mathbf{I} = \sigma(\mathbf{E} + \mathbf{E}^e).$$

The quantity denoted by $\mathbf{E}^e$ is the *impressed electric field strength*. In this connection eq. (18 a) shows that an energy flux may be present even when there are no temperature differences, provided that $\mathbf{I} \neq 0$. In other words, the transport of electricity is coupled with a transport of energy. This is contingent on the fact that the transport of electricity and that of energy have a common cause in the motion of electrons in the metal and the electron theory of metals confirms this assumption in all respects. In accordance with the present notation the energy transported per Coulomb is equal to $-(\Pi + \zeta/e)$.

We shall now proceed to consider thermoelectric phenomena and we shall begin by discussing the *Thomson effect*. This effect occurs in an electric conductor which carries a current and along which a temperature gradient is maintained and consists in the fact that so-called *Thomson heat* appears in addition to Joule heat, as is seen by substituting (18 b) into (17). The amount of Thomson heat per unit volume and time is equal to μ ($\mathbf{I} \cdot \operatorname{grad} T$) where μ (not to be confused with the μ from eq. (17)) is the *Thomson coefficient*. This additional quantity of heat can be positive or negative depending on the relative direction of $\mathbf{I}$ and grad T. It is customary to refer to this effect as being reversible because it changes sign with a change in the direction of $\mathbf{I}$ or grad T. This term is, however, a misnomer, because the Thomson effect constitutes only one aspect of the whole process, and, moreover, it is intimately interlocked with heat conduction and with the generation of Joule heat, both of them typically irreversible processes.

The existence of Thomson heat is implied in the preceding fundamental equations and this is readily seen when considering the accumulation of heat per unit volume and time, or $\rho\, \partial u/\partial t = (\mathbf{I} \cdot \mathbf{E}) - \operatorname{div} \mathbf{W} - \Phi \operatorname{div} \mathbf{I}$ with the substitution of $\mathbf{W}$ and $\mathbf{E}$ from eqs. (18 a) and (18 b) respectively. Since div $\mathbf{I} = 0$ (which is always true for direct current, and approximately so if the current varies but slowly) we obtain

$$\rho \frac{\partial u}{\partial t} = \operatorname{div}(\varkappa \operatorname{grad} T) + \left(\frac{\partial \Pi}{\partial T} - \varepsilon\right)(\mathbf{I} \cdot \operatorname{grad} T) + \frac{1}{\sigma} \mathbf{I}^2.$$

The first term on the right-hand side gives the accumulation of heat for heat conduction alone, the last term being the Joule heat. The second term has a form to be expected of Thomson heat. From the definition of the Thomson coefficient *the first Thomson relation* is deduced, i. e.

$$\mu = \frac{\partial \Pi}{\partial T} - \varepsilon. \tag{20}$$

It was first obtained by Thomson; speaking more precisely Thomson inferred the existence of an additional heat term from the fact that, generally speaking, $\partial \Pi / \partial T - \varepsilon \neq 0$ and that, otherwise the energies would not balance.

The coefficient Π is known as the *Peltier coefficient*. It is connected with the *Peltier effect*, i. e. with the positive or negative flow of heat at a boundary between two different metals. Considering the arrangement in Fig. 22 and assuming that the temperature remains constant throughout we can calculate that the flow of energy from left to right in metal I is given by: $-[\Pi_\mathrm{I} + (1/e)\zeta_\mathrm{I}]\,\mathrm{I}$, whereas that in metal II is given by: $-[\Pi_\mathrm{II} + (1/e)\zeta_\mathrm{II}]\,\mathrm{I}$, assuming a unit cross-sectional area of both conductors at the boundary.

Hence we obtain an accumulation of energy of a magnitude

$$\left[\Pi_\mathrm{II} - \Pi_\mathrm{I} + \frac{1}{e}(\zeta_\mathrm{II} - \zeta_\mathrm{I})\right]\mathrm{I}. \tag{21}$$

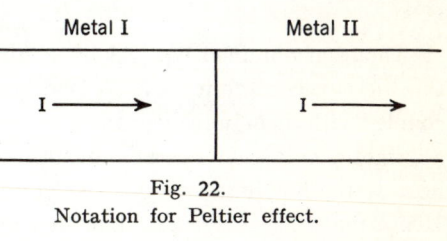

Fig. 22.
Notation for Peltier effect.

The quantity $1/e(\zeta_\mathrm{II} - \zeta_\mathrm{I})\,\mathrm{I}$ is used up in lifting the carriers of electricity through a potential difference from Φ_I to Φ_II across the boundary, so that only the quantity $(\Pi_\mathrm{II} - \Pi_\mathrm{I})\,\mathrm{I}$ is left over.

We can, finally, calculate the emf of a circuit which is composed of two different metals in which the metals are not kept at a constant temperature. Thus

$$\oint (\mathbf{E}^e \cdot d\mathbf{r}) = \oint \varepsilon\,(\mathrm{grad}\,T \cdot d\mathbf{r}) + \oint \left(\mathrm{grad}\,\frac{\zeta}{e} \cdot d\mathbf{r}\right) = \oint \varepsilon\,dT.$$

It is convenient to express the integral with the aid of two part integrals, each of which is taken over one metal. It T_1 and T_2 denote the temperatures of the junctions and ε_I and ε_II denote the absolute values of the emf's of the two metals respectively, we obtain

$$\oint (\mathbf{E}^e \cdot d\mathbf{r}) = \int_{T_1}^{T_2} \varepsilon_\mathrm{II}\,dT + \int_{T_2}^{T_1} \varepsilon_\mathrm{I}\,dT = \int_{T_1}^{T_2} (\varepsilon_\mathrm{II} - \varepsilon_\mathrm{I})\,dT. \tag{22}$$

This shows that the emf of a closed circuit depends only on the temperatures of the junctions and that it vanishes when both metals are identical (as in such a case $\varepsilon_I = \varepsilon_{II}$) and, finally, that for sufficiently small temperature differences it is approximately equal to $(\varepsilon_{II} - \varepsilon_I)(T_2 - T_1)$. It is necessary to remark here that from measurements of the Peltier heat as well as from measurements of the emf's it is possible to obtain only differences of Peltier coefficients or emf's for *pairs* of metals. Equations (20) and (23) can, therefore, be verified only with respect to two metals. On the other hand, the electron theory of metals is capable of defining Π and ε for a single metal.

Equation (20) shows that the three thermoelectric effects are coupled with each other. Onsager's reciprocal relations lead to another important equation connecting these quantities. According to it the matrix of coefficients in eq. (18) must be symmetrical which leads to *Thomson's second relation* $\beta = \gamma$ or, from eq. (18 c)

(23) $$\Pi = T\varepsilon.$$

Thomson obtained eq. (23) by a different line of reasoning. He separated the thermoelectric effects from heat conduction and Joule heat, which are coupled with them in reality, and considered that the thermocouple constituted a Carnot engine when operated in steady-state. In such a case only the Peltier heat at the hot junctions is considered as the heat absorbed by the system. Hence the efficiency of a cycle with a temperature difference dT between the source and the sink and with a quantity of work equal to the electrical energy $I \oint (\mathbf{E} \cdot d\mathbf{r})$ becomes

$$\frac{\Delta T}{T} = \frac{(\varepsilon_{II} - \varepsilon_I)\Delta T}{\Pi_{II} - \Pi_I}$$

which is a consequence of eq. (23).

There is little justification for such a separation of the so-called reversible and irreversible effects, in spite of the fact that it leads to a result which can be verified experimentally.[1] It is, therefore, very gratifying that the electron

---

[1] Thomson himself explicitly stated that such a separation involves a new hypothesis and that an experimental proof is required, because "Not only are the conditions prescribed in the second Law of the Dynamical Theory not completely fulfilled, but the part of the agency which does fulfil them is in all known circumstances of thermo-electric currents excessively small in proportion to agency inseparably accompanying it and essentially violating those conditions" (Trans. Roy. Soc. of Edinburgh, Vol. XXI, p. 128, 1 May 1854). In his careful analysis, Boltzmann showed that this hypothesis is untenable, thus fully confirming Thomson's misgivings (Sitzungsber. d. Akad. d. Wiss., Vienna, Math. Naturw. Klasse, II Div. **96**, 1258 (1888)).

theory of metals was able to confirm Thomson's second relation without the need for such additional 'ad hoc' hypotheses.[1] It provided, moreover, one of the first examples for the direct proof of an Onsager reciprocal relation and articulated the basic principle which underlies it, namely the principle of microscopic reversibility (i. e. the property of invariance of fundamental laws with respect to a change in the direction of time). This principle was later so imaginatively generalized by Onsager.

In this connection it is necessary to point to certain generalizations which are important when considering thermoelectric phenomena in anisotropic bodies and when an external magnetic field **B** is superimposed. It is particularly noteworthy that in the case of magnetic fields Onsager's principle must be modified. In the simplest case for the tensor of thermal conductivity we obtain

$$\varkappa_{ik}(\mathbf{B}) = \varkappa_{ki}(-\mathbf{B})$$

instead of eq. (15).

This change of sign is connected with the fact that the principle of microscopic reversibility applies only when the magnetic field is reversed simultaneously. Considering, for example, the equation of motion of an electron in a magnetic field, i. e. $m\,\dot{\mathbf{v}} = -e(\mathbf{v} \times \mathbf{B})$ it is noticed that it remains unaltered by the substitution $t, B \rightarrow -t, -\mathbf{B}$.

The process of *thermal diffusion*, and that of *thermal effusion* which was discovered by Dufour[2] and very convincingly demonstrated by Clausius and Waldmann[3] can be analysed by similar methods. An apparent complication occurs due to the fact that changes in volume and flow phenomena must be taken into account. The application of Onsager's principle leads to a relation connecting the two effects and that relation is confirmed experimentally. To-day use is made of this relation to obtain very accurate values of coefficients of thermal diffusion from measurements on thermal effusion.[4]

D. Internal transformations

In the preceding sections we have considered transport phenomena (the transport of energy and electricity) and now we propose to discuss irreversible processes in non-homogeneous matter, i. e. so-called internal transformations or relaxation phenomena which are not accompanied by transport phenomena.

---

[1] A. Sommerfeld, Zeitschr. f. Physik Vol. 47, pp. 1 and 43 (1928).
[2] Dufour, Ann. Physik, Vol. 28, p. 490 (1873).
[3] Kl. Clausius and L. Waldmann, Naturw. Vol. 30, p. 711 (1942).
[4] L. Waldmann, Z. f. Physik, Vol. 124, (1944) p. 30.

(If the internal transformations consist of chemical reactions in the usual sense they are also termed homogeneous reactions.) In this case the volume must be kept constant and the introduction of heat must be excluded, so that we put $dV = 0$ and $dU = 0$. We shall consider 1 gram of matter consisting of three components A_0, A_1 and A_2 which may take part in two reactions $A_0 \rightleftarrows A_1$ and $A_0 \rightleftarrows A_2$. All other possible cases involving an arbitrary number of components and arbitrary types of chemical reactions can be treated in the same way so that we may restrict ourselves to this simple scheme. With $dV = 0$ and $dU = 0$ eq. (2) of Sec. 14 becomes

$$T\,ds = -\mu_0\,dn_0 - \mu_1\,dn_1 - \mu_2\,dn_2.$$

The quantities μ_i represent the chemical potentials of the single components and they will be referred here to 1 gram instead of to 1 mol of substance, whereas the quantities n_0, n_1 and n_2 denote the quantity of each component respectively, all per 1 gram of mixture. Thus $n_0 + n_1 + n_2 = 1$ and only n_1 and n_2 are independent. Hence

(24) $$T\,ds = (\mu_0 - \mu_1)\,dn_1 + (\mu_0 - \mu_2)\,dn_2.$$

The rate of change of entropy with time or, since transport phenomena are excluded, the rate at which entropy is created becomes

(25) $$\rho\frac{ds}{dt} = \frac{\rho}{T}(\mu_0 - \mu_1)\frac{dn_1}{dt} + \frac{\rho}{T}(\mu_0 - \mu_2)\frac{dn_2}{dt}.$$

This expression is also a linear function of dn_1/dt and dn_2/dt and their coefficients describe the deviation from equilibrium. In equilibrium $ds \leqslant 0$, as shown by eq. (8.1), for every virtual change of state dn_1, dn_2 which implies $\mu_0 - \mu_1 = \mu_0 - \mu_2 = 0$.

For small deviations from thermodynamic equilibrium we may, again, express dn_1/dt and dn_2/dt in eq. (25) as linear functions of their coefficients:

(26) $$\begin{cases} \rho\dfrac{dn_1}{dt} = a_{11}(\mu_0 - \mu_1) + a_{12}(\mu_0 - \mu_2) \\ \rho\dfrac{dn_2}{dt} = a_{21}(\mu_0 - \mu_1) + a_{22}(\mu_0 - \mu_2) \end{cases}$$

where $a_{ik} = a_{ik}(T, v)$ and in accordance with Onsager's principle we have $a_{12} = a_{21}$. The connection with the principle of detailed equilibrium is discussed in the solution to Problem II 7.

To conclude this section we shall add some general remarks. It was necessary to impose the condition $dU = 0$ in order to exclude transport phenomena. A change in internal energy can take place only through the performance of work, i. e. through a change in volume or through an exchange of heat. On the other hand the assumption that only the two reactions $A_0 \rightleftarrows A_1$ and $A_0 \rightleftarrows A_2$ took place was superfluous; the argument will not change if the reaction $A_1 \rightleftarrows A_2$ is also admitted. This circumstance is due to the fact that only *independent* reactions are important and their number determines the number of independent n_i' s. In the case under consideration the reaction $A_1 \rightleftarrows A_2$ would not be independent being the difference between the reactions $A_0 \rightleftarrows A_1$ and $A_0 \rightleftarrows A_2$. Evidently any two reactions can be taken as the independent ones. Furthermore, the reactions as written down need not faithfully represent their molecular mechanism i. e. they are so-called elementary reactions. It is even permissible to choose arbitrary linear combinations of such elementary reactions with arbitrary numerical coefficients without in any way affecting the argument. This is a consequence of the fact that the thermodynamic theory of irreversible processes concerns itself only with the phenomenological aspects of the processes and does not consider their molecular mechanism.

At this point it is useful to recall the statement in eq. (8.9) which asserted that for a *reversible* process $\Sigma X_i dx_i = 0$. Equation (25) shows that the rate of entropy creation, θ, per time dt for a process involving the changes dx_i is given by $\rho \Sigma X_i dx_i$ and that for an *irreversible* process we always have

$$\sum_i X_i dx_i > 0.$$

E. General Relations

The discussion of more general irreversible processes involving the coupling of internal changes with transport phenomena exceeds the scope of these lectures. We shall confine ourselves to the remark that in each particular case the change in entropy of an isolated unit of mass considered along its path in the field of motion can be calculated with the aid of the conservation laws and Gibb's equation (14.2) and that it can always be written in the form[1]

(27)
$$\rho \frac{ds}{dt} + \text{div } \mathbf{S} = \theta$$

where $\mathbf{S}$ denotes the *entropy flux*. The entropy flux is a linear function of the energy flux and of the diffusion fluxes and electrical current density if

---

[1] This fact was first clearly formulated by G. Jaumann; his name has already been mentioned in another connection in Vol. III.

they exist in the system. It does not contain the convective entropy flux, i. e. the entropy transported by matter in motion because eq. (27) is valid for an observer who travels with the element of matter. The quantity θ is the *entropy created locally* and can always be written as

$$(28) \qquad \theta = \frac{1}{T} \sum_i K_i X_i.$$

The quantities X_i are generalized fluxes such as the energy flux (or its three components), the flux of momentum due to internal friction, i. e. its six components p_{ik} considered in Vol. II eq. (10.18), the quantities dn_i/dt considered in the preceding section etc. The coefficients K_i will be referred to as thermodynamic forces. They are, e. g. $-1/T$ grad T, T grad μ_i/T, $\partial v_i/\partial x_k$, the chemical potentials themselves or their differences (as in eq. (25)) etc. The energy dissipation term $T\theta$ is thus always a sum of products of fluxes and forces.

When the deviations from thermodynamic equilibrium are small the fluxes X_i are linear functions of the forces K_k, thus

$$(29) \qquad X_i = \sum_{k=1}^{n} a_{ik} K_k.$$

From this point onwards we are in a position to formulate *Onsager's reciprocal relations*. However, contrary to what might be expected from previous sections their general formulation cannot be taken to be

$$(30) \qquad a_{ik} = a_{ki}.$$

Casimir[1] was the first to point out that these relations are true only for cases when the forces K_i and K_k which are associated with the pair of subscripts i, k are both even or both odd functions of the flow velocities or of the molecular velocities. If, however, one of the forces is even and the other odd, we have to write with Casimir, that

$$(31) \qquad a_{ik} = -a_{ki}.$$

We cannot give here the general proof of eqs. (30) and (31). Following Onsager and Casimir it is possible to obtain these relations from the *principle*

---

[1] H. B. G. Casimir, Rev. Mod. Physics, Vol. 17, p. 343 (1945).

of microscopic reversibility, or, in special cases, e. g. with the aid of the kinetic theory of gases as applied to gaseous flows. In the latter case it would be necessary at least to consider gaseous mixtures if non-trivial Onsager-type relations are to be obtained. For a homogeneous gas the energy dissipation, as will also be shown in Chap. V, is given by

$$(32) \qquad T\theta = \frac{1}{2} \sum_i \sum_k p_{ik}\left(\frac{\partial v_i}{\partial x_k} + \frac{\partial v_k}{\partial x_i}\right) + \left(\mathbf{W}\cdot - \frac{1}{T}\operatorname{grad} T\right)$$

(for an explanation of the second term see Section B, and for the explanation of the first term see Vol. II, Sec. 10.18). The phenomenological relations in eq. (29) would represent the quantities W_x, W_y and W_z as linear functions of the terms $-(1/T)\,(\partial T/\partial x_i)$ and of the six terms $(\partial v_i/\partial x_k + \partial v_h/\partial x_i)$. The three phenomenological relations must remain covariant with respect to a rotation of the system of coordinates, since the gas is isotropic at all points, or, in other words, its form must be the same, with the same coefficients, when expressed with the aid of the components in the new system of coordinates.

It follows that the coefficients of the terms $(\partial v_i/\partial x_k + \partial v_k/\partial x_i)$ must vanish (a vector cannot depend linearly on a tensor if the continuum is isotropic). Further, it also follows from the condition of isotropy that the heat conduction tensor must reduce itself to a multiple of the unit tensor δ_{ik}. A similar argument involving the phenomenological assumptions for the p_{ik} leads to the conclusion that there can be no coupling with the temperature gradient. Thus almost all mixed phenomenological coefficients vanish, only some mixed coefficients in the relation between p_{ik} and $(\partial v_i/\partial x_k + \partial v_k/\partial x_i)$ remain. Thus, as already shown in Vol. II, Sec. 10.21, and as we shall show again in Chap. V of the present volume, we can write

$$(33) \qquad p_{ik} = \eta\left(\frac{\partial v_i}{\partial x_k} + \frac{\partial v_k}{\partial x_i}\right) + \left(\zeta - \frac{2}{3}\eta\right)\left(\frac{\partial v_1}{\partial x_1} + \frac{\partial v_2}{\partial x_2} + \frac{\partial v_3}{\partial x_3}\right)\delta_{ik}.$$

In this equation η denotes the ordinary viscosity and ζ denotes the volume, or bulk viscosity. If we consider the coefficient of $(\partial v_2/\partial x_2)$ in p_{11} and the coefficient of $(\partial v_1/\partial x_1)$ in p_{22} we find that they must be equal. In this manner it can be seen that all reciprocal relations follow from considerations of symmetry in the present example. However, insofar as mixtures of two gases are concerned, *one* non-trivial reciprocal relation can be deduced. It expresses the relation between thermal diffusion and thermal effusion.

F. Limitations of the Thermodynamic Theory of Irreversible Processes

All previous considerations were restricted to small deviations from thermodynamic equilibrium and it is now necessary to specify more precisely the degree of deviation from thermodynamic equilibrium which may be regarded as small. It is natural to postulate that the thermodynamic concepts of temperature as well as of thermodynamic functions still retain their meaning. However, a rigorous specification of the limits of applicability becomes possible only when a more general theory can be formulated. Such a theory would then contain the theory of irreversible processes as a special limiting case. The kinetic theory of gases satisfies these requirements for gases. It does, in fact, as we shall see in Chap. V, confirm the conservation laws of fluid dynamics, it leads to the expression for the dissipation function in eq. (32) and also yields the phenomenological hypotheses concerning the flow of heat and the tensor of viscous stresses. If it were desired to establish the range of validity of these equations it would be necessary to carry the expansions in Chap. V further by adding the next term in order of magnitude. This was done e. g. by Enskog.[1] This method leads to clear statements about the conditions under which the next term in order of magnitude can be neglected. In this manner it is found that the temperature variations along a mean free path must be small compared with the absolute temperature and that changes in velocity must be small compared with the velocity of sound, or compared with the mean molecular velocity of random thermal motion. Considering that under normal conditions the mean free path is of the order of 10^{-4} cm it must be concluded that these limitations still leave an extremely wide margin which is seldom transgressed, excepting shock wave phenomena. It is true that in the case of numerous other irreversible phenomena it is not now possible to indicate quantitatively and in a simple manner the field of validity of the respective formulations but the results with gases allow us to expect with confidence that in many other cases there exists a field of validity which is sufficient for many purposes.

---

[1] D. Enskog, Zeitschr. f. Physik, Vol. 54, p. 498 (1929).

CHAPTER III

THE ELEMENTARY KINETIC THEORY OF GASES

The beginnings of the kinetic theory of gases can be traced to Daniel Bernoulli. A derivation of the expression for the pressure of a gas from the change in momentum of the molecules impinging on its walls can be found in his book "Hydrodynamica." Strassbourg, 1738 (cf. Vol. II, Sec. 11). The further development of this theory was resumed in the middle of the 19th century: Krönig 1856, Clausius 1857, Maxwell 1860. Ludwig Boltzmann's papers in which Maxwell's law of velocity distribution was given its most general form stand at the peak of this development.

22. The equation of state of a perfect gas

Let us now center our attention on the collisions which a solid, flat (or continuously curved), smooth (and hence frictionless) wall experiences from the impact of the gas molecules. We shall notice that the graph of these collisions taken for any element of surface $d\sigma$ is represented by a curve $f(t)$ which possesses a very large number of sharp dents corresponding to the enormous number of collisions. The pressure (force per unit area) on the element is defined as the smoothed time average of this curve. The contribution of a single collision is equal to the change in the momentum of one molecule resulting from the impact and from the subsequent reflection. When impact occurs at a velocity c and in a direction forming the angle θ with the normal to the wall, the change in momentum is equal to

(1) $$2\,m\,c \times \cos\theta.$$

The factor 2 stems from the recoil experienced by the wall during the appearance of the reflected impulse (angle of reflection = angle of incidence, the magnitude of velocity $|\mathbf{v}| = c$ remains preserved owing to the absence of friction). Taking the three Cartesian components ξ, η, ζ of $\mathbf{v}$ and placing the ξ-axis normal to the element with its positive direction outwards, we can replace (1) by

(1 a) $$2\,m\,\xi, \quad \xi > 0.$$

In order to account for all collisions we construct an oblique cylinder over the element of area $d\sigma$, Fig. 23, inclining its axis at an angle θ with respect to the normal and making its length equal to c. In the interior of this cylinder we now mark all molecules whose velocities point towards the wall and have values which lie between c and $c + dc$, and directions confined between θ and $\theta + d\theta$ and ϕ and $\phi + d\phi$. The angles θ and ϕ are measured with respect to the normal to the elementary area and around it. All velocity vectors drawn from each such molecule intersect our elementary area; during a unit of time all these, and only these molecules will impinge on the element $d\sigma$ under consideration.

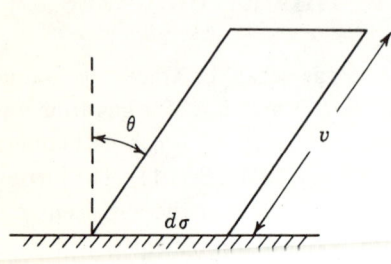

Fig. 23.
Illustrating the calculation of kinetic pressure.

The volume of the cylinder is equal to cross-sectional area × height $= d\sigma \times \xi$. Denoting the number of molecules per unit volume by n we find that the number contained within the cylinder is

(2) $$n \times \xi\, d\sigma.$$

Of these only such molecules count as will meet the wall and only such whose velocity lies in a given region $d\omega$ of the "velocity space." This region has, so far, been described by the quantities $dc, d\theta, d\phi$. The range of velocities has thus been described in a polar system of coordinates which is natural from the point of view of the element $d\omega$ in the physical space. It will, however, be more convenient from the point of view of the succeeding argument to express $d\omega$ in terms of the rectangular coordinates ξ, η, ζ and to put $d\omega = d\xi\, d\eta\, d\zeta$.

In any case, and independently of the choice of the system of coordinates, the number of colliding molecules is given by

(2 a) $$d\nu = \nu\, \xi\, d\sigma\, d\omega.$$

The symbol ν which must be carefully distinguished from n, denotes the density of molecules per unit volume and per unit velocity space $d\omega$. Evidently

$$n = \int \nu\, d\omega.$$

The corresponding change in momentum is found from (2 a) by multiplication by (1), or

(2 b) $$2\,\nu\, m\, \xi^2\, d\sigma\, d\omega.$$

The contribution of this group of collisions to the pressure is thus (division by $d\sigma$)

(2 c) $$dp = 2 \nu m \xi^2 d\omega, \quad \xi > 0.$$

The total pressure follows when we construct our cylinder for all possible velocity ranges:

(3) $$p = n m \overline{\xi^2}, \quad \overline{\xi^2} = \frac{2}{n} \int \nu \xi^2 d\omega, \quad \xi > 0;$$

the value $\overline{\xi^2}$ is obtained by averaging over half the velocity space $\xi > 0$. Since, however, to each molecule which travels towards the wall ($\xi > 0$) there corresponds one which travels away from it ($\xi < 0$) with equal probability we can extend the integration indicated in (3) over the whole of the velocity space and write

(3 a) $$p = n m \overline{\xi^2}, \quad \overline{\xi^2} = \frac{1}{n} \int \nu \xi^2 d\omega, \quad \text{with} \quad \xi \gtreqless 0,$$

instead of (3). The symbol $\overline{\xi^2}$ denotes the mean value of ξ^2 at a given point in space. A more rational representation, and one independent of the choice of the system of coordinates, is obtained by taking into account the fact that all directions are equivalent (isotropy of the velocity space). We then have

(3 b) $$\overline{\xi^2} = \overline{\eta^2} = \overline{\zeta^2} = \frac{1}{3}\overline{c^2}$$

because $v^2 = \xi^2 + \eta^2 + \zeta^2 = c^2$. Hence (3) yields

(4) $$p = \frac{n m}{3} \overline{c^2} = \frac{2}{3} n \overline{E_{tr}}, \quad \overline{E_{tr}} = \tfrac{1}{2} m \overline{c^2},$$

where $\overline{E_{tr}}$ denotes the kinetic energy of *translation*; rotational energy or that associated with internal motions need not be considered in the calculation of pressure.

Equation (4) contains more than the kinetic explanation of *pressure*; it also contains a kinetic definition of *temperature*. In order to see this we put

(4 a) $$n = \frac{N}{V},$$

where N denotes the total number of molecules in a volume V. Thus eq. (4) leads to

(4 b) $$p V = \frac{2}{3} N \overline{E_{tr}}.$$

Applying eq. (4 b) to one mol, we have $V = V_{mol}$ and $N = L$ = number of molecules per mol = Loschmidt-Avogadro number. Hence eq. (4 b) transforms into

(4 c) $$p V_{mol} = \frac{2}{3} L \overline{E}_{tr}.$$

Comparing this result with the equation of state of a perfect gas in its forms (3.11) and (3.11 a) it is seen that its right-hand side must be equal to RT. In this way

(5) $$\overline{E}_{tr} = \frac{3}{3} \frac{R}{L} T = \frac{3}{2} k T,$$

where k denotes the Boltzmann constant, already defined in Sec. 20, i. e. the gas constant reduced to a single molecule, R/L. In a three-dimensional space translation has three degrees of freedom. Consequently the contents of (5) (effecting simultaneously a change from subscript tr to subscript fr) can be expressed as follows: *The mean kinetic energy per degree of freedom is given by*

(6) $$E_{fr} = \tfrac{1}{2} k T.$$

This statement contains our (provisional)*kinetic definition of temperature.*

It should be noted here that eq. (5) contains the theory of the specific heats of a monatomic gas. Since its energy is wholly due to translation we can find u and c_v per mol of such a gas from eq. (5). Thus

(6 a) $$u = L \overline{E}_{tr} = \frac{3}{2} RT, \quad c_v = \frac{du}{dT} = \frac{3}{2} R \approx 3 \text{ kcal/kmol}.[1]$$

As $c_p - c_v = R$ eq. (6 a) determines c_p as well, and we have

(6 b) $$\frac{c_p}{c_v} = 1 + \frac{2}{3} = 1.66.$$

We now reiterate what we have already said in Sec. 4 C: The kinetic theory of gases is capable of filling the general framework of thermodynamics with actual numerical values which agree with experiment. In Chap. IV we shall revert to the values concerning polyatomic gases as discussed in Sec. 4 C.

In the preceding argument we have encountered the mean values of the squares $\overline{\xi^2}$, $\overline{c^2}$ etc. The linear mean $\overline{\xi}$ is, evidently, equal to zero, because the velocity space is isotropic and the positive semi-axis of ξ is statistically

---

[1] 1 kmol = 1000 mol = molar weight in kg.

indistinguishable from its negative semi-axis, or from any other semi-axis. The mean value of c which, obviously, differs from zero will be calculated in Sec. 23 B. The quantity

(7) $$(\overline{c^2})^{1/2}$$

is given directly from experimental results and constitutes a measure of the velocity of the gas. By way of example we shall now find its numerical value for the gases, H_2, He, O_2, N_2. As long as the gases remain perfect the mean velocity depends on temperature only, and not on pressure.

Denoting the mass of a molecule by m we can find from (5) that

$$\frac{m}{2}\overline{c^2} = \frac{3}{2}kT$$

or, multiplying by the Loschmidt-Avogadro number, that

$$\frac{Lm}{2}\overline{c^2} = \frac{3}{2}RT.$$

The product Lm is the mass μ of a mol and it is seen that the mean of the square of the molecular velocity $\overline{c^2} = 3RT/\mu$ is found from macroscopic measurements alone. For example for hydrogen we have $\mu = 2$ kg/kmol. With $T = 273$ K and the value (3.9) of R we obtain

$$\overline{c^2} = \frac{3}{2} \times 8.31 \times 273 \times 10^2 \, \text{m}^2\text{sec}^{-2}.$$

Hence

(8) $$\sqrt{\overline{c^2}} = 1.85 \text{ km/sec}.$$

Correspondingly for helium (monatomic, atomic weight 4, i. e. double that of the molar weight 2 of H_2) we find

$$\sqrt{\overline{c^2}} = \frac{1.85}{\sqrt{2}} \text{ km/sec} = 1.30 \text{ km/sec}.$$

The molar weight of oxygen is 16 times larger than that of hydrogen. Consequently it is necessary to divide the velocity (8) by $\sqrt{16} = 4$. In the case of nitrogen it is necessary to divide by $\sqrt{14} = 3.74$.

Even at 0 C the velocities of molecules appear to be extraordinarily high; with increasing temperatures their values increase somewhat, namely in proportion to $(1 + t/273)^{\frac{1}{2}}$, where t is the Celsius temperature. We can gain some insight into these relations if we consider that the velocity of sound cannot exceed the velocity of the molecules which propagate it. Hence it must be of the same order of magnitude, namely $c^2 = \dfrac{f+2}{f} \cdot \dfrac{RT}{\mu}$ [*cf.* Vol. II, eq. (13.17 a)]. The same is true about the velocity of compressed gases at the exit of an expansion nozzle.

The mean velocities of O_2, N_2, ... in atmospheric air *differ from each other* as is the case with any gaseous mixture. In accordance with one definition of temperature the *mean energies of translation are equal* to each other at *thermal equilibrium*. Consequently, on the average, the total energy of translation is distributed equally among the different components of a gaseous mixture and in proportion to their masses. The above constitutes the simplest example of the more general *law of the equipartition of energy*.

In the preceding argument we have considered the pressure on the wall only. However, all relevant statements can be seen to apply to the pressure in the interior of the gas if it is imagined that a small membrane is introduced there to measure the pressure. Hence the pressure inside the gas and including that at the walls appears to be independent of the point at which it is measured (*cf.* however, Sec. 26). This is due to the fact that external forces, e. g. gravitation, have been neglected. This influence on the statistical considerations concerning gases will be considered in Sec. 23 C.

23. The Maxwellian velocity distribution

In the preceding Section we have made a distinction between the physical space and the velocity space. We considered, e. g. in eq. (4 a), that the *physical space* is uniformly filled with molecules, in apparent agreement with macroscopic observation, it being implied that no external forces are acting on the molecules. We now turn our attention to the *velocity space*. It will be noted that in the preceding argument we only needed to know the mean values $\overline{\xi^2}, \ldots, \overline{c^2}$.

A. THE MAXWELLIAN DISTRIBUTION FOR A MONATOMIC GAS. PROOF OF 1860

If we select an arbitrary molecule of the gas we find that its velocity components have some arbitrary values, say ξ, η, ζ. We shall now consider the *probability* that the first component has a value which lies between ξ and $\xi + d\xi$ (a lamina confined between two planes at right angles to the ξ-axis), and denote it by

$$f(\xi)\,d\xi.$$

THE MAXWELLIAN VELOCITY DISTRIBUTION

The same can be said about the components η and ζ, the probability function being the same as before because there is no preferred direction in the space. It is not, however, evident, that the probable value for η is unaffected by a value which has been found for, say, ξ. As is well known this is the case with lotteries: Having won the great prize in one year we still have exactly the same chance of winning it next year. At first Maxwell assumed that this independence of probabilities was true in the theory of gases, but proved it later explicitly (*cf.* Sec. C and Chap. V).

The velocity vector **v** resulting from ξ, η, ζ belongs to the volume element $d\xi, d\eta, d\zeta$ of the velocity space (intersection of the three laminae ξ, η, ζ), and the probability that the tip of **v** will be found in this volume element is equal to

$$\tag{1} f(\xi)\,f(\eta)\,f(\zeta)\,d\xi\,d\eta\,d\zeta,$$

because of the lottery assumption. Taking into account the isotropy of all velocity directions and hence the fact that they are all equally probable we can introduce a new unknown function F which depends only on the velocity. Thus

$$\tag{1 a} F(c)\,d\omega, \qquad d\omega = d\xi\,d\eta\,d\zeta.$$

Comparing (1) with (1 a), we find that

$$\tag{2} F(\sqrt{\xi^2 + \eta^2 + \zeta^2}) = f(\xi)\,f(\eta)\,f(\zeta).$$

In order to determine the functions F and f from the preceding functional equation we can (purely formally) proceed as follows:

a) Logarithmic differentiation of (2) with respect to ξ:

$$\tag{3} \frac{\xi}{c}\frac{F'(c)}{F(c)} = \frac{f'(\xi)}{f(\xi)}.$$

b) Introduction of the abbreviations

$$\tag{3 a} \Phi(c) = \frac{1}{c}\frac{F'(c)}{F(c)}, \qquad \phi(\xi) = \frac{1}{\xi}\frac{f'(\xi)}{f(\xi)},$$

whence (3) becomes

$$\tag{3 b} \Phi(c) = \phi(\xi).$$

c) Differentiation with respect to η or ζ leads to

$$\tag{3 c} \Phi'(c) = 0, \qquad \Phi(c) = \text{const.}$$

Assuming[1] that the const $= -2\gamma$ we find that in view of (3 b) we also have

$$\phi(\xi) = -2\gamma,$$

so that according to (3 a)

(3 d) $$\frac{d \log f(\xi)}{d\xi} = -2\gamma\,\xi, \qquad \log f(\xi) = \alpha - \gamma\,\xi^2.$$

Putting $e^{-\alpha} = a$, we have

(4) $$f(\xi) = a\,e^{-\gamma\xi^2}.$$

It is remarkable as well as gratifying to find that the probability is given by the standard form of all statistical laws, namely by Gauss' error distribution function. The most probable value of the velocity component ξ is $\xi = 0$; the deviations from it distribute themselves symmetrically on either side and trace *Gauss' probability curve* (see Fig. 24a).

The constant a in (4) can be determined from the condition that it is absolutely certain that ξ has some value between $-\infty$ and $+\infty$. Hence

(5) $$\int_{-\infty}^{+\infty} f(\xi)\,d\xi = 1.$$

Making use of Laplace's integral, we have

(5 a) $$a \int_{-\infty}^{+\infty} e^{-\gamma\xi^2}\,d\xi = a \left(\frac{\pi}{\gamma}\right)^{\frac{1}{2}} = 1 \qquad \text{hence} \qquad a = \left(\frac{\gamma}{\pi}\right)^{\frac{1}{2}}.$$

In order to determine the value of γ we shall calculate the mean kinetic energy of the degree of freedom of the component ξ:

$$\frac{m}{2}\,\overline{\xi^2} = \frac{m}{2}\left(\frac{\gamma}{\pi}\right)^{\frac{1}{2}} \int_{-\infty}^{+\infty} \xi^2 e^{-\gamma\xi^2}\,d\xi = -\frac{m}{2}\left(\frac{\gamma}{\pi}\right)^{\frac{1}{2}} \frac{d}{d\gamma} \int_{-\infty}^{+\infty} e^{-\gamma\xi^2}\,d\xi =$$

$$= -\frac{m}{2}\left(\frac{\gamma}{\pi}\right)^{\frac{1}{2}} \frac{d}{d\gamma} \left(\frac{\pi}{\gamma}\right)^{\frac{1}{2}} = \frac{m}{4\gamma}.$$

---

[1] The negative sign is required in order to satisfy, for example, the following eq. (5); the factor 2 is introduced for convenience.

Equating this with $kT/2$ in accordance with eq. (22.6), we find

(6) $$\gamma = m/2kT.$$

Taking into account (5) and (6) we obtain the final form of eq. (4)

(7) $$f(\xi) = \left(\frac{m}{2\pi kT}\right)^{\frac{1}{2}} \times e^{-E_1/kT}\,; \qquad E_1 = \frac{m}{2}\xi^2.$$

The symbol E_1 denotes here the kinetic energy of the component ξ which corresponds to the point of the velocity space under consideration. Analogous equations apply to the remaining components η, ζ. The new distribution function $F(c)$ can be obtained at once with the aid of eq. (2). Thus

(8) $$F(c) = \left(\frac{m}{2\pi kT}\right)^{\frac{3}{2}} \times e^{-E/kT}, \qquad E = \tfrac{1}{2}m\,(\xi^2 + \eta^2 + \zeta^2).$$

The symbol E denotes now the kinetic energy of the translation which results from the components ξ, η, ζ.

B. Numerical values and experimental results

If instead of the velocity **v** we are interested in the distribution of the absolute value of the velocity, denoted by $c = $ celeritas, we consider the spherical shell described about the origin by the radii c and $c + dc$. Its probability will be denoted by

$$\phi(c)\,dc$$

and it is equal to the volume $4\pi c^2\,dc$ within the spherical shell times the value (8) of $F(c)$. Thus we obtain

(9) $$\phi(c) = 4\pi c^2 \left(\frac{m}{2\pi kT}\right)^{\frac{3}{2}} \times e^{-E/kT}, \qquad E = \tfrac{1}{2}mc^2.$$

In this way we are led to a distribution function which is no longer Gaussian, as shown in Fig. 24, and which ceases to be symmetrical with respect to the most probable value. For large values of c it decreases to zero exponentially, in the same way as the previous curve, but for small values of c it tends to zero only quadratically; for $c < 0$, obviously, ϕ remains undefined.

The maximum of the curve is found from $\phi'(c) = 0$ and according to (9) we have

(10) $$c_w = \left(\frac{2kT}{m}\right)^{\frac{1}{2}} = \text{most probable velocity.}$$

It is different from the root mean square. $\sqrt{\overline{c^2}}$ as well as from the linear mean

(10 a) $$\overline{c} = \int_0^\infty c\,\phi(c)\,dc.$$

These three velocities satisfy the ratios:

(11) $$c_w : \overline{c} : \sqrt{\overline{c^2}} = 1 : 1.13 : 1.22,$$

as shown in Problem III.2.

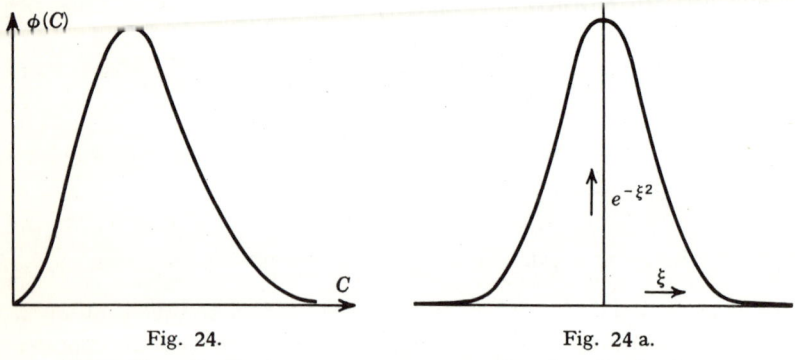

Fig. 24.
Maxwellian distribution.

Fig. 24 a.
Gaussian distribution.

A qualitative confirmation of Maxwell's distribution is obtained by observing the *broadening of spectral lines* of a luminous gas with increasing temperature. This is due to the *Doppler effect*. If ν_0 denotes the natural frequency of the luminous particle (atom or molecule) and $\lambda_0 = c_L/\nu_0$ (we shall provisionally denote the velocity of light by c_L in order to distinguish it from the preceding c) then the observer who is looking, say, in the x-direction, perceives the wavelength λ_0 only of such particles whose velocity is almost zero in the ξ-direction. Generally speaking, he will observe a wavelength $\lambda_0 + \Delta\lambda$. According to Vol. IV, Sec. 11 (and neglecting relativistic corrections), we have $\Delta\lambda/\lambda_0 = \xi/c_L$, so that

(12) $$\Delta\lambda = \xi/\nu_0.$$

All particles with equal ξ's contribute towards the specific intensity, J, of the spectrogram in equal degrees at a positive or negative distance, $\Delta\lambda$, from its center. Their number is determined by the distribution function, $f(\xi)$, from eq. (4). We are justified in assuming that all particles are excited with equal strength and that their intensities (not amplitudes) are superimposed owing to the lack of coherence in the emission. Consequently, the observed intensity becomes directly proportional to $f(\xi)$ where ξ is to be taken equal to $\nu_0 \Delta \lambda$ in accordance with (12). In this way we obtain from (4) that

$$(13) \qquad J = J_0 \exp\left[-\gamma(\nu_0 \Delta \lambda)^2\right].$$

J_0 denotes the specific intensity in the center of the spectrogram; in accordance with eq. (6) γ is inversely proportional to T and determines the width of the spectral line. Its width at half intensity is given by $J = J_0/2$, or

$$\Delta \lambda_H = \frac{1}{\nu_0}\left(\frac{\log 2}{\gamma}\right)^{\frac{1}{2}} = \frac{1}{\nu_0}\left(\frac{kT}{m} 2\log 2\right)^{\frac{1}{2}}.$$

It is seen that the form of the spectral line given by (13) constitutes a direct image of Gauss' error distribution curve and hence also of Maxwell's distribution function.

The first measurements on the shape and half-width of *emission spectra* at varying temperature and atomic weight m are due to Michelson.[1] The shape of Fraunhofer's absorption spectra is of fundamental importance in astrophysics. In this connection, in addition to the Doppler effect, it is necessary to take into account the broadening due to pressure (damping due to collisions), whereas the natural breadth of the lines (*cf.* Vol. III, Sec. 36) becomes insignificant by comparison. Otto Stern[2] was the first to succeed in directly verifying the Maxwellian distribution when he made use of his method of atomic beams.

C. GENERAL REMARKS ON THE ENERGY DISTRIBUTION. THE BOLTZMANN FACTOR

In the preceding Section we have made use of the original, and somewhat primitive, first proof advanced by Maxwell. We shall justify the lottery assumption in Chap. IV when we shall make use of a much more general and essentially simpler "combinatorial method," based on classical mechanics. It will be shown that the more precise method leads to Maxwell's distribution function for the case of monatomic gases in the absence of external forces. The restriction to monatomic gases was stressed in the title of Sec. A: the restriction to gases with no external forces acting on them was implied when, e. g. in eq. (4 a), we assumed that the density was uniform throughout the physical space. This would not be true, for example, of a gas in a gravitational field.

---

[1] Phil. Mag. **34**, 280 (1892).
[2] Zeitschr. f. Phys. **3**, 417 (1920).

Anticipating the results given in Chap. IV we now proceed to generalize our argument to include a polyatomic gas in an external field of force possessing a potential Φ. We shall proceed from eq. (8) but we must now replace the translational energy E by the total energy of a particle

(14) $$\varepsilon = E_{tr} + E_{rot} + \ldots + \Phi.$$

In addition to the translational energy, which is the only form applicable to monatomic gases, we have to consider the rotational energy E_{rot} as well as the energy of internal motions of particles (vibrational energy etc.) indicated as ... in the above equation, as well as the potential energy Φ in the field of forces. Furthermore, we shall extend the scope of an element $d\omega = d\xi\, d\eta\, d\zeta$ of the velocity space and we shall introduce the element $d\Omega$ of the "phase space" to be defined later. Thus instead of (8) we obtain

(15) $$F\, d\Omega = A\, e^{-\varepsilon/kT}\, d\Omega.$$

The constant A introduced here is determined by the normalizing condition

(15 a) $$\int F\, d\Omega = 1$$

in the same way as the constant a was normalized in eq. (4). Concerning the meaning of $d\Omega$ we shall remark here only that in the case of a monatomic gas $d\Omega = d\tau \times m^3\, d\omega$; $d\tau$ denotes the element of volume $dx\, dy\, dz$ in the physical space; the factor m^3 is due to the fact that in the phase space we shall introduce the momenta $m\xi, m\eta, m\zeta$ instead of the velocities ξ, η, ζ themselves. We now center our attention on the factor

(16) $$e^{-\Phi/kT}$$

contained in (15). Since we assume that only Φ depends on the space coordinates x, y, z, the value of this factor indicates the probability that a gas molecule may be found in a cell at x, y, z. Thus it permits us to calculate the spatial density distribution ρ of the particles. If ρ_0 denotes the density at a reference level of the potential, we can put generally that

(17) $$\frac{\rho}{\rho_0} = e^{-\Phi/kT}.$$

In the simplest case of a gravitational field, for which $\Phi = mgz$, eq. (17) transforms into the *barometric formula* cf. Vol. II, Sec. 7, eq. (15 a) or (15 c); J. Perrin's experiments on models of the atmosphere, see *ibid*, can be regarded as a macroscopic confirmation of the Boltzmann factor.

24. Brownian motion

The oscillatory motion performed by the smallest particles (specks of dust, colloidal particles), suspended in a liquid or a gas and observable with the aid of a microscope, was described in 1826 by the botanist Robert Brown. Its nature remained a puzzle for a long time. A final clarification was given in a paper by Einstein published in the memorable year 1905. Even in 1906 the critical Röntgen endeavored to counter the assertion that Brownian motion might after all be due to the energy of the lighting system of the microscope by a series of suitable control experiments.

Brownian motion comes under the general heading of *fluctuations*, i. e. departures from *thermodynamic equilibrium*. In line with the elementary character of this volume of our Lectures we shall confine ourselves here to a special derivation due originally to Langevin[1] because it leads to Einstein's result in a very simple way.

In Sec. 33 of Vol. IV we have proved the following theorem: If we place a very large number N of unit vectors of entirely arbitrary directions in a plane, then the resultant vector is equal to $\sqrt{N}$. In the present case we are interested in the collisions undergone by the colloidal particle owing to the thermal agitation of its surroundings. Complete directional isotropy of collisions is statistically assured and the number of collisions is proportional to the observation time t. The distances traversed by particles between two collisions (in actual fact reference is made to the projections of the paths **r** of a particle onto the focal plane of the microscope of which the observed zig-zag is composed) do not constitute unit vectors; they are small distances which fluctuate about a mean value whose magnitude, in turn, depends on the properties of the surrounding fluid as well as on the particle whose motion (so-called "random walk") is being observed. The resultant translation **r** of a particle can be calculated from eq. (33.4) in Vol. IV by interchanging S with **r** and by replacing the unit vector by $\mathbf{r}_i$. The mean value is then

(1) $$\overline{\mathbf{r}^2} = \sum \overline{\mathbf{r}_i^2} = \overline{\mathbf{r}_i^2} \times N = P t.$$

In order to determine the factor of proportionality in this equation we shall make reference to the equation of motion of the particle:

(2) $$M\ddot{\mathbf{r}} = \mathbf{K}(t) - C\dot{\mathbf{r}}.$$

M denotes the mass of the colloidal particle, **r** is the vectorial displacement of its center of mass from a fixed initial point 0; $\mathbf{K}(t)$ is the force which varies

---
[1] Comptes rendus 1908, p. 530.

in jumps both as regards magnitude and direction, and which transfers the collisions to M; the last term in the equation represents the frictional resistance assumed proportional to the velocity $\dot{r}$ as in Stokes' hypothesis; this assumption implies that the surrounding medium is regarded as a continuum which, obviously, is permissible only if the particle is many times larger than the molecular structure of the fluid; this is a reasonable assumption in the circumstances. Assuming that the particle is a sphere of radius a and that the viscosity of the fluid is η, we can assume that

(2 a) $$C = 6\pi\eta a$$

as given in eq. (35.20) of Vol. II Taking the scalar product with $\mathbf{r}$, we find from eq. (2) that

(3) $$M(\mathbf{r}\cdot\ddot{\mathbf{r}}) = (\mathbf{r}\cdot\mathbf{K}) - C(\mathbf{r}\cdot\dot{\mathbf{r}}).$$

where

$$\mathbf{r}\cdot\dot{\mathbf{r}} = \frac{1}{2}\frac{d}{dt}(\mathbf{r}^2).$$

The product $(\mathbf{r}\cdot\mathbf{K})$ is known as the "virial of force $\mathbf{K}$," the term being used in the study of the mechanics of material points; its usefulness in the kinetic theory of gases was first recognized by Clausius. We now apply the elementary transformation used in connection with the virial theorem:

$$\mathbf{r}\cdot\ddot{\mathbf{r}} = \frac{d}{dt}(\mathbf{r}\cdot\dot{\mathbf{r}}) - (\dot{\mathbf{r}}\cdot\dot{\mathbf{r}}) = \frac{1}{2}\frac{d^2}{dt^2}(\mathbf{r}^2) - \mathbf{v}^2;$$

thus eq. (3) transforms to

(4) $$\left(\frac{1}{2}M\frac{d^2}{dt^2} + \frac{1}{2}C\frac{d}{dt}\right)\mathbf{r}^2 - M\mathbf{v}^2 = (\mathbf{r}\cdot\mathbf{K}).$$

We now integrate this equation with respect to time from 0 to t and divide all terms by t. During the time interval t the product $(\mathbf{r}\cdot\mathbf{K})$, i. e. the projection of the rapidly varying force on the direction of $\mathbf{r}$, changes sign many times. Dividing by the large value of t (large compared with the interval which corresponds to a change in sign of $\mathbf{r}\cdot\mathbf{K}$) we may expect that

(4 a) $$\frac{1}{t}\int_0^t (\mathbf{r}\cdot\mathbf{K})\,dt = 0.$$

With $r_0 = 0$ (for the initial position of the particle) the integrated eq. (4) becomes

$$(5) \qquad \frac{M}{2t}\frac{dr^2}{dt} + \frac{C}{2t}r^2 = \frac{1}{t}\int_0^t M\mathbf{v}^2\,dt.$$

The right-hand side contains the temporal mean of double the kinetic energy of the particle. According to the law of equipartition the mean value of the kinetic energy averaged over a large number of particles is equal to kT (2 degrees of freedom in the two-dimensional motion under consideration; in the case of linear motion it would be necessary to write $\frac{1}{2}kT$ instead). We can now make use of our result in Sec. 23 B concerning gaseous mixtures: The mean velocities of our colloidal particles are much smaller than those of the molecules in the surroundings owing to the much larger mass of the former but the mean kinetic energy is equal for both and can, therefore, be expressed in terms of the absolute temperature of the surrounding fluid, as indicated. Denoting the mean for the larger aggregate by a bar over the symbol (transition from a single particle to a certain aggregate of particles), we obtain

$$(5\text{ a}) \qquad \frac{1}{t}\int_0^t \frac{1}{2}M\overline{\mathbf{v}^2}\,dt = kT.$$

We shall verify presently that the first term on the left-hand side of (5) decreases exponentially with t so that we are able provisionally to neglect it. Performing the transition to an aggregate of particles on the left-hand side of (5) and taking into account eqs. (2 a) and (5 a), we conclude that

$$(6) \qquad \overline{r^2} = \frac{2kT}{3\pi\eta a}t.$$

This contains our rough estimate in eq. (1) together with the evaluation of the factor P which appeared in it.

When observations are made only on one-dimensional translations of the particle, e. g. on those in the x-direction, eq. (6) is replaced by

$$(6\text{ a}) \qquad \overline{x^2} = \frac{kT}{3\pi\eta a}t$$

in accordance with the principle of equipartition. This is *Einstein's equation* which has been confirmed experimentally in numerous ways and which was used, for example, to determine the Boltzmann constant, k, or the Loschmidt-Avogadro number $L = R/k$.

We shall now supplement our derivation with a more accurate integration of eq. (5), performing the transition to the aggregate before the integration. Putting $\overline{r^2} = u$ in eq. (4) and integrating once we obtain

$$\dot{u} + \frac{C}{M} u = \frac{4kT}{M} t. \tag{7}$$

The integral of the associated homogeneous equation is

$$u_1 = A\, e^{-Ct/M} \tag{7 a}$$

and it is easy to find that

$$u_2 = \frac{4kT}{C}\left(t - \frac{M}{C}\right) \tag{7 b}$$

constitutes a particular solution of the non-homogeneous equation. Now $M = \frac{4\pi}{3} \rho a^3$ where ρ denotes the density of the particle and hence, according to (2 a), we have

$$\frac{M}{C} = \frac{2}{9} \cdot \frac{\rho a^2}{\eta}.$$

Assuming $a = 10^{-4}$ cm (limit of visibility), $\eta \approx 10^{-2}\, \frac{\text{g}}{\text{cm sec}}$ (water) and $\rho \approx 1\, \text{g} \times \text{cm}^{-3}$ (the particle floats in water), we find that

$$\frac{M}{C} = \frac{2}{9} \times 10^{-6} \text{ sec.} \tag{7 c}$$

The presence of M/C in the brackets in (7 b) denotes an unmeasurably small shift in the zero of the time scale. Its presence in the exponent of (7 a) represents a very fast decay of any initial disturbance A that may have been present. Thus our assumption that $u = u_1 + u_2$ simplifies to

$$u \approx u_2 \approx \frac{4kT}{C} t$$

which is identical with (6).

It is evident that observations performed on a single particle will deviate considerably from the mean values in (6) or (6 a); furthermore, as is easily demonstrated, the scatter will follow Gauss' error distribution curve so that, consequently, it is necessary to take mean values averaged over large numbers of single observations in order to obtain an experimental verification.

The behavior of a torsional microbalance constitutes an extremely instructive variant of Brownian motion. The investigation of the fluctuations of a microbalance was first suggested by M. von Smoluchowski who also gave the relevant theory. The method was improved by E. Kappler[1] to such an extent that it could be used to determine the Loschmidt-Avogadro number to within 1 per cent.

In the case of Brownian motion the observation is concerned with the quadratic mean ($\overline{r^2}$ or $\overline{x^2}$) of a displacement; in the case of a microbalance the relevant quantity is given by the quadratic mean $\overline{\phi^2}$ of the angular displacement.

The following remarks may suffice to give a description of circumstances encountered during an experiment: A thin mirror of about 1 mm^2 in area is suspended from a quartz strand several μ in diameter. The torsional fluctuations caused by the impacts from air molecules are registered on a photographic film with the aid of reflected light. It is, of course, necessary to maintain a constant temperature and to insure freedom from vibrations. The "directional force," i. e. the elastic constant D of the quartz strand, is determined in the usual way by observing the free oscillations of the system when provided with an additional mass. In order to exclude radiometric influences it is desirable to keep the pressure either very low (e. g. 1/100 mm Hg), or comparatively high (e. g. 1 atm). The duration of one film recording was about 10 hrs.

When making a theoretical analysis of these fluctuations it is necessary to note that the mirror possesses not only the kinetic energy

(8 a) $\qquad \tfrac{1}{2} I \dot{\phi}^2, \qquad I =$ moment of inertia[2]

but also the potential energy

(8 b) $\qquad \tfrac{1}{2} D \phi^2, \qquad D =$ directional force (elastic constant).

Since the time-averaged values of kinetic and potential energy are equal, they have each to be ascribed the statistical mean energy $\tfrac{1}{2} k T$, corresponding to one degree of freedom. On taking mean values we find from eqs. (8 a) and (8 b) respectively that

(9 a) $$\overline{\dot{\phi}^2} = \frac{kT}{I},$$

(9 b) $$\overline{\phi^2} = \frac{kT}{D}.$$

---

[1] Ann. d. Phys. 11, 233 (1931); *cf.* Naturw. 649 and 666 (1939).
[2] *Cf.* Sec. 31, remark following eq. (7).

The statistical deviations from these two mean values have been registered by Kappler; both plot exactly along a Gaussian error distribution curve. The distribution of $\dot\phi^2$ around the mean value (9 a) can be directly denoted as a "Maxwellian distribution" for the angular velocity.

It is realized that Kappler's method furnishes an elementary and reasonably accurate way of determining the Loschmidt-Avogadro number L, because the equation $R = L k$ permits us to calculate the Loschmidt-Avogadro number when k is known.

It is evident that no analogy to eq. (9 b) occurs in the consideration of the Brownian motion of a freely floating particle, because the particle is not restricted to a definite position of equilibrium. Equation (9 a) is replaced by our preceding eq. (5 a). Moreover the angular displacement ϕ (measured with respect to a fixed axis) of a Brownian particle satisfies a relation which is analogous to Einstein's equation (6 a), as already shown by Perrin, in which it suffices to make a suitable change in Stokes' frictional constant (2 a) (cf. Vol. II, eq. (35.21)).

In order to complete the dynamical analysis of the torsional balance experiment we again follow Langevin's method. In analogy with eq. (2) we can write the equation of motion as

(10) $$I \ddot\phi = \mathsf{M}(t) - C \dot\phi - D \phi,$$

where $\mathsf{M}(t)$ denotes the torque due to the molecular collisions with the mirror; $C \dot\phi$ denotes Stokes' frictional moment in the surrounding air (or at low pressure); $D \phi$ is the elastic couple of the quartz strand (this term was absent in eq. (2)). Multiplying by ϕ and applying the virial theorem we have

(11) $$\left(\frac{1}{2} I \frac{d^2}{dt^2} + \frac{1}{2} C \frac{d}{dt}\right) \phi^2 - I \dot\phi^2 + D \phi^2 = \phi \mathsf{M}(t),$$

instead of (4). Integrating with respect to t and dividing by t we find that the right-hand side vanishes. The last two terms on the left-hand side cancel each other in accordance with eqs. (9 a, b), when the mean over the aggregate is taken. Hence eq. (11) reduces to

$$I \frac{du}{dt} + C u = 0, \qquad u = \overline{\dot\phi^2}$$

because the constant of integration is zero in view of the fact that the mean value of the moment is equal to zero. Repeated integration gives

(11 a) $$u = u_0 e^{-C t/I}.$$

The decay of an initial angular displacement (or velocity) indicated by eq. (11 a) was investigated by Kappler. By reducing the pressure of the air we can cause the value of the frictional constant C to decrease to an arbitrarily small value and thus it is possible to increase the "time of decay" I/C to an order of magnitude of several seconds. In this way it was possible to obtain an experimental verification of the relation in eq. (9 a) for the torsional balance.

The same could not be achieved in the case of a freely floating particle, eq. (5 a), because of the order of magnitude of M/C calculated in (7 c). The latter also indicates the time interval during which the motion can be regarded as being linear for all intents and purposes (i. e. without significant changes in direction). Because of this order of magnitude the motion of a Brownian particle observed by eye is in fact seen to be a mean of a very large number of displacements.

This is the reason why the very careful measurements performed by Franz Exner (1900) led to values of *velocity* which were smaller by many orders of magnitude than those implied in eq. (5 a). The existence of these measurements proved to be an obstacle to the acceptance of the view that Brownian motion is essentially of a molecular nature. The latter view gained universal acceptance only after Einstein deduced theoretically in 1905 the expression for a quantity which could be measured directly, namely one for the *mean square of displacement*.

Reverting to the torsional balance we conclude with a remark of general *interest: Thermal fluctuations which can be studied quantitatively on the example of a torsional balance set an insurmountable limit to the sensitivity of all indicating instruments*; this principle was first stated in relation to ultra-sensitive galvanometers by Ising, 1926.

25. Statistical considerations on paramagnetic substances

In order to give a statistical derivation of Langevin's function which has been used in Sec. 10 in anticipation of this proof, it is sufficient to make a simple application of the Boltzmann factor from eq. (23.15).

We suppose that the paramagnetic body is made of single, independent elementary magnets which are free to rotate and which are arranged in a disorderly manner. We shall denote their magnetic moments by $\mathbf{m} = p\,l\,\mathbf{e}$ (p = magnetic pole strength, l = distance between poles, $\mathbf{e}$ = unit vector). This model is sufficient to deduce not only the properties of paramagnetic gases (O_2, NO, ...) and liquids, but also of those of solid salts, both when classical considerations are applied as in Section A, and when quantum mechanics is used, as in Section B. From the classical point of view all directions of the magnetic axes are permissible and equally probable. From the point of view of quantum mechanics only several discrete orientations of the magnetic axis with respect to the external magnetic field are admitted. In any case the probability of a given orientation is determined by the interaction between the directed force of the external field and thermal agitation, i. e. the temperature of the system.

A. The classical Langevin function

Let θ denote the angle between the axis of the elementary magnet and the direction of the external field. The latter will be denoted by B (denoted by $\mu_0 H$ in Sec. 19). The moment acting between them is

$$D = m\, B \sin \theta$$

and tends to *decrease* θ, that is to render **m** parallel to B. The work of *increasing* θ against the field is

$$D\, d\theta = m\, B \sin \theta\, d\theta = -m\, B\, d \cos \theta.$$

On turning by $d\theta$ the change in the potential energy Φ of our elementary magnet is of equal magnitude. Thus we have

(1) $$d\Phi = -m\, B\, d \cos \theta, \qquad \Phi = -m\, B \cos \theta.$$

Φ increases as θ is increased in the same way as the gravitational potential $\Phi = m\, g\, z$ increases as the distance z from the surface of the earth is increased. Our present Φ is so normalized as to place its minimum $\Phi = -m\, B$ at the stable position $\theta = 0$ and its maximum at the unstable position $\theta = \pi$.

From the classical point of view the a *priori* probability dW of a given orientation of our elementary magnet is *the same* for equal ranges of angles $d\omega = \sin \theta\, d\theta\, d\phi$. It becomes *different* for different temperatures when multiplied by the Boltzmann factor, namely

(2) $$dW = A\, e^{-\Phi/kT}\, d\omega.$$

This exhibits the opposing influences of temperature and field. At high temperatures all directions have roughly the same probability, as it was a *priori*, because then $\exp(-\Phi/kT) \approx 1$. At low temperatures the stable orientation $\theta = 0$, $\Phi = \Phi_{min}$ outweighs all other possibilities. The coefficient A

introduced in (2) can best be evaluated from the condition that the probability of finding the magnet at *any* direction is equal to 1. It follows that

$$\int dW = 1 = A \int_0^{2\pi} d\phi \int_0^{\pi} e^{-\Phi/kT} \sin\theta\, d\theta$$

so that

(2 a) $$1/A = 2\pi \int_0^{\pi} e^{-\Phi/kT} \sin\theta\, d\theta.$$

Substituting (2) and integrating with respect to ϕ, we have

(3) $$dW = \frac{e^{-\Phi/kT} \sin\theta\, d\theta}{\int e^{-\Phi/kT} \sin\theta\, d\theta}.$$

We now proceed to calculate the mean value of the component of **m** in the direction of the field and we shall denote it by $\overline{m}$. With dW from (3) we have

(4) $$\overline{m} = \int_0^{\pi} m \cos\theta\, dW.$$

When the number of elementary magnets is very large only this component and only its probable value enter into the calculation because the components at right angles to the field lie in different azimuthal planes ϕ in an irregular manner and cancel each other. Multiplying the mean value $\overline{m}$ by the number n of elementary magnets per mol we obtain a quantity M which in Sec. 19 was called the magnetization:

$$M = n\overline{m}.$$

On the other hand $M_\infty = nm$ (uniform orientation of all elementary magnets in a sufficiently strong field), eq. (19.7), denotes the *magnetization at saturation*. Introducing the abbreviations

(5) $$x = \cos\theta, \qquad \alpha = \frac{mB}{kT}$$

we find that eq. (4) yields

(6) $$\frac{M}{M_\infty} = \int_{-1}^{+1} e^{\alpha x} x\, dx \bigg/ \int_{-1}^{+1} e^{\alpha x}\, dx.$$

Performing the integration indicated in the denominator, we have

$$\frac{1}{\alpha}(e^\alpha - e^{-\alpha}) = \frac{2}{\alpha} \sinh \alpha.$$

The numerator is the derivative of the denominator with respect to α, i. e. equal to

$$\frac{2}{\alpha} \cosh \alpha - \frac{2}{\alpha^2} \sinh \alpha.$$

Hence it follows from (6) that

(7) $$\frac{M}{M_\infty} = \frac{\cosh \alpha}{\sinh \alpha} - \frac{1}{\alpha}.$$

The right-hand side represents the Langevin function from eq. (19.5 b); it is easy to see that our definition of α in (5) agrees with that given in (19.5 a). Our present eq. (7) fills the gap left open in Sec. 19 B and completes the classical theory of paramagnetic substances as well as Weiss' theory of ferromagnetic substances which followed in Sec. 19 C.

B. Modification of Langevin's Function with the Aid of Quantum Mechanics

According to the point of view of quantum mechanics the assemblage of all possible orientations of the elementary magnets does *not* constitute *a continuum*, $-1 \leqslant \cos \theta \leqslant +1$, but is restricted to certain *discrete* values of $\cos \theta$ ("angular quantization"). Their number is determined by the spectroscopic character of the atom (or molecule) in its ground state; it is equal to 2, 3, 4, ..., depending on whether the ground state corresponds to a doublet, triplet, quadruplet, ... (The ground state is *not* associated with a magnetic moment in the case of a singlet system; the atom does *not* then constitute an elementary magnet.)

The principles of quantum mechanics lead to the following rule, which we cannot, obviously, justify here: Let r denote the multiplicity of the term system ($r = 2$ doublet, $r = 3$ triplet, ...); put $r = 2j + 1$ (j = azimuthal quantum number) and $\cos \theta = s/j$ (so that $|s| \leqslant j$ because $|\cos \theta| \leqslant 1$). The rule states that only such values of s are admitted as differ from each other by unity. They are seen tabulated below.

$r =$	2	3	4	5
$j =$	1/2	1	3/2	2
$s =$	$\pm 1/2$	$\pm 1; 0$	$\pm 3/2; \pm 1/2$	$\pm 2; \pm 1; 0.$

In the case of a doublet system, for example, only the two orientations which correspond to $\cos\theta = \pm 1$, i. e. parallel and antiparallel to the field, are permissible. In a system composed of triplets the orientation at right angles to the field, corresponding to $\cos\theta = 0$, is also permitted; there are four possible orientations in a system composed of quadruplets, namely those given by $\cos\theta = \pm 1$, $\cos\theta = \pm 1/3$ etc. It is assumed *a priori* that such orientations are associated with equal probabilities. When calculating the mean value $\overline{m}$ of the component in the field direction it is necessary again to take into account the Boltzmann factor. Generally speaking the integral in (4) is now replaced by the sum of r terms. Introducing the abbreviations in (5), we have for $r = 2$

$$(8) \qquad \frac{\overline{m}}{m} = (e^\alpha - e^{-\alpha})/(e^\alpha + e^{-\alpha}) = \tanh\alpha,$$

and for $r = 3$

$$(8\text{ a}) \qquad \frac{\overline{m}}{m} = (e^\alpha - e^{-\alpha})/(e^\alpha + 1 + e^{-\alpha}) = \frac{2\sinh\alpha}{1 - 2\cosh\alpha}.$$

Similarly for $r = 4$

$$(8\text{ b}) \qquad \frac{\overline{m}}{m} = \left(e + \frac{1}{3}e^{\alpha/3} - \frac{1}{3}e^{-\alpha/3} - e^{-\alpha}\right)/(e^\alpha + e^{\alpha/3} + e^{-\alpha/3} + e^{-\alpha}).$$

Depending on the spectroscopic nature of the state of the atom it is necessary to replace Langevin's function by the right-hand side of one of the preceding equations. The classical Langevin function appears as a limit for $r \to \infty$. We shall denote the latter by $L_\infty(\alpha)$, denoting the preceding modified forms by $L_2(\alpha)$, $L_3(\alpha)$, ... The function L_2 was introduced by W. Lenz[1] in 1920. We shall now compare it with Langevin's function

$$(9) \qquad L_2 = \frac{\sinh\alpha}{\cosh\alpha}, \quad L_\infty = \frac{\cosh\alpha}{\sinh\alpha} - \frac{1}{\alpha}.$$

On a graph the curves L_3, L_4, ... will be seen to fall between the limiting curves (9).

This is exhibited, in particular, by the slope of the tangent at the origin. According to (9) we have (see also eq. (19.6))

$$L_2'(0) = 1, \qquad L'_\infty(0) = \frac{1}{3}.$$

---

[1] Phys. Zeitschr. **21**, 613 (1920).

whereas (8 a, b) give

$$L_3'(0) = \frac{2}{3}, \qquad L_4'(0) = \frac{5}{9}, \ldots .$$

The slope is steepest for L_2 and decreases in steps until it reaches $1/3$ for L_∞.

Since, according to (19.11 a), the *Curie temperature* Θ from eq. (19.11 a) depends on $L'(0)$, its value is also modified by quantum mechanics as compared with the classical value. The same applies to the *Curie constant* whose classical value (19.7 a) should be replaced by, for example,

$$C = \frac{\mu_0 M^2{}_\infty}{R}$$

in the case of a doublet.

Additional modifications introduced by quantum mechanics have already been described at the end of Sec. 19 C and in connection with Fig. 19. We shall not pursue this point here but we wish to emphasize that the theory of para- and ferromagnetic phenomena, being of statistical origin, often illustrates the difference between classical statistical mechanics and quantum mechanics.

26. The statistical significance of the constants in van der Waals' equation

In Sec. 9 we introduced the constants a and b of the van der Waals equations in a purely phenomenological way giving only a brief account of their physical significance. In his dissertation, van der Waals fully established their statistical origin and his derivation was subsequently simplified by Boltzmann in his "Vorlesungen über Gastheorie," 1898. In our description we shall follow that given in Chapter A of F. Sauter's paper[1] which contains a modern account of the essential points of Boltzmann's method. The circumstance that only the simplest statements of the elementary theory of gases are required for the purpose, as stressed by Sauter, allows us to fill the gap left open in Sec. 9 already at this stage and without having to draw on the general methods described in Chap. IV.

A. The volume of a molecule and the constant b

When considering perfect gases it is assumed that the volume of a molecule is equal to zero. This assumption is justified under certain conditions (not too low a temperature, and sufficiently high rarefaction). Generally speaking, and in connection with real gases in particular, it is necessary to assume that

---

[1] Ann. d. Physik (6) **6**, 59 (1949).

the presence of a molecule whose (very small) volume is equal to v_0 prevents other molecules, whose volumes are also equal to v_0, from penetrating into its sphere of influence. It suffices here to assume that v_0 corresponds to a rigid sphere. The sphere of influence can thus be defined as that sphere whose surface can be reached by the *centers* of the other, impinging molecules, without causing the molecular volumes to overlap. It is evident that this sphere of influence has double the radius of a molecule and hence its volume is $8 v_0$.

Let us now consider a unit of volume within the gas and let n_i denote the number of molecules present in it (molecular density). The portion of space barred to an additional molecule is then $8 n_i v_0$, so that

(1) $$\frac{v_i}{v_1} = 1 - 8 n_i v_0,$$

where v_1 denotes the unit volume; it has been included in the denominator for reasons of dimensional consistency.

Let us now consider an element of area, $d\sigma$, inside the gas. It is seen that eq. (1) is satisfied on both sides of it. If, however, $d\sigma$ denotes an *element* of a *solid wall* then only half of the molecules present will constitute a bar to penetration and it will be due to the molecules present on the side of the element exposed to the gas. Thus we must make a distinction between n_i and n_w (molecular density near a wall). The volume open to the penetration of an additional molecule is larger than (1) and is equal to

(2) $$\frac{v_w}{v_1} = 1 - 4 n_w v_0.$$

The ratio v_w/v_i represents simultaneously the ratio of the probabilities of finding an additional molecule near the wall as compared with the interior of the gas (the probability of being able to place a molecule at one point as compared with the other). This ratio is equal to n_w/n_i so that comparing (2) with (1), we have

(3) $$\frac{v_w}{v_i} = \frac{n_w}{n_i} = \frac{1 - 4 n_w v_0}{1 - 8 n_i v_0}.$$

Solving for n_w we find that

(3 a) $$n_w = \frac{n_i}{1 - 4 n_i v_0}.$$

The number of molecules, n, and hence the density of a gas is slightly higher near a wall than in the bulk. This is the real reason for the correction term b in the van der Waals equation, as shown by Sauter who followed Boltzmann in this respect.

In order to show this we refer once more to the calculation of pressure given in Sec. 22. We have considered then a cylinder on base $d\sigma$ and finite height ξ. We can, however, decrease the height ξ as much as we please, and so we can confine the whole cylinder to the neighborhood of the wall, if we simultaneously allow the base to increase in proportion. Thus eq. (22.4) remains valid, except that n must be replaced by n_w. Taking into account the definition of temperature in eq. (22.5) as well as eq. (3 a), we have

$$(4) \qquad p = n_w \times k T = \frac{n_i k T}{1 - 4 n_i v_0}.$$

In order to reduce this equation to the form due to van der Waals and given in Sec. 9 we first apply it to one mol of a gas. The number of all molecules is then equal to the Loschmidt-Avogadro number L and the number of molecules in a unit of volume $n_i = L/v$, where v now denotes the molar volume. It follows from eq. (4) that

$$(4\text{ a}) \qquad p = \frac{L k T/v}{1 - 4 L v_0/v} = \frac{R T}{v - 4 L v_0}.$$

On comparing with eq. (9.1) in which we must put $a = 0$, because cohesion forces have so far been left out of account, we find

$$(5) \qquad b = 4 L v_0;$$

the constant b is equal to four times the total volume of all molecules present in a mol. This is the physical significance of the constant b; it was deduced already by van der Waals.

B. The van der Waals cohesion forces and the constant a

In the present argument we shall neglect the volume of each molecule but we shall take into account the short-range cohesion forces acting between neighboring molecules and given by $f(r)$; for example in the case of noble gases these forces decrease approximately in proportion to r^{-7} as r increases. We may now put $n_i = n$ as if no cohesion forces were present, because in the *bulk of the gas* all forces acting on one molecule are evenly distributed and cancel each other. However, near *a boundary* conditions are different because the forces of cohesion exert a pull towards the interior of the gas ("upwards") because the molecules which would exert a pull in the opposite direction ("downwards") do not exist. There is thus a kind of sedimentation in the upward direction, i. e. away from the wall, which leads to the normal value n outside the boundary zone but generates a value $n_w < n$ in the neighborhood of the wall. Qualitatively speaking, the influence of the forces of cohesion consists in creating a density distribution near the boundary which differs from the normal, in a way similar to the influence of molecular volumes.

In order to obtain a quantitative expression it is necessary to recall the barometric formula (23.16). In deducing it, account was taken of the gravitational potential i. e. of the force of gravity which was independent of position and which was directed downwards. Its value was

(6) $$\Phi = m\,g\,z \quad \text{or} \quad -\frac{\partial \Phi}{\partial z} = -m\,g.$$

In the present case, however, the force is not constant; it is equal to zero above the boundary zone (mutual compensation of cohesion forces), and it is directed upwards within it. We imagine that the sphere of influence of cohesion forces is represented by a spherical surface centered at the point under consideration and we use the symbol $\gamma(z)$ to denote the influence from the segment of the sphere intercepted by the wall. The "upward pull" will be proportional to this well-defined quantity $\gamma(z)$. Furthermore, in the present case we are not considering a molecule in a given field of forces, as was the case with the gravitational field, but we wish to enquire into the interaction between the molecule which experiences the pull upwards and the surrounding neighboring molecules. This interaction depends on the density with which the neighboring molecules would fill the segment $\gamma(z)$ in the absence of the wall. It is thus seen to be proportional to the number of molecules n; since our present considerations are directed towards the establishing of a correction term, the variation of n with space coordinates may be neglected. Hence, as distinct from (6), we put

(7) $$-\frac{\partial \Phi}{\partial z} = n\,\gamma(z); \qquad \Phi = n \int_z^\infty \gamma(z)\,dz,$$

where the coefficient of proportionality has been included in $\gamma(z)$. In this equation Φ is normalized to give $\Phi = 0$ for $z = \infty$, which is different from the condition $\Phi = 0$ at $z = 0$ in the case of gravity. Hence at the wall

$$\Phi(0) = n\,\overline{\gamma}, \qquad \overline{\gamma} = \int_0^\infty \gamma(z)\,dz,$$

and the preceding barometric formula

$$\frac{n_z}{n_0} = e^{-\Phi/kT}$$

transforms into

(8) $$\frac{n_w}{n} = e^{-\Phi(0)/kT} = e^{-n\bar{\gamma}/kT}$$

at $z = 0$.

Applying eq. (4), or Bernoulli's construction, we find that the pressure is given by

(9) $$p = n_w k T = n k T \, e^{-n\bar{\gamma}/kT}.$$

Expanding into a series for large values of kT, and neglecting higher-order terms we have

(9 a) $$p \approx n k T - n^2 \bar{\gamma}.$$

We apply this equation once more to one mol putting $n = L/v$ with $v = v_{mol}$. Hence[1]

(10) $$p = \frac{RT}{v} - \frac{L^2 \bar{\gamma}}{v^2}.$$

---

[1] At this point one might ask whether eq. (9) does not lead to a pressure correction term which is more precise than that in the van der Waals equation. When the full expression is retained the van der Waals equation is replaced by that due to Dieterici

$$p = \frac{RT}{v-b} e^{-a/vRT}.$$

The critical values (cf. Sec. 9 A) implied are:

$$v_{cr} = 2b, \quad R T_{cr} = \frac{a}{4b}, \quad p_{cr} = \frac{a}{4b^2} e^{-2},$$

and the *critical constant* is

$$K = \frac{R T_{cr}}{p_{cr} v_{cr}} = \frac{1}{2} e^2 = 3.69$$

(instead of $K = 8/3 = 2.67$ given by the van der Waals equation). Empirical values for He, A and Xe are $K = 3.31$; 3.42, and 3.57, respectively.

The first term is of the same form as for a perfect gas, as expected, because the volume of the molecules has been neglected. On comparing with the van der Waals equation the second term explains the physical significance of a in eq. (9.1):

$$a = L^2 \bar{\gamma}. \tag{11}$$

The constant a in the van der Waals equation is equal to L^2 times a certain cohesive action at the boundary which vanishes in the bulk of the gas. It may be worth mentioning that an artificial wall is created when a manometer is introduced into the interior of the gas and so it measures the "cohesion pressure" $p_a = a/v^2$ in addition to the normal pressure.

Sauter (*l. c.*) deduced an expression for the cohesion pressure from the statistical behavior of the whole collection of molecules and not, as we have done it in the preceding argument, from the behavior of a single isolated molecule, which experiences the pull of the rest. Consequently, Sauter's derivation is more satisfactory but also more tedious. In addition Sauter proved that the nature of the wall and the supposed existence of long-range cohesion forces between wall and gas have no influence on the pressure being measured.

27. The problem of the mean free path

The concept of the mean free path was introduced by Clausius as early as 1858. It represents the expected length l which is traversed by any molecule under consideration and moving within the conglomeration of molecules which constitute the gas, during a time interval between an arbitrary initial position and the first collision with another molecule. In the case of perfect gases whose molecules have been assumed to be shrunk to points we would have $l = \infty$. Consequently, it is necessary to take into account the finite volume of a molecule, v_0, as in Sec. 26. We have found previously that this volume entered into the van der Waals equation in the form of the constant b which was in turn proportional to $4 v_0$. It will be shown that in the present case the final equation will contain the cross-section of v_0 multiplied by the factor 4. As in the preceding argument, the present considerations are restricted to rigid, spherical molecules.

We shall, further, restrict the calculation to the case when the velocity of the impinging molecule is large with respect to that of the molecule suffering the impact. The "mean free path" which was calculated already by Maxwell on the assumption that the molecules satisfy the Maxwellian distribution is of the same order of magnitude (coefficient $\sqrt{2}$ in the denominator). It appears to us that the present restriction is justified because in the exact theory, *cf.* Sec. 44, it is not necessary to know the mean free path as it can be replaced by an argument in a series expansion which must then be properly defined depending on the particular problem under consideration.

It will be remarked at this point that the problems to be considered in the following Sections B and C are concerned with *irreversible processes* and hence transgress the scope of ordinary thermodynamics and statistical mechanics (*cf.* here Sec. 21 and Chap. V).

A. Calculation of the Mean Free Path in One Special Case

When the impinging particles move very fast it is possible to assume that the slow particles suffering the impact are at rest. We now attach the volume of the latter to the former, as was done in Sec. 26 A, and construct an "effective cross-section." The radius of the effective cross-section (the sum of the radii r_1 and r_2 of the two molecules under consideration) will be denoted by s (in the present case $r_1 = r_2 = r$ and $s = 2r$). The volume swept by the sphere of influence per unit time has the form of a circular cylinder of cross-section πs^2 and height c. When the number of molecules is n, the interior of the circular cylinder will contain on the average

$$n \pi s^2 c$$

molecules suffering impact, for these can now be regarded as points, as they have been deprived of their volume. The preceding expression represents also the number of collisions per unit time, denoted by ν. Since ν collisions are associated with the path traversed per unit time we find that the mean free path for a single collision is

$$(1) \qquad l = \frac{c}{\nu} = \frac{1}{n \pi s^2}.$$

In the case of rigid spherical molecules of effective cross-section πs^2, l denotes precisely the smallest thickness of layer for which the effective cross-section of the molecules would cover the whole area if there were no intersections.

We now proceed to calculate the order of magnitude of l. In order to do this we introduce a reference length l_1 defined by the condition that a cube of side l_1 contains on the average just *one* molecule. Thus $n = 1/l_1^3$ and according to (1), we have

$$(2) \qquad \frac{l}{l_1} = \frac{1}{\pi} \left(\frac{l_1}{s}\right)^2.$$

The numerical value of the radius of a hydrogen atom (first Bohr orbit) is about 0.5×10^{-8} cm. Estimating that the radius of a hydrogen molecule is equal to the double of that for an atom, we find that the radius s of the sphere of influence is equal to 2×10^{-8} cm. The number of molecules n should be taken per unit of volume and can be found from the Loschmidt-Avogadro number. At 0 C and 1 at pressure it is $2.8 \times 10^{19}/\text{cm}^3$. Hence

(2 a) $$l_1^3 = \frac{1}{28} \times 10^{-18} \text{ cm}^3, \qquad \frac{l_1}{s} = \frac{1/3 \times 10^{-6}}{2 \times 10^{-8}} = \frac{50}{3},$$

and according to (2)

(2 b) $$\frac{l}{l_1} = \frac{10^4}{36\pi} \approx 100.$$

Consequently the following proportion can be set up:

(3) $$l : l_1 : s = 10^4 : 10^2 : 6.$$

Under the present assumptions the mean free path l is 100 times larger than the mean distance l_1 between molecules; at atmospheric pressure it is, in turn, much larger than the radius s of the cross-section, or, as we may also say, than the molecular diameter.

According to Avogadro's law the length l_1 is the same for all perfect gases; furthermore the diameters of molecules do not differ materially from each other. Consequently the proportion (3) is also approximately valid for O_2 and N_2.

We can conclude, further, that the molecules of a perfect gas suffer enormously large numbers of collisions on moving over a distance of several cm from an initial position. This shows that the unexpectedly large values of $\sqrt{\overline{c^2}}$ found in Sec. 22 are confined to ultra-microscopic distances. We can also see from eq. (1) (l being inversely proportional to pressure) that the concept of a mean free path becomes illusory when applied to very high vacua: The molecules are in a position to reach the walls of the vessel without first colliding with other molecules. This limiting case of the kinetic theory of gases was clarified both experimentally and theoretically in detailed papers due to M. Knudsen. In the opposite limiting case (liquefaction, touching of molecules) we have $l \approx s$ and the concept of the mean free path loses its sense once more.

B. Viscosity

We now propose to consider the case when the molecules of a gas possess a *molar* (macroscopic) velocity in addition to their molecular velocity. We assume that the former is directed along the x-axis and that it increases linearly in the direction of y. In the phenomenological science of fluid dynamics, *cf.* Vol. II, Sec. 10, it is assumed that a small element of area at right angles to y is acted upon by a shearing stress in the x-direction given by

$$\sigma_{yz} = \pm \eta \frac{\partial u}{\partial y}. \tag{4}$$

It decelerates the velocity of the upper (faster) layers and accelerates that of the lower (slower) layers (see the respective sketch in Sec. 10 of Vol. II). It is now our task to deduce the same result with the aid of kinetic theory and to calculate the value of the coefficient η (viscosity). The proof consists in setting up a balance equation for the *transfer of momentum*.

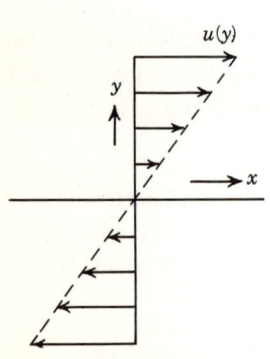

Fig. 25.
Mean velocity in Couette flow.

Let m denote the mass of a molecule, u—the molar, and ξ—the molecular velocity component in the direction of x. The total momentum of the molecules in the direction of x is thus:

$$g_x = m(u + \xi), \tag{5}$$

where u depends on y, Fig. 25. The probability distribution of the values of ξ is independent of y if, as we shall assume, the temperature does not vary from point to point. The cross-section at $y = 0$ under consideration is continually traversed by slower molecules coming from the lower half space and entering the upper region; faster molecules traverse it in the opposite direction. We shall now inquire into the point from which the faster and slower molecules, respectively, arrive. This depends on the last preceding collision in the half space from which they come. In order to determine the probable distance at which this collision took place we produce the mean free path l starting at $y = 0$ and direct it along the path of the molecule. Denoting the angle between this direction and the axis of ordinates by θ we find that the abscissa of the position of the last collision is equal to $-l \cos \theta$ for a molecule coming from below. When it arrives from above, the abscissa is equal to $+ l \cos \theta$, *cf.* Figs. 26 and 26 a.

27.7 THE PROBLEM OF THE MEAN FREE PATH

We now proceed to calculate the molar momentum $g = m u$ which is lost by the upper half-space $y > 0$ owing to its having lost the molecules which have passed our cross-section at $y = 0$ per unit area. It is not necessary to take into account the molecular momentum $m \xi$ because ξ is independent of y and its change is, therefore, cancelled by the molecules arriving from below.

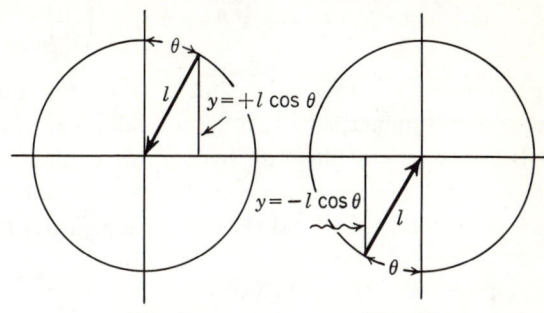

Fig. 26.   Fig. 26 a.
Particles from above   Particles from below
arriving from a distance l (= radius of circle).

We begin by considering the group of molecules which move upwards with a velocity between c and $c + dc$ and whose directions lie between θ and $\theta + d\theta$. Since they come from below, Fig. 26 a, their contribution is

(5 a) $$g \uparrow = m u(-l \cos \theta) = m u (0) - m l \cos \theta \left(\frac{\partial u}{\partial y}\right)_0.$$

(Since $u(y)$ is linear there are no higher-order terms in the Taylor expansion for u.) Correspondingly the group of molecules travelling upwards (θ = angle with the negative semi-axis of ordinates), Fig. 26 a:

(5 b) $$g \downarrow = m u (+ l \cos \theta) = m u (0) + m l \cos \theta \left(\frac{\partial u}{\partial y}\right)_0.$$

The difference of the two is

(6) $$g \uparrow - g \downarrow = -2 m l \cos \theta \left(\frac{\partial u}{\partial y}\right)_0.$$

This value must now be multiplied by the number of molecules dv, equal for the two groups, which cross our section at $y = 0$ per unit time and area. According to eqs. (22.9 a) and (23.9) the number of molecules which arrive from the volume in Fig. 23 per unit time and area is equal to

(7) $$dv = \frac{n}{4 \pi c^2} \phi(c) \times c \cos \theta \times \sin \theta \, d\theta \, d\phi \times c^2 d c,$$

where the normal to the surface is taken in the ζ-direction (also the polar axis). The total change in momentum due to the molecules in the volume shown in Fig. 23 is obtained by multiplying eq. (6) by this expression. Integrating with respect to the azimuth ϕ and rearranging, we have

$$- n\,m \times l\,c\,\phi(c)\,dc \times \cos^2\theta \sin\theta\,d\theta \times \left(\frac{\partial u}{\partial y}\right)_0.$$

The values of $g\uparrow$ and $g\downarrow$ have been multiplied by the same factor because the Gaussian distribution is symmetrical. In order to evaluate the total momentum transferred, it is necessary to integrate over all velocities c and over all directions θ from which molecules may come. Integration over θ is thus extended from 0 to $\tfrac{1}{2}\pi$ and then total change in momentum becomes

$$\bar{g} = \int (g\uparrow - g\downarrow)\,dv = -m\,n\left(\frac{\partial u}{\partial y}\right)_0 \int_0^\infty l\,c\,\phi(c)\,dc \cdot \int_0^{\pi/2} \cos^2\theta \sin\theta\,d\theta$$

(8)
$$= -\frac{m\,n}{3}\left(\frac{\partial u}{\partial y}\right)_0 \int_0^\infty l\,c\,\phi(c)\,dc.$$

This equation takes into account the fact that, strictly speaking, l depends on c. It would be possible to determine this relation, the calculation being somewhat complicated, on the assumption that the molecules obey Maxwell's distribution law. It is, however, necessary to note that the problems under consideration do not involve states of equilibrium so that, strictly speaking, the distribution law is not Maxwellian, cf. Sec. D. We shall, therefore, abstain from further investigating the integral in (8) replacing l by its mean value, for example by $l/\sqrt{2}$ (see here the introduction to the present Section). This allows us to place it in front of the integral, so that (8) yields

(8 a)
$$\bar{g} = -\frac{m\,n\,l}{3}\left(\frac{\partial u}{\partial y}\right)_0 \int_0^\infty c\,\phi(c)\,dc.$$

Making use of eq. (23 10 a) and replacing the integral in (8 a) by $\bar{c}$, we find

(9)
$$\bar{g} = -\frac{m\,n\,l\,\bar{c}}{3}\left(\frac{\partial u}{\partial y}\right)_0.$$

Evaluating the change in momentum for the negative half space we obtain the same formula, except that the sign is reversed. Thus

(9 a) $$\overline{g} = + \frac{m n l \overline{c}}{3} \left(\frac{\partial u}{\partial y}\right)_0.$$

Comparing (9) and (9 a) with (4) we see that the viscosity is given by

(10) $$\eta = \frac{m n l \overline{c}}{3}.$$

It will be noticed that $m n$ represents the mass per unit volume, i. e. the density ρ of the gas. The preceding equation (10) was first deduced by Maxwell in 1860. It only seemingly implies a dependence on pressure, because the product of

(11) $$\rho = m n \quad \text{and} \quad l \approx \frac{1}{\sqrt{2} \, n \pi \, s^2}$$

is independent of the number of molecules n, which cancels in the numerator and denominator. The remaining quantity is independent of the mass and volume of the particles. This *lack of dependence* of viscosity on *pressure* was at first considered paradoxical; it has, however, been verified by Kundt and Warburg in 1875 down to pressures of the order of 1/60 at. The equation is also limited in its application at the high pressure range because the gas then approaches its point of liquefaction when the concept of a mean free path ceases to have a meaning, as is also the case for very low pressures.

The *dependence on temperature* defined by the factor $\overline{c}$ in eq. (10) is also of great interest. In actual fact $\overline{c}$ (as well as $(\overline{c^2})^{\frac{1}{2}}$ increases in proportion to the square root of absolute temperature. Whereas the viscosity of liquids (e. g. of oils) decreases fast with temperature, *the viscosity of a gas increases in proportion to* $\sqrt{T}$.

C. Thermal conductivity

The preceding considerations, concerning the transfer of momentum g, apply to the transfer of any quantity G which may be transported by the molecules. In particular, we are interested in the transport of energy, E, when we consider a gas which is not in thermal equilibrium, and whose temperature increases linearly along the axis of ordinates. We shall once more use Figs. 26 and 26 a to illustrate the argument.

The energy E is given by eq. (23.14) in which the potential Φ may now be omitted. Denoting the number of degrees of freedom by f (translation, rotation, and internal vibrations) and making use of the law of equipartition of energy we find that each degree of freedom is associated with the energy $\frac{1}{2} k T$, so that the total energy becomes

$$E = \tfrac{1}{2} f k T$$

or, in terms of y

$$E = \tfrac{1}{2} f k \left[T(0) + y \left(\frac{\partial T}{\partial y} \right)_0 \right].$$

Thus the quantity G being transferred is given by

$$G \uparrow - G \downarrow = -2 \cdot \tfrac{1}{2} \cdot f k l \cos \theta \left(\frac{\partial T}{\partial y} \right)_0$$

instead of (6) ($m u$ has been replaced by $\tfrac{1}{2} f k T$). Instead of (8 a), we obtain

$$\overline{G} = -\frac{1}{3} \frac{n f k l}{2} \left(\frac{\partial T}{\partial y} \right)_0 \cdot \int_0^\infty c\, \phi(c)\, dc = -\frac{n l}{3} \cdot \frac{f k \overline{c}}{2} \left(\frac{\partial T}{\partial y} \right)_0.$$

Comparing this expression with Fourier's phenomenological assumption about the conduction of heat along a corresponding temperature gradient (Q = quantity of heat transferred, $\varkappa$ = thermal conductivity)

$$Q = -\varkappa \left(\frac{\partial T}{\partial y} \right)_0,$$

we obtain the following interpretation of $\varkappa$ from kinetic theory

(12) $$\varkappa = \frac{n\, l\, f k \overline{c}}{3 \quad 2}.$$

It is seen that $\varkappa$ is *independent of pressure* in the same way as η from eq. (10) but *varies with temperature* (owing to the factor $\overline{c}$). The lack of dependence of thermal conductivity on pressure was demonstrated experimentally by Kundt and Warburg (when the temperature is not too low) at the same time when they performed experiments on viscosity.

The value of the ratio $\varkappa/\eta$ which results from the preceding considerations is also of great interest, because it does not contain the questionable quantities $\overline{c}$ and l. Combining (12) with (10) we obtain

$$\frac{\varkappa}{\eta} = \frac{f k}{2 m}.$$

Multiplying the numerator and the denominator by the Loschmidt-Avogadro number L per mol, we obtain

$$\frac{\varkappa}{\eta} = \frac{\frac{1}{2} f R}{\mu}$$

(because $kL = R =$ universal gas constant and $mL = \mu =$ molecular weight). It is seen from eq. (4.13) that the numerator contains the molar specific heat at constant volume, c_v, of the gas. This quantity can be *measured directly* as distinct from the original quantitites $\bar{c}$ and l which appear in the expressions for $\varkappa$ and η. The resulting equation

(13) $$\frac{\varkappa}{\eta} = \frac{c_v}{\mu}$$

in which c_v/μ represents the specific heat. In a more accurate derivation it would still contain a numerical factor on the right-hand side. The numerical value of the constant for rigid, spherical molecules (Enskog) is 2.52, becoming rather smaller for polyatomic molecules. It may be worth noting that eq. (13) is to a certain extent reminiscent of the *Wiedemann-Franz law* in the theory of metals (*cf.* Sec. 45).

(13 a) $$\frac{\varkappa}{\sigma} = \frac{2}{3} \left(\frac{k}{e}\right)^2 T,$$

($\sigma =$ electrical conductivity, $e =$ electron charge). The analogy consists in the fact that the questionable quantity l (as well as the number of free electrons, n) has been canceled in the quotient (13 a). The two equations differ in that the temperature dependence has been suppressed in (13) as distinct from (13 a).

D. SOME GENERAL REMARKS ON THE PROBLEMS ASSOCIATED WITH THE CONCEPT OF THE MEAN FREE PATH

It has been mentioned on several occasions that the assumption of a Maxwellian distribution in connection with non-equilibrium states constitutes a rough approximation. The essential task in problems of this kind is to *determine the velocity distribution in terms of space and time*. The importance of this problem was recognized quite early by Boltzmann; it was subsequently formulated by Hilbert. When solving it, it is found that a certain length l, which can be interpreted as the mean free path, enters naturally into the equations and that it can be defined only in terms of the velocity distribution.

It is by no means evident that it will be equal for problems of internal friction and heat conduction, as assumed hitherto. We shall revert to this point in Sec. 44.

The relations become particularly difficult in problems involving diffusion, but we do not propose to pursue this matter here. We feel, however, forced to consider a difficulty which is contained right in the foundations of probability and which becomes particularly acute in problems involving the mean free path. We shall explain its nature with reference to the throwing of dice by inquiring into the number of throws which are required in order to make it probable for a prescribed numeral, for example 1, to show up for the first time. This very imprecisely formulated question is traditionally answered by introducing the concept of "mathematical expectation" defined as

$$(14) \qquad \overline{x} = \sum_{x=1}^{\omega} x \, w_x$$

where w_x denotes the probability that the prescribed number will first occur at the x'th throw. In the case of dice w_x is known uniquely so that the sum in (4) can be calculated without difficulty. The computation is shown carried out in Problem III. 4 and it is found that the result is 6. This number can be said to be equal to the "mean free path of the numeral one" in a series of throws.

However, it is possible to imagine that this series has been extended rearwards and we can inquire as to when the numeral one most probably showed up for the last time before the present series of throws had been instituted. The mathematical expectation for this event is again given by eq. (14) and is, consequently, also equal to 6. Thus it appears that there is a contradiction between the "mean free path 6" and the interval of 12 throws between the last appearance of unity in the series extended rearwards and its first appearance in the series of throws following. It must be realized that such a contradiction is to be expected owing to the lack of regularity in chance and is conditioned by the arbitrary cut which was introduced into the sequence of events by selecting a definite start for the series of throws.

This cut finds its corollary in the kinetic theory of gases in the arbitrary cut which we introduced at $y = 0$ in Figs. 26 and 26 a; in a similar manner it is possible to argue that the collisions in the upper and lower half-spaces respectively correspond to the occurrence of the unity before and after the cut. Consequently it should not be surprising that the path covered by a molecule between two such collisions is equal to twice the distance which, in accordance with (14), we would define as the mathematical probability for the occurrence of a single collision; its approximate value was given in eq. (1).

CHAPTER IV

GENERAL STATISTICAL MECHANICS: COMBINATORIAL METHOD.

It has been shown in Chap. III that statistical considerations concerning states of equilibrium lead to simple general laws, the Maxwell-Boltzmann velocity distribution law, the equation of state of a perfect gas, and the values of its specific heats. On the other hand, towards the end of Chap. III we were forced to conclude that the considerations of irreversible processes and of any arguments involving the use of the mean free path required the introduction of fairly arbitrary assumptions and cumbersome modes of reasoning. Surveying the problem from a higher vantage point we are compelled to suppose that states of equilibrium can be treated in a simple and comprehensive manner, whereas it is necessary to employ more complex methods in order to obtain satisfactory solutions of problems associated with irreversible processes. A sketch of the latter will be given in Chap. V. The present Chap. IV rests on the work of Ludwig Boltzmann (1847—1906) and is astonishingly simple. In any case, the simplicity of the theory must not be judged from the point of view of elementary and trivial intuition but must be assessed from the point of view of higher mathematical lucidity. The concept of *entropy*, this fundamental pillar of thermodynamics which was conspicuously absent in the elementary Chap. III will only now be given its due place and in the form of *Boltzmann's principle* it will turn out to be the arithmetical consequence of a simple procedure of a combinatorial character.

28. Liouville's theorem, Γ-space and μ-space

Before applying any calculations involving probability it is necessary to make a decision about "events of equal probability". In the case of dice the equal probability attached to the numbers 1 to 6 is warranted by their geometrical shape and by the homogeneity of the material of which they are made. In card games an attempt is made to give an equal chance to each player by careful shuffling.

A. The Multidimensional Γ-space (Phase Space)

In the kinetic theory of gases it is expected that any initial regularities in the physical or velocity space will be quickly smoothed out owing to the enormous number of particles and to their intense motion. It is expected that a state of complete "molecular chaos" prevails so that is is possible to apply the calculus of probability to it. The gas is taken to be a mechanical system of F degrees of freedom. Let N be the number of molecules present, all being assumed identical and each possessing f degrees of freedom, so that $F = f N$. The system will be described by Hamilton's canonical equations given in Vol. I, eq. (41.4),

$$(1) \qquad \dot{p}_k = -\frac{\partial H}{\partial q_k}, \qquad \dot{q}_k = +\frac{\partial H}{\partial p_k}, \qquad H = H(p_1, \ldots, p_F, q_1, \ldots, q_F)$$

where the q_k's denote the space coordinates, and the p_k's represent the associated momentum coordinates of the molecules. It is assumed that the p's and q's are numbered (cf. e. g. Sec. 12 of Vol. I), so that a given molecule is associated with many values of k. The Hamiltonian function H represents the total energy of the system and is represented in terms of p and q; the molecular interactions during collisions and the repulsion caused by the walls is assumed included in H. It is derived from Lagrange's function

$$L = L(q_1, \ldots, q_F, \dot{q}_1, \ldots, \dot{q}_F)$$

and is

$$H = \sum_k p_k \dot{q}_k - L, \qquad p_k = \frac{\partial L}{\partial \dot{q}_k},$$

(see e.g. Vol. I, Sec. 41, eq. (1)).

The space of $2F$ dimensions defined by the p, q will be called the Γ-space or the phase space.[1] The instantaneous state of a system is represented by a point in that space, the point describing a path with the course of time. We do not, however, consider one single such Γ point but a large number of them, in particular, such as lie within a small spatial element

$$(2) \qquad \Delta \Omega = \Delta p \Delta q \qquad \text{with } \begin{cases} \Delta p = \Delta p_1 \ldots \Delta p_F \\ \Delta q = \Delta q_1 \ldots \Delta q_F. \end{cases}$$

---

[1] Following P. and T. Ehrenfest in their now standard Sec. 32 of the Mathem. Enzykl. Vol. IV, 4, which deals with the axiomatic foundations of the theory of gases. The letter Γ stands for "gas".

We further imagine that the curves emanating from it are marked in some way. They represent gases of identical molecular species and identical numbers of molecules, differing only somewhat as regards their micro-states; they cannot be distinguished macroscopically when they possess equal (or almost equal) energies.[1] The spatial element $\Delta \Omega$ changes its shape with time in a multiplicity of ways. We do assert, however, that it retains its volume in the process.

B. Liouville's theorem

In order to arrange the proof in the simplest way possible we shall recall the results of ordinary kinematics given in Sec. 1 of Vol. II. The relative change of volume (volumetric dilatation) was shown to be given by

$$(3) \qquad \Theta = \frac{\partial \xi}{\partial x} + \frac{\partial \eta}{\partial y} + \frac{\partial \zeta}{\partial z},$$

where x, y, z denoted arbitrary Cartesian coordinates fixed in space, ξ denoted the translation of a point of a body in the direction of x; η, ζ denote translations in the directions y and z respectively. Let it now be imagined that this formula is extended to many dimensions, x and y corresponding to our p_1 and q_1. The symbols ξ and η now denote the translations $\dot{p}_1 \, dt$ and $\dot{q}_1 \, dt$ in the Γ-space and $\partial \xi / \partial x + \partial \eta / \partial y$ transforms into

$$(4) \qquad \left(\frac{\partial \dot{p}_1}{\partial p_1} + \frac{\partial \dot{q}_1}{\partial q_1} \right) dt \quad \text{or, in view of (1) into} \quad \left(-\frac{\partial}{\partial p_1} \frac{\partial H}{\partial q_1} + \frac{\partial}{\partial q_1} \frac{\partial H}{\partial p_1} \right) dt = 0.$$

In a similar manner pairs of succeeding terms of the sum in (3) imagined extended to the many dimensions under consideration can be combined to give zero. Such partial sums have no independent physical meaning because they represent projections on a coordinate plane placed in an arbitrary manner. However, the sum of them all, our dilatation Θ, is an invariant of the Γ-space. It is clear that its geometrical meaning is given by $\dfrac{d\Delta \Omega}{dt} / \Delta \Omega$. It follows that

$$(5) \qquad \frac{d\Delta \Omega}{dt} = 0, \qquad \Delta \Omega = \text{const}$$

which is the statement of Liouville's theorem.

---

[1] When it is prescribed that the energies are exactly equal, the number of dimensions of the Γ-space is reduced from $2F$ to $2F - 1$ owing to the existence of the energy equation $H(p, q) = \text{const}$; when the energies are approximately equal, $H_1 < H < H_2$, attention is centered on a shell-like portion of the space of $2F$ dimensions. The latter case occurs, for example, when the system is in thermal equilibrium with its surroundings.

Equation (5) can also be expressed as follows: If we imagine that the phase points of different micro-states are distributed over the Γ-space with uniform density, then their motion is like that of an *incompressible fluid*. Interpreting the density as a measure of the probability of finding a phase point in an element of the phase space we can conclude that the assumption that equal elements $\Delta\Omega$ enclose regions of equal probability is true at all times if it was true at any instant.

Consequently, it is possible to replace the initial conditions of mechanics by the statistical assumption that equal phase elements in the Γ-space are associated with equal probabilities. It corresponds to the assumption that the faces of dice are associated with equal probabilities. In the last resort the question as to whether or not this is true, i. e. whether the sides of dice and equal phase elements are, in fact, associated with equal probabilities, is a matter to be decided in the light of experience. The preceding argument could only demonstrate that the fundamental hypothesis of statistical mechanics is compatible with the equations of motion. Generally speaking, such statistical rules replace the initial conditions of the mechanics of systems whereas the equations of motion remain unaltered.

In conclusion we wish to stress once more that considerations involving the phase space are not related to a single gas but to a very large group of states of the gas under consideration or of an arbitrary system, which we imagine placed one beside the other and whose behavior is surveyed at the same time. Their states differ only microscopically; macroscopically they are alike.

C. Equality of probability for the perfect gas

We now proceed to introduce the simple representation of the *phase space which is associated with a single molecule, or the "μ-space"*, according to the terminology introduced by P. and T. Ehrenfest. This has only $2f$ dimensions instead of the $2Nf$ of the Γ-space. It is thus six-dimensional in the case of a monatomic gas ($f = 3$), and ten-dimensional in the case of a diatomic gas composed of two rigid molecules ($f = 5$ owing to the two additional rotational degrees of freedom) etc.

It should be realized that the transition from the Γ- to the μ-space is possible only in cases when the mechanical system described by $H(p, q)$ possesses very special properties. So far we have assumed it to be quite general in its properties (arbitrary interactions between the molecules were admitted, e. g. repulsive forces at impact or cohesive forces at larger separations), but now we shall stipulate that the system corresponds to a *perfect gas* (volume of molecule $v_0 = 0$, hence no collisions, infinite mean free free path). Consequently the molecules are seen to move independently of each other; their phase spaces are separated and identical. H is the sum of the Hamiltonian functions corresponding to the individual molecules; it contains no terms which involve the coordinates of several molecules simultaneously so that during the motion the contributions of the individual molecules remain separate. From the fact that the volume elements in the Γ-space are constant and from the fact that the $\Delta\Omega_\Gamma$'s have equal probabilities (statistical hypothesis and Liouville's theorem) it is possible to conclude that, under the restrictions under consideration, the $\Delta\Omega_\mu$'s also have *equal probabilities*. This equality of probability forms the basis of *Boltzmann's combinatorial method* which will occupy our attention in this Section. The method *a priori* attaches equal probability to the elements

(6) $$\Delta\Omega = \Delta p\,\Delta q \quad \text{with} \quad \begin{cases} \Delta p = \Delta p_1 \ldots \Delta p_f \\ \Delta q = \Delta q_1 \ldots \Delta q_f. \end{cases}$$

The simplificiation of this formula compared with (2) should be noted: The subscripts $1, 2, \ldots, f$ in (6) refer to the same molecule, whereas in (2) the indices $1, 2, \ldots, F$ were enumerated through all the molecules.

When deriving eq. (2) we have assumed that the element $\Delta\Omega$ in the Γ-space was "small" and in (4) we applied the rules of differential calculus to the differences Δp, Δq. We have not, however, decided *how small* these elements should be. When describing the method, Boltzmann stresses repeatedly that it is necessary to stipulate finite elements Δp, Δq to make sure that the phase elements enclose numerous molecules; he does, however, always pass to the limits $\Delta p \to 0$, $\Delta q \to 0$ in the final results. In fact the answer to the question of "how small" these are to be was only given by the *quantum theory*.

In anticipation we wish to make the following remark: The product $p_k\,q_k$ (and hence the product $\Delta p_k\,\Delta q_k$) has the dimension of *an action*. This is quite clear as far as Cartesian coordinates are concerned ($[q] = $ cm; $[p] = [m\,\dot q] = $ g cm sec^{-1}); it is, however, true of coordinates q with arbitrary dimensions, as seen from the relation $p = \partial L/\partial \dot q$ already quoted: Since L has the dimension *erg* it follows, in fact, that

$$\text{Dimension of } p_k\,q_k = \text{erg sec} = \text{action}.$$

Quantum theory teaches that the action is composed of elementary quanta in a discontinuous way. We shall see that Planck's constant h ("quantum of action") puts a *lower limit* to the size of $\Delta p_k\,\Delta q_k$ for every pair of coordinates p_k, q_k:

(7) $$\Delta p_k \cdot \Delta q_k = h.$$

We shall make use of this result here. According to (6) the magnitude of an elementary cell of the phase space for a molecule of f degrees of freedom is

(8) $$\Delta \Omega = h^f.$$

Two states p, q of a molecule whose representations lie within the same (properly bounded) elementary cell cannot be distinguished statistically. The cells $\Delta \Omega$ defined in this way constitute the *units* with which is is necessary to work when applying the combinatorial method. They correspond to the six events of equal probability in the game of dice.

To conclude, we must pose one more difficult question: Are these elementary cells large enough to contain, as required by Boltzmann, a large number of molecules? Normally this is not the case. Earlier reasoning was based on sufficiently large phase cells so that, generally speaking, there was no difficulty associated with too small numbers. This is, however, not *a priori* possible when the size of an elementary cell is determined by quantum mechanics. However, with the aid of a simple transformation it is possible to show that, nevertheless, a large number of elementary cells can be combined into a higher entity (Sec. 29 C) thus creating conditions in which the assumptions which are essential for the application of Boltzmann's method are satisfied. Thus the small size of the elementary cells constitutes no serious obstacle.

Darwin and Fowler have developed a different method which allows us to calculate the mean values even in cases when the numbers of molecules per cell are small. The results agree with those due to Boltzmann, but their field of applicability is larger because variations in energy from cell to cell impose limitations on the process of combining a large number of cells into one big cell. We shall refrain here from describing the derivation due to Darwin and Fowler,[1] but we shall use this method to advantage in another connection (*cf.* Sec. 37).

The succeeding argument will be concerned mostly with the μ-space. However, the inability to discern between equal particles, demanded by quantum mechanics, will force us to revert to the Γ-space. It will be found that the formulae which are applicable in this case are quite similar to those which we are about to derive for the μ-space. They can, moreover, be derived in a similar manner, but we shall justify them in Sec. 36 with the aid of methods which can be traced to W. Gibbs, to whom, together with L. Boltzmann, we owe the existence of statistical mechanics. It will turn out that Gibbs' method is quite independent of the problems connected with the smallness of cells.

---

[1] Reference should be made to: R. H. Fowler, 'Statistical Mechanics', Cambridge 1929 or to the small, but eminently readable booklet by M. Born: "Natural Philosophy of Cause and Chance", Oxford, 1949.

29. Boltzmann's principle

Boltzmann's principle interprets entropy in terms of the probability of states and expresses it in the terse formula:

(1) $$S = k \log W.$$

So it stands carved out on Boltzmann's memorial in the Central Cemetery in Vienna, floating in the clouds over his majestic bust.

It is immaterial that Boltzmann himself never wrote down the equation in this form. This was first done by Planck, e. g. in the First Edition of his "Vorlesungen über die Theorie der Wärmestrahlung," 1906. The constant, k, was also introduced by Planck and not by Boltzmann. Boltzmann only referred to the proportionality between S and the logarithm of the probability of a state. The designation of "Boltzmann's principle" was advocated by Einstein for the reverse of (1), namely

(1 a) $$W = e^{S/k}$$

in which S was considered to be known empirically, the quantity W being the unknown for which an expression was sought. According to it the "second part of the Second Law" signifies, as recognized already by Helmholtz, a transition from an artificial state of order to a more probable state of disorder.

The right-hand side of (1), particularly in the case of non-homogeneous systems, must be augmented by a constant, i. e. by a quantity which is independent of the parameters of state but is related only to the numbers of mols of the components. In general, eq. (1) has the form

(1 b) $$S = k \log W + \text{const.}$$

In 1877 in connection with eq. (1) (which he had not written down) Boltzmann remarked,[1] still quite vaguely, that: "It might even be possible to calculate the probability of the different states from the ratio of the number of the various distributions, and this might lead to an interesting calculation of thermal equilibrium". Shortly afterwards[2] he added: "I do not think that one is justified in accepting this result without reservations as something evident, at least not until an exact definition of what we are to understand by the term 'most probable distribution' has been given." In the same paper Boltzmann intimates that the existence of Liouville's theorem constitutes a necessary limitation for the parameters of state which are to be chosen.

---

[1] Vienna Academy, No. 39 in "Gesammelte Werke", p. 121.
[2] *Ibid*, No. 42, p. 193.

A. Permutability as a Measure of the Probability of a State

Let us now consider a perfect gas; we assume that it consists of N molecules, that it is enclosed in a volume V, and that its total energy is U. If the number of degrees of freedom of a single molecule is f, then the phase space has $2f$ dimensions. We subdivide it into M cells,

$$1, 2, \ldots, i, \ldots, M.$$

In view of the fact that the cells are finite, as given in eq. (28.8), and in view of the prescribed finite values of U and V, M is a finite, if extremely large, number. At first, we distribute the molecules over the cells in a completely arbitrary fashion, and denote the numbers of molecules in the cells by

(2) $$n_1, n_2, \ldots, n_i, \ldots, n_M,$$

where, naturally,

(2 a) $$\sum_i n_i = N.$$

Any distribution n_i determines a definite microstate of the gas. We now propose to inquire into the number of ways in which one specific microstate can be realized with N molecules distributed over M cells, and denote this number as the thermodynamic probability (or weight), W, of the state.[1] It is given by the permutability[2]

(3) $$W = \frac{N!}{n_1! \, n_2! \, \ldots \, n_M!}.$$

In order to clarify the concept we shall first consider a case when M and N are small.

---

[1] It is essential to note that the thermodynamic probability, being an integer, has not been normalized to 1. Thus W depends on the size of the elementary cells.

[2] The term permutability (Germ. Permutabilität) was suggested by Boltzmann on p. 191 of the paper last cited. It is more graphic than the normally employed German term "Komplexion."

(It denotes the number of combinations of N elements taken $n_1, n_2, \ldots$ at a time—*Transl.*)

Example: $N = 2, \quad M = 2;$

a) $\quad n_1 = 1, \quad n_2 = 1, \quad W = \dfrac{2!}{1!\,1!} = 2,$

b) $\quad n_1 = 2, \quad n_2 = 0, \quad W = \dfrac{2!}{2!\,0!} = 1,$

c) $\quad n_1 = 0, \quad n_2 = 2, \quad W = \dfrac{2!}{0!\,2!} = 1.$

In connection with a) we can put the first molecule either in cell 1 or in cell 2; the placement of the second molecule is then determined; hence $W = 2$. In connection with b) and c) there is no freedom of choice; hence $W = 1$.

Although eq. (3) is known from the elementary theory of combinatorials to represent the polynomial coefficient, we shall, for completeness, prove it by induction from $N - 1$ to N. Thus we assume that eq. (3) has been proved to be true up to $N - 1$ molecules. In order to obtain W_N from the known expression for W_{N-1} it is necessary to consider one of the following W_{N-1} arrangements:

$n_1 - 1, \quad n_2, \ldots \quad n_M, \quad W_{N-1}^{(1)} = \dfrac{(N-1)!}{(n_1-1)!\,n_2!\ldots} = \dfrac{(N-1)!\,n_1}{n_1!\ldots n_M!},$

$n_1, \quad n_2 - 1, \ldots n_M, \quad W_{N-1}^{(2)} = \dfrac{(N-1)!}{n_1!\,(n_2-1)!\ldots} = \dfrac{(N-1)!\,n_2}{n_1!\ldots n_M!},$

$\ldots\ldots\ldots\ldots\ldots\ldots\ldots\ldots\ldots\ldots\ldots\ldots\ldots\ldots\ldots\ldots\ldots\ldots$

$n_1, \quad n_2, \ldots n_M - 1, \quad W_{N-1}^{(N)} = \dfrac{(N-1)!}{n_1!\ldots(n_M-1)!} = \dfrac{(N-1)!\,n_M}{n_1!\ldots n_M!}.$

Each of these arrangements leads to the required arrangement (2) if the Nth molecule is properly placed. Thus we obtain W_N if we add up the values of $W_{N-1}^{(i)}$ on the right-hand sides. In this way we obtain

$$W_N = \sum_{i=1}^{M} W_{N-1}^{(i)} = \dfrac{(N-1)!}{n_1!\ldots n_M!}(n_1 + n_2 + \ldots + n_M) = \dfrac{N!}{n_1!\ldots n_M!},$$

which is identical with (3) in view of (2 a).

The use of Stirling's formula suggests itself in order to simplify the expression for W. In our case (since N is very large) it is sufficient to use the approximation

(4) $$N! = \left(\dfrac{N}{e}\right)^N.$$

A slightly more accurate estimate would give

(4 a) $$N! = (2\pi N)^{\frac{1}{2}} \cdot \left(\frac{N}{e}\right)^N.$$

We may note the elementary derivation of eq. (4) (logarithmic curve replacing an inscribed step-like series):

$$\log N! = \log 1 + \log 2 + \ldots + \log N \approx \int_1^N \log x \, dx$$

$$= [x(\log x - 1)]_{x=N} - [x(\log x - 1)]_{x=1} = N(\log N - 1) + 1 \approx N \log \frac{N}{e}.$$

Taking the *numerus logarithmi* we are led to eq. (4).

We shall use the approximation in (4) to represent the numbers n_i as well as N which implies that they are assumed to be large too. This is undoubtedly permissible if the series of n_i's includes any large numbers, because the small numbers are then negligible and can be replaced by unities. However, the method fails when all n_i's are small numbers. Unfortunately this is exactly the case where Boltzmann's method is applicable and if the argument is associated with the phase elements h^f. It is, for example, possible to show that in the case of a perfect gas under normal conditions out of 30,000 quantum cells, in round figures, at most only one contains a molecule at all (*cf.* Sec. 37). It is, in this case, most unlikely that two molecules will be contained in one cell. Fortunately, as mentioned at the end of Sec. 28, this presents no serious obstacle. We shall, therefore, disregard this difficulty at first, and perform the calculation as if the n_i's included large numbers. At the conclusion of the argument we shall investigate the changes which must be effected when a large number of elementary cells are combined into one big cell, so that the n_i's really include large numbers.

Following Boltzmann we substitute for $n_i!$ the approximations

(4 b) $$n_i! = \left(\frac{n_i}{e}\right)^{n_i}, \qquad \log n_i! = n_i (\log n_i - 1)$$

into eq. (3), and obtain

(5) $$\log W = \text{const} - \sum_{i=1}^{M} n_i \log n_i.$$

29. 8a   BOLTZMANN'S PRINCIPLE   217

The constant in (5) includes all terms which are independent of the n_i's; hence we have

(5 a) $$\text{const} = N (\log N - 1) + \sum_{i=1}^{M} n_i = N \log N.$$

B. The maximum of probability as a measure of entropy

We now ask for the "most frequent arrangement" of molecules, or, in other words, we shall try to calculate the values of the n_i's for which W becomes a *maximum*. In order to do this we shall apply a virtual variation δn_i to n_i, taking into account that, according to the condition in eq. (2 a), we must have

(6) $$\sum \delta n_i = 0.$$

Hence, we obtain from eq. (5) that

$$\delta \log W = - \sum \delta n_i (\log n_i + 1).$$

In view of condition (6) the unity in the brackets can be omitted. As long as the n_i's only have to satisfy condition (6), the criterion of maximum probability becomes

(7) $$\delta \log W = - \sum \delta n_i \log n_i = 0.$$

This means, in agreement with our statistical hypothesis, that all the n_i's must be equal, because in view of (6), eq. (7) can be satisfied only if $n_1 = n_2 = \ldots$

It is, however, necessary to subject the n_i's to another condition. Since the total energy is prescribed (see the beginning of this section), it follows that

(8) $$U = \sum n_i \varepsilon_i,$$

where U denotes the sum taken over all the cells, and ε_i is the total energy of the molecules in a cell whose coordinates in the phase space are p_i, q_i. The value of ε_i changes from cell to cell, but it must be considered fixed within a cell by quantum theory. When n_i is varied in (8) at constant values of U and ε_i, we obtain

(8 a) $$\sum \varepsilon_i \delta n_i = 0.$$

In order to satisfy the two conditions (6) and (8 a) simultaneously it is best to make use of the elegant method of Lagrange's multipliers, as explained in Vol. I, eq. (12.5). Thus (7) is replaced by

(9) $$\delta \log W = - \sum \delta n_i (\log n_i + \alpha + \beta \varepsilon_i) = 0.$$

Since the multipliers α and β are yet to be fixed, we may regard the n_i's as independent. In this manner eq. (9) leads to:

(10) $$\log n_i = -\alpha - \beta \varepsilon_i, \qquad n_i = e^{-\alpha} \cdot e^{-\beta \varepsilon_i}.$$

Substituting this value of $\log n_i$ into (5), we obtain the maximum

(11) $$\log W_{max} = \text{const} + \alpha \sum n_i + \beta \sum n_i \varepsilon_i.$$

Here α and β refer to the state of the whole system, and not to that of a single cell so that we were justified in placing them in front of the summation sign which is extended over i.

Making use of (2 a) and (8) and inserting the value of the constant from (5 a), we obtain simply

(12) $$\log W_{max} = N \log N + \alpha N + \beta U.$$

It will be shown later that the maximum of W which we have just calculated is extremely large as compared with all states associated with only a slightly different value of n_i. For this reason we are justified in identifying the state of maximum probability with the "real state" which we would expect to find experimentally. If this is so, then, according to Boltzmann's principle (1), eq. (12) represents the value of S/k. It follows at once from (11) that for a process during which N remains constant but during which the energy U is changed by external interaction, we must have

(13) $$dS/k = N\, d\alpha + U\, d\beta + \beta\, dU.$$

The changes in α and β are related to each other by the condition

(14) $$N = \sum n_i = e^{-\alpha} \sum e^{-\beta \varepsilon_i} = \text{const}.$$

The sum in eq. (14) is the so-called *partition function* (in German *Zustandssumme*):

(15) $$Z_0 = \sum e^{-\beta \varepsilon_i}.$$

More precisely, the sum in eq. (14) refers to the partition function in the μ-space. Its importance stems from the circumstance that all thermodynamic properties can be derived from it (*cf.* also eq. (33.14)). According to (14), we have $\alpha = \log(Z_0/N)$. Thus eq. (12) becomes:

$$(12') \qquad \frac{1}{k} S = \log W_{max} = N \log Z_0 + \beta U$$

and the arrangement of molecules and the energy follow from eqs. (8) and (10), respectively, and are

$$(16) \qquad n_i = e^{-\alpha} e^{-\beta \varepsilon_i} = -\frac{N}{\beta} \frac{\partial \log Z_0}{\partial \varepsilon_i}$$

$$U = e^{-\alpha} \sum \varepsilon_i e^{-\beta \varepsilon_i} = -N \frac{\partial \log Z_0}{\partial \beta}.$$

C. The combining of elementary cells

There is no doubt about the fact that the approximation in Section B is inadmissible. The results, however, are correct. It is easy to show that they apply at least to cases when the values of energy differ little from cell to cell. It is then possible to regard them as being constant over a large region and to combine a large number of cells, say $\varkappa$, into a higher entity, a *macro-cell*, as we shall call it.

Let $N_1, N_2, \ldots, N_m$ denote the numbers of molecules per macro-cell. In analogy with eqs. (2 a) and (8), we may write

$$(17) \qquad \sum_j N_j = N, \qquad \sum_j N_j \bar{\varepsilon}_j = U.$$

The summation $\sum_j$ extends over all macro-cells and $\bar{\varepsilon}_j$ denotes the mean energies in them. The thermodynamic probability of a given distribution of molecules over the macro-cells is obtained by summing up all elementary probabilities associated with the arrangements of molecules which lead to the same distribution over the macro-cells. Hence

$$(18) \qquad W' = \frac{N!}{N_1! \ldots N_m!} \sum_{(n)}' \frac{N_1!}{n_{11}! \ldots n_{1\varkappa}!} \cdot \frac{N_2!}{n_{21}! \ldots n_{2\varkappa}!} \cdots$$

Mathematically this means that the sum must be taken over all arrangements of N_1, N_2 etc. Consequently the sum changes into a product of sums

$$W' = \frac{N!}{N_1! \ldots N_m!} \prod_j \sum \frac{N_j!}{n_{ji}! \ldots n_{j\varkappa}!}$$

which can all be calculated with the aid of the binomial theorem. The factors are: $\varkappa^{N_1}, \varkappa^{N_2}, \ldots$, and we have:

$$(19) \qquad W' = \frac{N!}{N_1! \ldots N_m!} \varkappa^N.$$

This equation differs from (3) only by the term $\varkappa^N$. It is now possible to apply Boltzmann's methods to W' because we can choose $\varkappa$ large enough for the N_j's to become large numbers. In analogy with eq. (10), we obtain

$$(20) \qquad N_j = e^{-\alpha - \beta \bar{\varepsilon}_j}.$$

Furthermore, eq. (12) is replaced by

$$(21) \qquad \log W'_{max} = N \log \varkappa N + \alpha N + \beta U$$

and (14) is replaced by

$$(22) \qquad N = e^{-\alpha} \sum_j e^{-\beta \bar{\varepsilon}_j}.$$

We can now write down an expression for the partition function (15) in the μ-space

$$(23) \qquad Z_0 = \varkappa \sum_j e^{-\beta \bar{\varepsilon}_j},$$

and it follows from (22) that

$$e^\alpha = \frac{Z_0}{\varkappa N}, \qquad \alpha = \log Z_0 - \log \varkappa N.$$

Accordingly, we are once more led to eq. (12')

$$(21') \qquad \log W'_{max} = N \log Z_0 + \beta U = \log W_{max},$$

i. e. a value which is independent of $\varkappa$. The same can be said about the remaining quantities, and the results are seen to be the same as for $\varkappa = 1$. This time, however, the justification for them is more sound.

30. Comparison with thermodynamics

A. Constant volume process

If $V = $ const, the division of the phase space into elementary cells remains unaltered and it follows from (29.14) by logarithmic differentiation that

(1) $$0 = -d\alpha - d\beta \frac{\sum \varepsilon_i e^{-\beta \varepsilon_i}}{\sum e^{-\beta \varepsilon_i}} = -d\alpha - d\beta \frac{U}{N}.$$

Thus eq. (29,13) reduces to

(2) $$dS = k\beta\, dU.$$

Now, in accordance with the Second Law (*cf*. Sec. 6), we have

(2 a) $$dS = dQ_{rev}/T = dU/T.$$

On comparing (2) with (2 a), we obtain

(3) $$\beta = \frac{1}{kT}.$$

We shall see later that this fundamental relation remains true also in cases B and C.

B. General process performed by a gas in the absence of external forces

We assume that the process consists not only of a change in energy, dU, but also of a change in volume, dV. Thus in addition to the phase cells 1 to M we shall have cells $M+1$ to M' (i. e. if dV is positive; if dV is negative cells $M'+1$ to M will disappear). The corresponding change of $\sum\limits_{i=1}^{M}$ will be denoted by

(4) $$d\Sigma = \sum_{i=M+1}^{M'} \quad \left(\text{or} = -\sum_{i=M'+1}^{M}\right)$$

where, in the additional terms in the sum, the α's and β's which are associated with the whole system, and not with the single phase elements, retain the same values in the same way as in the original terms. Differentiating eq. (29.14) logarithmically we obtain, instead of (1), that

(4 a) $$-d\alpha - d\beta \frac{U}{N} + \frac{d\Sigma}{\Sigma} = 0.$$

In the absence of external forces we shall have

(4 b)
$$\frac{d\Sigma}{\Sigma} = \frac{dV}{V}$$

as will be proved shortly. Substitution of (4 a, b) into (29.13) yields

$$dS = k\beta\, dU + kN\frac{dV}{V}.$$

On comparing with the statement of the Second Law

$$dS = \frac{dQ_{rev}}{T} = \frac{dU + p\, dV}{T}$$

which now differs from (2 a) we obtain, in addition to (3), the *equation of state of a perfect gas*

(5)
$$\frac{p}{T} = \frac{kN}{V} \qquad \text{as well as} \qquad pv = RT.$$

The latter is obtained when the number of molecules, N, is chosen equal to the Loschmidt-Avogadro number, i. e. when V is made equal to the molar volume, v.

In order to prove eq. (4 b) we are obliged to examine more closely the structure of the phase element $\Delta\Omega$. If x, y, z denote the coordinates of a molecule (e. g. of its center of gravity) in the physical space, we may put

(5 a)
$$\Delta\Omega = \Delta\tau\,\Delta\Omega', \qquad \Delta\tau = \Delta x\,\Delta y\,\Delta z$$

where $\Delta\Omega'$ denotes the volume element of all momentum coordinates of the molecule, as well as of the space coordinates of its internal degrees of freedom, if they exist. Our statement in Sec. 28 that the $\Delta\Omega$'s are equal to each other, being all equal to h^f, can now be supplemented by the statement that the $\Delta\tau$'s can also be chosen equal, because

$$\frac{\partial \dot{x}}{\partial x} + \frac{\partial \dot{y}}{\partial y} + \frac{\partial \dot{z}}{\partial z} = \frac{\partial^2 H}{\partial x\,\partial p_x} + + = \left(\frac{\partial K_x}{\partial p_x} + +\right) = 0$$

since the external force $\boldsymbol{K}$ has been assumed to be equal to zero.

In this case the values of energy ε_i are independent of x, y, z. Thus in the summation with respect to i in eq. (29.14) for each $\Delta\Omega'$ there are as many equal terms as there are space cells, $\Delta\tau$. This number is $V/\Delta\tau$. Consequently, eq. (29.14) may be replaced by

(6)
$$N = \sum_i n_i = e^{-\alpha}\frac{V}{\Delta\tau}\sum_j e^{-\beta\varepsilon_j}$$

where the subscript j denotes now the summation over the phase elements $\Delta \Omega_j'$ only. By logarithmic differentiation of (6), subject to the requirement that N is constant, and for varying V, α, and β (the summation over j remains unaffected thereby), it follows that

$$(6\text{ a})\qquad -d\alpha - d\beta\, \frac{U}{N} + \frac{dV}{V} = 0.$$

The additional term dV/V in this equation corresponds to the term $d\Sigma/\Sigma$ in eq. (4 a) and hence eq. (4 b) is seen to have been proved.

C. A GAS IN A FIELD OF FORCES; THE BOLTZMANN FACTOR

It was found in (5) that the pressure was uniform throughout the gas. This, however, is true only when the energy ε is independent of the space coordinates. When external forces are present (we assume that they possess a single-valued potential $\Phi(x, y, z)$, since otherwise no state of equilibrium could exist, cf. Vol. II, end of Sec. 7), we assume

$$(7)\qquad \varepsilon = \Phi(x, y, z) + \varepsilon',$$

where ε' denotes that portion of the energy of a molecule (inclusive of rotational energy, etc.) which is independent of x, y, z. The problem is now seen to depend on the change in volume dV (i. e. on how we insert the volume elements $\Delta \tau$).

In the present case we shall restrict ourselves to an isochoric process as in Sec. A. Since the external forces are independent of the process, and since $V = \text{const}$, we obtain

$$(8)\qquad \beta = \frac{1}{kT}$$

in exactly the same way as in eqs. (1) to (3); here k is a constant of the system, and independent of the space coordinates, in spite of the fact that Φ does depend on them. *Hence T is independent of the space coordinates.* In particular this is also true of the atmosphere in the gravitational field, on condition that it is in thermal equilibrium. (Some meteorologists have doubted this in the past.)

On the other hand, however, the pressure and density depend on the space coordinates. The latter follows directly from eq. (29.10). The number of particles contained in a cell $\Delta \tau$ becomes

$$(9)\qquad n = e^{-\alpha - \Phi/kT} \cdot \sum e^{-\varepsilon_j/kT}$$

if the value of ε_j is taken from (7) and if the summation is extended over the phase space excluding the cells in the physical space. Multiplying by the mass of the molecule, m, and dividing by $\Delta\tau$, we obtain the *density* ρ. Comparing with the density ρ_0 at the reference level of Φ, we obtain

$$(10) \qquad \frac{\rho}{\rho_0} = e^{-\Phi/kT}.$$

This is the Boltzmann factor from eqs. (23.16) and (23.17) of which use has already been made on several occasions.

D. The Maxwell-Boltzmann velocity distribution law

Maxwell's velocity distribution law for a monatomic gas with zero external forces can be deduced in an equally simple manner. We consider a definite cell $\Delta\omega_j$ of the momentum space and an element $\Delta\tau$ of the physical space (which is arbitrary when no forces are acting). The momentum cell is specified by the coordinates

$$p_x = m\xi, \qquad p_y = m\eta, \qquad p_z = m\zeta,$$

where ξ, η, ζ denote the velocity components, as before.

In the case of a monatomic gas, when the inner degrees of freedom are absent, the cell is associated with the energy

$$\varepsilon = \frac{m}{2}(\xi^2 + \eta^2 + \zeta^2).$$

Substituting the universal relation $\beta = 1/kT$ into eq. (29.10), we obtain

$$(11) \qquad n_i = e^{-\alpha} \exp\left\{-\frac{m}{2}(\xi^2 + \eta^2 + \zeta^2)/kT\right\}.$$

Summation over all elements of space, yields

$$(11\text{ a}) \qquad n_j = \sum_i{}' n_i = \frac{V}{\Delta\tau} e^{-\alpha} \exp\left\{-\frac{m}{2}(\xi^2 + \eta^2 + \zeta^2)/kT\right\}.$$

Putting

$$(11\text{ b}) \qquad \frac{n_j}{N} = F_i \Delta\omega_1,$$

we see that $F_j \Delta\omega_j$ denotes the probability that an atom selected arbitrarily from among the N present will belong to the momentum cell $\Delta\omega_j$ or, in other words, that its velocity will be ξ, η, ζ.

Since $N = \Sigma n_j$ eqs. (11 a) and (11 b) give

(12) $$F = \frac{\exp\{\ \}_j}{\sum_i \exp\{\ \}_i \Delta \omega_i}.$$

The factor $\Delta \omega_j$ with which F_j was multiplied in eq. (11 a) has been here taken into the denominator and each individual term of the sum has been multiplied by it. (Since the phase cells $\Delta \Omega_j$ are equal, and since the $\Delta \tau$'s from eq. (5 a) have been assumed equal, it follows that the momentum cells $\Delta \omega_j$ are equal.)

In order to evaluate the denominator in (12), which, as it is easy to see, is closely related to our partition function Z_0 in (29.15), we go over to the limit $\Delta \omega_j \to 0$. The denominator then becomes equal to:

(12 a) $$m^3 \int_{-\infty}^{+\infty} e^{-m\xi^2/2kT} d\xi \cdot \int_{-\infty}^{+\infty} e^{-m\eta^2/2kT} d\eta \cdot \int_{-\infty}^{+\infty} e^{-m\zeta^2/2kT} d\zeta.$$

Performing the integrations indicated in (23.5 a), we find that

(12 b) $$m^3 \left(\frac{2\pi k T}{m}\right)^{3/2} = (2\pi m k T)^{3/2}.$$

Consequently, it follows from (12) that

(13) $$F = (2\pi m k T)^{-3/2} \exp\left\{-\frac{m}{2}(\xi^2 + \eta^2 + \zeta^2)/kT\right\}.$$

This expression is identical with that in (23.8) with the only formal difference that our present F refers to an element of momentum space, whereas the previous one referred to one in the velocity space; this explains the occurrence of the factor $m^{3/2}$ in the numerator of (23.8) as against its presence in the denominator of (13).

We can now state the following conclusions: It is seen that the path followed in the derivation of Maxwell's velocity distribution law proves to be a *royal* one. It leads, moreover, directly to Boltzmann's generalization for polyatomic molecules which was formulated at the end of Sec. 23. In fact the preceding derivation can be applied without change to a polyatomic gas possessing internal degrees of freedom as well as to a gas in a potential field because the respective additional factors will cancel each other in the numerator and in the denominator of eq. (12).

E. Gaseous mixtures

We may restrict ourselves to the case of two gases accommodated in the same volume V and possessing a total energy U; let the masses of their molecules be m_1, m_2, the numbers of molecules being N_1, N_2. We introduce a phase space for each of the gases; the numbers of molecules per cell in the phase space will be denoted by n_{i1} and n_{j2} respectively.

According to (29.3) the permutabilities are

$$W_1 = \frac{N_1!}{\Pi \, n_{i1}!}, \qquad W_2 = \frac{N_2!}{\Pi \, n_{j2}!}.$$

Since both distributions are independent of each other, the thermodynamic probability for the mixture is

$$W = W_1 \times W_2.$$

Applying Stirling's formula, eq. (29.5) becomes

(14) $$\log W = \text{const} - \left(\sum_i n_{i1} \log n_{i1} + \sum_j n_{j2} \log n_{j2} \right).$$

The calculation of the maximum of W is subject to the following three conditions:

(15) $$\sum n_{i1} = N_1, \quad \sum n_{j2} = N_2, \quad \sum n_{i1} \varepsilon_{i1} + \sum n_{j2} \varepsilon_{j2} = U.$$

In order to satisfy them it is necessary to introduce three Lagrangian multipliers α_1, α_2, β. From (14), we have

$$\delta \log W = - \sum_i \delta n_{i1} (\log n_{i1} + \alpha_1 + \beta \varepsilon_{i1}) - \sum \delta n_{j2} (\log n_{j2} + \alpha_2 + \beta \varepsilon_{j2})$$

and consequently:

$$\log n_{i1} = -\alpha_1 - \beta \varepsilon_{i1}, \qquad \log n_{j2} = -\alpha_2 - \beta \varepsilon_{j2}.$$

Substitution into (14) yields

$$\log W_{max} = \text{const} - \alpha_1 \sum_i n_{i1} - \alpha_2 \sum_j n_{j2} - \beta \left(\sum_i n_{i1} \varepsilon_{i1} + \sum_j n_{j2} \varepsilon_{j2} \right),$$

and in view of (15), we have

(16) $$\log W_{max} = \text{const} - \alpha_1 N_1 - \alpha_2 N_2 - \beta U.$$

This is Boltzmann's principle for the representation of the entropy of a mixture. It follows once more that

(17) $$\beta = \frac{1}{kT}$$

as before. This means that the *temperature of the mixture is uniform* and changes of volume satisfy the *equation of state of a mixture* in the form:

(18) $$p = p_1 + p_2, \quad p_1 = \frac{kN_1 T}{V}, \quad p_2 = \frac{kN_2 T}{V},$$

where p_1 and p_2 denote the partial pressures exerted by each of the components, on the assumption that it alone fills the volume V.

The velocity distributions are also superimposed on each other, as are the pressures, and each separately retains the Maxwellian form.

31. Specific heat and energy of rigid molecules

In spite of the fact that the concept of a rigid molecule, just as the concept of a rigid body in mechanics, is undoubtedly unacceptable from the physical point of view, it is useful to perform a careful study of the thermal behavior of a gas consisting of rigid molecules because such a study will, in particular, determine for us the limits of validity of classical statistical mechanics.

The difficulties attendant on such a task were mentioned by *Lord Kelvin* in his Baltimore Lectures in 1884, Appendix B, when he referred to them as the *Nineteenth Century Clouds over the Dynamical Theory of Heat*; they led him to the conclusion, most revolutionary at the time, that the principle of equipartition would have to be abandoned. We shall see in Secs. 33 to 35 that the physics of the Twentieth Century, the quantum theory in particular, was able to throw brilliant light on all hitherto dark regions of statistical mechanics.

A. The monatomic gas

In view of its lack of a definite structure, the assumption that a monatomic molecule is rigid presents, as yet, no difficulties. We have found in Sec. 22, eqs. (6 a) and (6 b), that the molar energy and the molar specific heats of a monatomic gas are given by

(1) $$u = \frac{3}{2} RT, \quad c_v = \frac{3}{2} R, \quad \frac{c_p}{c_v} = \frac{5}{3}$$

and it only now remains to show how its entropy, already known to us from thermodynamics, fits into the statistical method of expression.

According to Boltzmann's principle, eq. (29.12) becomes

(2) $$S = kN \log N + \alpha k N + \beta k U.$$

The last term on the right-hand side of this equation is a constant and equals $3/2\, k N$ (because $k = 1/T$, and $U = 3/2\, N k T$); it can, therefore, be combined with the first term. Thus eq. (2) simplifies to

(2 a) $$S = kN\left(\frac{3}{2} + \alpha + \log N\right).$$

The value of α for a monatomic gas can be deduced from the argument in Sec. 30 D. We begin with

(3) $$N = \sum n_i = e^{-\alpha}\frac{V}{\Delta \tau}\sum_j \exp\{\ \}_j.$$

The term in the brackets $\{\}$ has the same meaning as in (30.12) and, as before, the summation over j extends only over the momentum cells. The factor $V/\Delta \tau$ denotes the number of space cells contained in the volume V, and hence the multiplicity with which each term j must be counted in order to supplement the original summation over i with that over j. Multiplying both sides of (3) by

$$\Delta \Omega = \Delta \tau \cdot \Delta \omega$$

and putting $\Delta \omega$ as a factor in the summation sign $\underset{j}{\Sigma}$, in accordance with the remark to (30.12), we obtain

(4) $$e^\alpha = V\sum_j \exp\{\ \}_j \Delta \omega_j / N \Delta \Omega.$$

The sum in the above equation is identical with the denominator in (30.12); hence, according to (30.12 a/12 b) it is equal to

$$(2\pi m k T)^{3/2},$$

and we conclude from (4) that

(4 a) $$\alpha = \log V + \frac{3}{2}\log T + \frac{3}{2}\log(2\pi m k) - \log(N \Delta \Omega).$$

SPECIFIC HEAT AND ENERGY OF RIGID MOLECULES

Substituting this value of α into (2 a), we obtain

$$\text{(5)} \qquad S = kN\left(\log V + \frac{3}{2}\log T\right) + C.$$

This is the thermodynamic entropy equation, known to us from Sec. 5, eq. (10), except that it refers here to N particles of a monatomic gas, rather than to one gram.

It must, however, be realized that our present result by far transcends that obtained in thermodynamics because the constant C now has a definite numerical value. It is, namely, equal to the product of kN and the constant from eq. (4 a), augmented by the constant in eq. (2 a):

$$\text{(5 a)} \qquad C = \frac{3}{2} kN\left(1 + \log(2\pi m k) - \frac{2}{3}\log \Delta\Omega\right).$$

According to Boltzmann's theory, $\Delta\Omega$ is undefined, but according to quantum theory we have $\Delta\Omega = h^f$. Thus for a monatomic gas we put $\Delta\Omega = h^3$, and we obtain from eq. (5 a) that

$$\text{(5 b)} \qquad C = kN \log \frac{(2\pi m k)^{3/2} e^{3/2}}{h^3}.$$

Substituting this value into (5) and simplifying, we have

$$\text{(5 c)} \qquad S_{transl} = kN \log\left\{V\left[(2\pi m k T)^{3/2} e^{3/2}/h^3\right]\right\}$$

or per mol,

$$\text{(6)} \qquad s_{transl} = R \log\left\{v\left[(2\pi m k T)^{3/2} e^{3/2}/h^3\right]\right\}.$$

The subscript attached to s indicates that the validity of this expression is not restricted to monatomic gases; it also represents the contribution of the translatory motion of a polyatomic molecule to the entropy of the gas. It is obvious that an extrapolation to $T = 0$ is not permissible because the ideal gas state ceases to exist at $T = 0$. Thus there is no contradiction between eq. (6) and Nernst's Third Law.

Equation (6) was first deduced by Sackur.[1] At about the same time, and independently, Tetrode[2] established an equation which formally differed only slightly from it. We shall return to this famous Sackur-Tetrode equation in Sec. 37 A.

---

[1] O. Sackur, Ann. d. Phys. **36**, 958 (1911); **40**, 67 (1913).
[2] H. Tetrode, Ann. d. Phys. **38**, 434 (1912).

At this stage we must draw attention to two points. First, the constant in the expression for entropy contains the finite volume of the phase cells. We are not allowed arbitrarily to choose the size of a phase element, as was still done by Boltzmann. Moreover, the constant in the expression for entropy determines its size. Unfortunately, this also implies that the assumption that single cells contain large numbers of molecules, n, can never be satisfied, because of the small value of a quantum of action (*cf.* here Sec. 29 C).

Secondly, the entropy must be proportional to the number of molecules N. According to (5 c) this will be the case when we subtract from it the value $N \log N$. Since the volume contained in the logarithm in eq. (5 c) is proportional to the number of molecules at constant temperature, the entropy would increase more than proportionately, if the term $N \log N$ had not been subtracted from it. However, this difficulty can only be removed with the aid of quantum theory. At this stage reference should be made to the derivation of Tetrode's equation in Sec. 37 A.

B. Gas composed of diatomic molecules

We imagine a rigid diatomic molecule in the form of a dumb-bell: The two atoms are assumed to be point-masses which are connected by a massless link of length l.[1] In addition to the two atomic masses, m_1, m_2, the model is assumed to possess two equal moments of inertia about axes at right angles to the line connecting m_1 and m_2; the moment of inertia about the axis joining m_1 to m_2 is assumed to be equal to zero. The same is true about any linear arrangement of atoms, such as occurs, for example, in the case of CO_2.

As is well known, the model possesses 5 degrees of freedom corresponding to the coordinate of the center of gravity x, y, z and the two angular coordinates θ, ψ which describe the position of the axis of the dumb-bell with respect to an arbitrary reference position. The third angle, ϕ, which measures the angular deflection about the axis of the dumb-bell does not count because it corresponds to a zero angular momentum. According to eq. (35.12) in Vol. I, the kinetic energy of rotation is

(7) $$\varepsilon_{rot} = \frac{1}{2} I \left(\dot{\theta}^2 + \sin^2 \theta \times \dot{\psi}^2 \right),$$

because here $C = 0$ and $A = I$. The symbol I denotes the moment of inertia of the body and it should be noted that the symbol Θ was used in Vol. I.

---

[1] Quantum theory demonstrates that we may here neglect the finite extent of atomic nuclei and of the electrons.

31. 9 SPECIFIC HEAT AND ENERGY OF RIGID MOLECULES

This change in notation is made necessary because we shall need the symbol Θ to denote the characteristic temperature of the rotator in Sec. 33.

The momentum coordinates are thus given by

(7 a) $$p_\theta = \frac{\partial \varepsilon_{rot}}{\partial \dot\theta} = I \dot\theta, \quad p_\psi = \frac{\partial \varepsilon_{rot}}{\partial \dot\psi} = I \sin^2 \theta \times \dot\psi.$$

The phase space has 10 dimensions. The phase element can be written as

(7 b) $$\Delta\Omega = \Delta x \Delta y \Delta z \cdot m^3 \Delta \xi \Delta \eta \Delta \zeta \cdot \Delta \theta \Delta \psi \cdot \Delta p_\theta \Delta p_\psi.$$

For the following argument it is far more convenient to transform ε_{rot} into a sum of squares with constant coefficients. As is well known, this can be achieved by introducing the angular velocities about two mutually perpendicular axes, both being at right angles to the axis of the dumb-bell:

(7 c) $$\omega_1 = \dot\theta = p_\theta/I, \quad \omega_2 = \sin\theta \times \dot\psi = p_\psi/I\sin\theta.$$

Here ω_1, ω_2 represent "non-holonomic" velocities, see Vol. I, Sec. 35.4; the quantities $I\omega_1$, $I\omega_2$ were designated as "momentoids" by Boltzmann, but we prefer to call them "impulsoids." The phase element of the impulsoid space of the $I\omega_1$, $I\omega_2$ differs from that of the momentum space of the p_θ, p_ψ by the functional determinant:

(7 d) $$\frac{\Delta p_\theta \Delta p_\psi}{\Delta(I\omega_1)\Delta(I\omega_2)} = \begin{vmatrix} \frac{\partial p_\theta}{\partial(I\omega_1)}, & \frac{\partial p_\theta}{\partial(I\omega_2)} \\ \frac{\partial p_\psi}{\partial(I\omega_1)}, & \frac{\partial p_\psi}{\partial(I\omega_2)} \end{vmatrix} = \begin{vmatrix} 1 & 0 \\ 0 & \sin\theta \end{vmatrix} = \sin\theta.$$

The transformation (7 c) changes (7 b) into

(8) $$\Delta\Omega = \Delta x \Delta y \Delta z \cdot m^3 \Delta\xi\Delta\eta\Delta\zeta \cdot \sin\theta \Delta\theta\Delta\psi \cdot \Delta\omega_1 \Delta\omega_2,$$

and (7) assumes the following form, commonly used in mechanics:

(8 a) $$\varepsilon_{rot} = \frac{1}{2} I(\omega_1^2 + \omega_2^2).$$

The extension of the phase space from 6 to 10 dimensions does not affect the universally valid results derived in Sec. 29, i. e. neither the meaning of $\beta = 1/kT$, or the equation of state, or the Boltzmann factor. However, the additional degrees of freedom demand their share of energy, each of a value of $\frac{1}{2}RT$, at least in accordance with the classical calculation, to be given here. Hence eqs. (1) must be changed to

(9) $$u = \frac{5}{2}RT, \quad c_v = \frac{5}{2}R, \quad \frac{c_p}{c_v} = 1 + \frac{2}{5} = \frac{7}{5}.$$

In order to prove it we express the molar energy in the form

(10) $$\frac{u}{L} = -\frac{\partial}{\partial \beta} \log \sum e^{-\beta \varepsilon_i}$$

in accordance with the universal formula (29.16). We change from the summation over i, which extends over the physical and the momentum spaces, to the summation over j which is only concerned with the momentum space. In analogy with (3) every term must be multiplied by the number of cells in the physical space. This is now equal to:

$$\frac{v \cdot 4\pi}{\varDelta \tau \cdot \varDelta \sigma} \text{ with } \begin{cases} \varDelta \tau = \varDelta x \varDelta y \varDelta z, \quad \varDelta \sigma = \sin\theta \varDelta \theta \varDelta \psi, \\ 4\pi = \int \sin\theta \varDelta \theta \varDelta \psi = \text{surface of unit sphere.} \end{cases}$$

We can now extend the numerator and the denominator of the preceding fraction with the aid of the five-dimensional $\varDelta \omega$ of the momentum space, so that in the denominator we obtain $\varDelta \Omega = \varDelta \tau \varDelta \sigma \varDelta \omega = h^5$, and $\varDelta \omega$ in the numerator can be taken into the $\sum_j$ as $\varDelta \omega_j$. Thus the partition function Z_0 for (10) becomes

(11) $$Z_0 = \frac{4\pi v}{\varDelta \Omega} \sum e^{-\beta \varepsilon_j} \varDelta \omega_j.$$

The contribution of the rotation to the distribution function can be factored out because the energies are additive. With $Z_0 = Z_{transl} \times Z_{rot}$ we have

$$Z_{rot} = \frac{4\pi}{h^2} \sum_j \exp\left(-\frac{1}{2}\beta I(\omega_1^2 + \omega_2^2)\right) \varDelta \omega_{j \text{ (rot)}}$$

$$= \frac{4\pi I^2}{h^2} \int \exp\left(-\frac{1}{2}\beta I(\omega_1^2 + \omega_2^2)\right) d\omega_1 d\omega_2.$$

These are two Laplace integrals, each of which is equal to $\sqrt{\frac{2\pi}{\beta I}}$. Hence

(12) $$Z_{rot} = \frac{8\pi^2 I}{h^2 \beta}.$$

According to (29.16), the contribution to the energy is

(13) $$U_{rot} = -N \frac{\partial \log Z_{rot}}{\partial \beta} = \frac{N}{\beta} = N k T.$$

Thus the energy per mol, including the contribution from translation is

(13 a) $$u = \frac{5}{2} R T, \quad c_v = \frac{5}{2} R, \quad \text{Q. E. D.}$$

31. 16   SPECIFIC HEAT AND ENERGY OF RIGID MOLECULES

The entropy of a diatomic gas must be changed accordingly. It is seen from eq. (29.12′) that the contribution from rotation per mol is equal to

$$s_{rot} = R \log T + R \log \frac{8\pi^2 \mathrm{e} k I}{h^2}. \tag{14}$$

The total entropy of one mol becomes:

$$s = R \log v + \frac{5}{2} R \log T + \text{const} \tag{15}$$

in complete agreement with the thermodynamic equation (5.10)

C. The polyatomic gas and Kelvin's clouds

Assuming a rigid molecule of a general structure in which the atoms are not arranged along a single line, we have to consider an additional degree of freedom: There are now three angular coordinates, θ, ψ, ϕ and three angular velocities, ω_1, ω_2, ω_3 about the principal axes of the ellipsoid of inertia; we shall assume that the principal moments of inertia are I_1, I_2, I_3. The calculation which led us to eq. (12), will now give

$$\left(\frac{2\pi}{\beta m}\right)^{3/2} \left(\frac{2\pi}{\beta I_1}\right)^{1/2} \left(\frac{2\pi}{\beta I_2}\right)^{1/2} \left(\frac{2\pi}{\beta I_3}\right)^{1/2}$$

and instead of (9), we now have

$$u = \frac{6}{2} R T, \quad c_v = 3 R \approx 6 \frac{\text{cal}}{\text{deg mol}}, \quad \frac{c_p}{c_v} = \frac{4}{3}. \tag{16}$$

It is very satisfactory to note that simple rule from Sec. 4 C, namely that $c_v = 5/2\,R$, or that $c_v = 3\,R$, now becomes comprehensible. However, on closer inspection the above rule turns out to be too simple. Consider, for example, the angular model of a molecule of water. According to spectroscopic results, the valency bonds which connect O with the two atoms of H form an angle $\gamma > \tfrac{1}{2}\pi$ (band spectrum of water vapor). The three moments of inertia are different from each other and from zero; if we consider steam to be an approximately perfect gas, we find that c_v is equal to 6 cal/deg mol independently of the moments of inertia and of temperature. The same would be true about molecules with more obtuse angles, but in the limiting case $\gamma = \pi$ the arrangement would become linear and c_v would jump to the diatomic value $5/2\,R \approx 5$ cal/deg mol. The discontinuity of values 3,5, and 6 cal/deg mol which reflects the 3,5, and 6 degrees of freedom of mono- dia- and polyatomic molecules, signifies one of the clouds which obscure the kinetic theory of gases.

However, there exists an even darker cloud. Induced by Nernst's representation of gas degeneration, Eucken performed measurements on the molar specific heat of H_2 at decreasing temperature. He discovered that it continuously decreases from 5 cal/deg mol and becomes equal to 3 cal/deg mol at 80 K. *The rotational degrees of freedom have died out*, they have become, as we sometimes say, *frozen*, and H_2 *has become monatomic*. Paraphrasing a quotation from Schiller,[1] the author stated in 1911, during a scientific congress in Karlsruhe, that: "Degrees of freedom should be weighted, not counted." Quantum theory shows how this is to be achieved.

32. The specific heat of vibrating molecules and of solid bodies

We now drop the physically untenable hypothesis of rigid molecules and take into account the fact that atoms in a molecule are capable of performing small vibrations about their position of equilibrium when, in addition to their kinetic energy, they also posses potential energy. The same is true about the atoms in the giant molecule of a solid body.

A. The diatomic molecule

The force with which the two atoms of such a molecule act on each other coincides with the axis of the arrangement, irrespective of the origin of the force, i. e. irrespective of whether it is of an electrical, polar or homeopolar nature. Let r_1, r_2 denote the two amplitudes of the atoms, measured outwards with respect to the position of equilibrium. In view of the equality of the *actio* and *reactio*, the two amplitudes are seen to be coupled. For the sake of simplicity we shall assume a quasi-elastic bond and hence a harmonic vibration. We put $r = r_1 + r_2$ and calculate the potential and kinetic energies of the coupled system. Evidently

(1) $$E_{pot} = \frac{C}{2} r^2, \quad E_{kin} = \frac{M}{2} \dot{r}^2.$$

The "reduced" mass of the two atoms (*cf.* e. g. Vol. IV, Problem III. 1) of mass m_1, m_2 respectively, is given by

(1 a) $$M = \frac{m_1 m_2}{m_1 + m_2}.$$

---

[1] Demetrius, 1st Act, end of Scene 1.

The elastic constant of the link has been denoted here by C. The rotary motion is superimposed on the vibration along the link. Strictly speaking, the rotation is not completely independent of the vibration, because the moment of inertia varies with the varying distance between the atoms. However, this influence may be neglected in a first approximation because the amplitude is small compared with the distance between the atoms (at $T = 2000$ K and for H_2 molecules it is about 10 per cent).

The phase space of a vibrating diatomic molecule must, thus, be enlarged, as compared with that of a rigid molecule, namely from 10 to 12 dimensions. The phase element $\Delta \Omega$ in (31.8) is to be supplemented with the factors Δr and $M \Delta \dot{r}$. This, however, constitutes no additional difficulty, since the additional energy terms (1) are quadratic in form. The previous integral (31.12) only becomes multiplied by the factors:

(2 a) $$\int_{-\infty}^{+\infty} \exp\left(-\frac{1}{2} \beta C r^2\right) dr = \sqrt{\frac{2\pi}{\beta C}},$$

(2 b) $$\int_{-\infty}^{+\infty} \exp\left(-\frac{1}{2} M \dot{r}^2\right) d\dot{r} = \sqrt{\frac{2\pi}{\beta M}},$$

if the changes in the moment of inertia are neglected. Since we are only interested in the dependence on β, the thus extended eq. (31.13) gives at once

(3) $$u = \frac{7}{2} R T, \quad c_v = \frac{7}{2} R \approx 7 \frac{\text{cal}}{\text{deg mol}}, \quad \frac{c_p}{c_v} = 1 + \frac{2}{7} = 1.29.$$

The preceding argument leads to an important remark and to an even more important question:

1. The potential energy in (2a) appears on an equal footing with the kinetic energy in (2b); according to (3) each contributes $\frac{1}{2} R T$ per mol to the energy distribution.

2. Why does the vibration remain unexcited in the case of air and other diatomic gases under ordinary conditions? If this were not so we should observe the smaller value 1.29 for c_p/c_v instead of 1.4.

B. Polyatomic Gases

We shall recall here a general proposition from the science of vibrations (*cf.* e. g. Vol. VI, Sec. 25): The number of free vibrations which any mechanical system can perform about a stable position of equilibrium is equal to the number of degrees of freedom of the system less the number of degrees of freedom for translations and rotations. Regarding each of these vibrations we can make the same statement as we have made about the linear vibration of a diatomic gas; each of them would have to increase the value of c_v by RT so that for increasingly complex systems of molecules c_v would increase without limit. Consequently, the value of c_p/c_v would approach unity. Why then, do we observe in the case of organic molecules an average value of c_p/c_v in the neighborhood of 1.33 (*cf.* Sec. 4 C) instead of this value, that is a value which corresponds only to active translational and rotational degrees of freedom?

C. The Solid Body and the Dulong-Petit Rule

The analysis of the structure of crystals reveals that every crystal is composed of atoms arranged in a lattice. Because of the mutual links between atoms, and disregarding translations and rotations of the whole body, we see that the atoms only vibrate about their positions of equilibrium. Since every atom possesses 3 degrees of freedom, a lattice composed of N atoms possesses $3N$ degrees of freedom, and deducting the 6 degrees of freedom of the rigid body, we are left with $3N - 6$ degrees of freedom for vibrational modes and an equal number of independent oscillators. In this connection it is necessary to remember that it is always possible to regard the motion of a number of *coupled* oscillators, such as, for example, that of the atoms, as the sum of an equal number of independent normal vibrations,[1] (*cf.* Vol. VI, Sec. 25), whose potential energies must be taken into account in the same way as in A.

Thus at thermodynamic equilibrium there is associated with each oscillator an average energy kT. Hence the molar specific heat of a solid body (in the small, every solid body is a crystal) becomes $c_v = 3R \approx 6$ cal/deg mol independently of the temperature. This is the very well known *rule* due to *Dulong and Petit*. However, this rule contradicts Nernst's Third Law (*cf.* Sec. 12.3), according to which $c_v \to 0$ for $T \to 0$, and it does not agree with experiment at lower temperatures. In fact, in the case of hard substances (diamond, carborundum) a decrease in c_p (and in c_v) is observed even at room temperatures.

---

[1] A normal mode occurs when all atoms vibrate simultaneously in a characteristic manner. Two sympathetic pendulums oscillating in parallel or in counter-stroke, or the natural vibrations of a string, may be recalled as examples.

33. The quantization of vibrational energy

Planck's discovery of the quantum of action, h, has induced us from the beginning to regard the phase elements $\Delta\Omega$ as being constant and to define the corresponding values of energy, ε_i, as a discontinuous series. It turned out that this was very important for the constant in the entropy equation. This series can be treated like a continuum only in cases when the difference between two succeeding levels of energy is vanishingly small in comparison with the equipartition energy kT, i. e. when

(1) $$\varepsilon_{i+1} - \varepsilon_i \ll kT.$$

The transition from the partition function expressed as a sum

(2) $$Z_0 = \sum_i e^{-\varepsilon_i/kT}$$

to the integral (cf. e. g. eq. (30.12 a))

(2 a) $$\frac{1}{\Delta\omega}\int e^{-\varepsilon/kT}\,d\omega$$

is justified only in such cases.

A. The linear oscillator

In what follows we shall replace the subscript i by the more common subscript n and we shall denote the natural frequency of the oscillator by ν. According to the original assumption made by Planck in 1900, we then have

(3) $$\varepsilon_n = n \cdot h\nu.$$

According to his suggestion made in 1911 we put

(3 a) $$\varepsilon_n = \left(n + \frac{1}{2}\right)h\nu,$$

which agrees with the final result in quantum mechanics. In both assumptions condition (1) implies

(4) $$h\nu \ll kT.$$

Introducing the *characteristic temperature*

(4 a) $$\Theta = \frac{h\nu}{k},$$

we find that (4) reduces to

(5) $$\Theta \ll T.$$

The preceding calculations with the aid of the integral partition function instead of the sum, is justified only if the above condition is satisfied. We shall prove that the specific heat vanishes in the reverse case, i. e. when $T \ll \Theta$.

The value of ν, and thus the value of Θ can be obtained from spectroscopical data (infrared rotational spectra), and so, for example, for HCl, for which the spectra have been investigated particularly thoroughly, we have $\Theta \approx 4000 \, \mathrm{K}$.

We now proceed to perform the calculation for arbitrary values of $T \lessgtr \Theta$ making, at first, use of assumption (3). From eq. (29.16) we find the molar energy is given by

(6) $$u = -L \frac{\partial}{\partial \beta} \log \sum_{n=0}^{\infty} e^{-\beta n h \nu}.$$

The geometrical series which appears in the above equation is convergent for any values of $\beta > 0$ and is equal to

$$1/(1 - e^{-\beta h \nu}).$$

Hence according to (6) we have

(6 a) $$u = L h \nu \frac{e^{-\beta h \nu}}{1 - e^{-\beta h \nu}}.$$

Substituting $\beta = 1/kT$ and $\Theta = h\nu/k$, we obtain

(7) $$u = \frac{R\Theta}{e^{\Theta/T} - 1}$$

(8) $$c_v = \frac{du}{dT} = \frac{R\Theta^2/T^2}{(e^{\Theta/T} - 1)^2} e^{\Theta/T}.$$

Thus in the two limiting cases we have

(9)

$T \gg \Theta$	$T \ll \Theta$
$u = RT$	$u = R\Theta\, e^{-\Theta/T} \to 0$
$c_v = R$	$c_v = \dfrac{R\Theta^2}{T^2} e^{-\Theta/T} \to 0$

In the first limiting case we have complete participation of the vibrational degree of freedom in equipartition, in the same way as in Sec. 32 A. In the second case we have no appreciable excitation of vibrations. The transition between these two limiting cases is shown in Figs. 27 and 27a; for example, for $T = \Theta$, we have according to (6) and (7) that

$$u = \frac{R\Theta}{e-1} = 0.58\, R\Theta, \qquad c_v = \frac{R\,e}{(e-1)^2} = 0.92\, R.$$

The above two diagrams settle the question which we have posed at the end of Sec. 32 A. The vibrational degree of freedom is now "weighted" according to temperature and not simply "counted" as in the classical theory. It now becomes clear that at normal temperatures and with values of Θ of the order of several thousand degrees, *the vibrational degree of freedom does not take part in equipartition.*

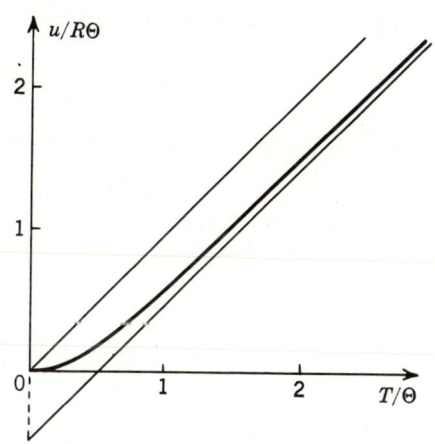

Fig. 27.

Molar energy of quantum mechanical oscillators as a function of temperature; (units: $R\Theta, \Theta$).

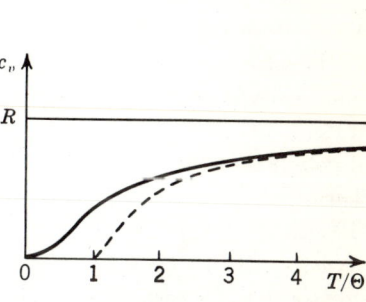

Fig. 27 a.

Molar specific heat of quantum mechanical oscillators as a function of temperature; (units: R, Θ).

It is, however, necessary to include two additional remarks:

1. If the correct quantum equation (3a) is substituted for the older equation (3) we find that (7) is replaced by

(10) $$u = \frac{1}{2} R\Theta + \frac{R\Theta}{e^{\Theta/T} - 1}$$

which is obtained easily from (6) by substituting $n + \tfrac{1}{2}$ for n. It is seen that (7) is increased by the "zero-point energy" $\tfrac{1}{2} R\Theta$; its existence has been confirmed by numerous experiments at very low temperatures. The curve representing u in Fig. 27 becomes thus shifted upwards by a certain distance (equal for all values of T).

2. We are, naturally, most interested in the relations for symmetrical molecules, such as H_2, N_2, O_2, ... Unfortunately, the preceding formulae do not rigorously apply to such molecules but we are forecd to refrain from considering the reasons for it, because this would lead us too deeply into the minutiae of wave mechanics. We shall content ourselves with stating that the "freezing" of the vibrational degrees of freedom at $T \ll \Theta$ is unaffected by these additional considerations.

For molecules possessing more vibrational modes it is, of course, necessary to consider a whole series of natural frequencies, ν_1, ν_2, ... whose influences are superimposed on one another. Instead of Fig. 27 a we obtain a curve for c_v which exhibits several steps, the vertical distance from step to step increasing by R each time. It is clear that at ordinary temperatures only the first step need be taken into account.

B. The solid body

The inconsistency of the theory with Nernst's Third Law which we were forced to record in Sec. 32 C resolves itself if we consider that every atom of the solid body is an independent oscillator and that each must be "quantized" separately. Thus for $T \to 0$ the specific heat per mol is given by the right-hand columns of (9): *The molar specific heat tends to zero exponentially.*

This application of the theory of quanta was suggested by Einstein (Ann. der Physik, Vol. 22) as early as 1907. However, the experiments carried out in Nernst's institute showed a weaker than exponential decrease towards zero. The reason for it is easy to perceive: The atoms do not vibrate independently but in larger or smaller groups, depending on the temperature (*cf.* Sec. 32 C and also Sec. 35).

C. Generalization to arbitrary quantum states

We define the general molecular partition function:

$$(11) \qquad Z_0 = \sum_{n=0}^{\infty} g_n \, e^{-\varepsilon_n k/T}.$$

The equation (11) differs from our previous definition by the "weighting factor", g_n. It serves to collect different quantum states of equal energy into one term. In the case of two unequal atoms vibrating we had $g_n = 1$, but already in the case of two equal atoms g_n varies from term to term (*cf.* remark 2 in Sec. A).

In the same way as in eq. (6) we now derive an expression for the total energy U from Z_0 but instead of one mol we consider a system composed of an arbitrary number of molecules, N. At the same time we shall replace the differentiation with respect to β by one with respect to T in that we write

$$-\frac{d}{d\beta} = -\frac{dT}{d\beta}\frac{d}{dT} = kT^2\frac{d}{dT}.$$

Equation (6) now transforms into

(12) $$U = NkT^2\frac{d\log Z_0}{dT}.$$

An expression for S is obtained from (29.12′)

(13) $$S = Nk\log Z_0 + \frac{U}{T}.$$

In order to prove it by the *methods of thermodynamics* we now vary T keeping V, as well as the remaining parameters, constant. Thus we obtain from (13) that

(13 a) $$dS = Nk\frac{dZ_0}{Z_0} - \frac{U}{T^2}dT + \frac{dU}{T}.$$

The second term on the right hand side can be calculated from (12):

$$-Nk\frac{d\log Z_0}{dT}dT = -Nk\frac{dZ_0}{Z_0},$$

and it is seen that it cancels the first term on the right-hand side of (13a). Hence we have

$$dS = \frac{dU}{T}$$

as we must have for an isochoric process. Thus the validity of (13) has been proved, except for an additive constant whose value will be adjusted in accordance with Nernst's Third Law. Combining (12) with (13), we obtain

(13 b) $$S = Nk\frac{d(T\log Z_0)}{dT}.$$

The expression for free energy $F = U - TS$ which follows directly from (13) turns out to be particularly simple:

(14) $$F = -NkT\log Z_0.$$

34. The quantization of rotational energy

The levels of rotational energy which are admissible by quantum theory also form a discontinuous, step-like series. For the simplest atomic model of a rotator (diatomic molecule), Planck's energy levels of a linear oscillator given in eq. (33.3) are now replaced by the expression

(1) $$\varepsilon_n = n(n+1)\frac{\hbar^2}{2I}, \qquad n = 0, 1, 2, \ldots, \qquad \hbar = \frac{h}{2\pi}.$$

The proof of (1) given in wave mechanics is based on the theory of spherical harmonics, but its detailed consideration would exceed the scope of these lectures.

We shall employ here the partition function from eq. (33.11) formally. In our case the weighting factor is given by

(2) $$g_n = 2n+1.$$

This also is a consequence of the theory of spherical harmonics, according to which, apart from zonal harmonics (dependence on θ alone), there are in addition $2n$ associated surface harmonics (dependence on θ and ϕ). These rotational states which differ from each other in the physical space, correspond to the same energy (1). Every energy level in the series (1) is, as we say, $(2n+1)$-fold degenerated. Thus the weighting factor corresponds to a summation over the physical space, which in this case must be carried out together with that over the momentum space.

Consequently, and in accordance with (33.11), we obtain the partition function

(3) $$Z_{rot} = \sum_{n=0}^{\infty} (2n+1)\, e^{-n(n+1)q} = 1 + 3\, e^{-2q} + 5\, e^{-6q} + \ldots$$

where q denotes the contraction

(3 a) $$q = \frac{\hbar^2}{2IkT} = \frac{\Theta}{T} \qquad \text{with} \qquad \Theta = \frac{\hbar^2}{2Ik}.$$

We now proceed to calculate Θ for H_2 as this constitutes a particularly interesting case. From spectroscopic data (multiple line spectra) we have[1]

$$I = 0.46 \times 10^{-40}\, \text{g cm}^2.$$

---

[1] Using the equation $I = 2m_H (a/2)^2$, we can calculate the distance between the two atoms of hydrogen, as it is known that $m_H = 1.67 \times 10^{-24}$ g. Thus $a = 0.74 \times 10^{-8}$ cm = = 0.74 Å, i. e. the same order of magnitude as the cross-section of a hydrogen atom. This remark serves to render the above value for I plausible from the point of view of its order of magnitude.

Making use of the values which were given at the end of Sec. 20 (eq. (20.43)):

$$\hbar = 1.06 \times 10^{-27} \text{ erg sec}, \quad k = 1.38 \times 10^{-16} \text{ erg/deg}$$

and of the definition of Θ in (3a), we calculate that

(4) $$\Theta \approx 80 \text{ K}.$$

This result leads to the following conclusions: For $T \ll 80$ K the second term in the series (3) is already exponentially small compared with the first. Thus

$$Z_{rot} \approx 1, \quad \log Z_{rot} \approx 0.$$

According to eq. (33.12) this expresses the fact that the contributions from rotational energy to the energy, U, and to the specific heat, c_v, vanish. The specific heat is reduced to the contribution from translation

(5) $$c_v = \frac{3}{2} R.$$

Hydrogen has become, so to say, "monatomic". This agrees with Eucken's discovery which was described at the end of Sec. 31 as the darkest cloud threatening the development of classical statistical mechanics,[1] particularly with regard to the numerical value of Θ.

We now proceed to consider the second limiting case when $T \gg \Theta$, i. e. when $q \ll 1$. The series (3) converges very slowly because its terms decrease to zero only for extremely large values of n. According to its analytical character, it belongs to the group of theta functions; the latter occur in the theory of elliptic functions and in heat conduction problems, see Vol. VI Sec. 15. The use of an analogue to the "transformation formula of the theta function", Vol. VI, eq. (15.8) would in the case of our eq. (3) also lead to a more exact and a more convenient formula for numerical computation. However, at this stage it suffices to perform an approximate estimation of the limit for $T \gg \Theta$, as the more accurate calculation is considered in Problem IV. 6.

Since q is small, we may write

(6) $$p = n(n+1)q$$

---

[1] The rotational energy of the electron shell and that due to the atomic nucleus remain unexcited for the same reason. The value of Θ becomes extremely large (100,000 to 10 million deg in round numbers), owing to the small mass of an electron, in the first case, and owing to the small radius in the second.

which is almost continuous and which varies from 0 to ∞ as n increases. The difference Δp between p_n and p_{n-1}

(6 a) $$\Delta p = p_n - p_{n-1} = n(n+1)q - n(n-1)q = 2nq$$

is a small number for all finite values of n for which the terms of (3) give a significant contribution. Thus in the following argument we shall denote it by dp. Consequently, without committing an appreciable error,[1] we can write $2n + 1 = dp/q$. Thus,

(7) $$Z_{rot} = \frac{1}{q} \int_0^\infty e^{-p}\,dp = \frac{1}{q} = \frac{T}{\Theta},$$

and it follows from eq. (33.12) that

(7 a) $$U = NkT^2 \frac{d}{dT} \log \frac{T}{\Theta} = NkT \quad \text{and hence} \quad \frac{dU}{dT} = Nk.$$

The last result demonstrates that the contribution of rotation to the molar heat for $T \gg \Theta$ is exactly equal to R. This contribution is represented graphically in Figs. 28 and 28a. They demonstrate clearly, just as was the case with the vibrational energy, that the rotational degree of freedom dies out gradually as we pass to $T < \Theta$; the full amount of $R = 2$ cal/deg mol is attained asymptotically only above $T = \Theta$.

In the preceding argument we have purposely considered "hydrogen" because of Eucken's discovery. Strictly speaking we should have referred to the semi-heavy hydrogen HD (D = deuterium), because in the case of the ordinary hydrogen H_2 (as well as in the case of the heavy hydrogen D_2) there are complications, as mentioned in remark 2 on p. 240. However, even in the case of atoms composed of two identical molecules, the general qualitative features of the behavior of the rotational heat c_v remain unchanged in essence.

We shall refrain here from discussing the problems in quantum mechanics which are connected with the rotational energy of a general *polyatomic* molecule. Even in the special case of a symmetrical molecule, such as NH_3, CH_4, …, there appears a new degree of freedom, namely rotation about the axis of symmetry. The scale of energy (1) must now be extended by a term which contains the moment of inertia about this axis and a new index of summation. The single summation in the partition function is now replaced by a double summation.

---

[1] The magnitude of the error can be estimated with the aid of Euler's summation formula.

At sufficiently high temperatures the new degree of freedom causes the molar heat to increase by $\frac{1}{2} R$, but decays at lower temperatures depending on the magnitude of the temperature Θ' which is characteristic for this degree of freedom. In this way the polyatomic molar heat $c_v = 3 R$ changes continuously until it reaches the diatomic value $c_v = 5/2 R$. For large values of Θ' (small moment of inertia about the axis of symmetry) the polyatomic molecule behaves like a *diatomic* one at ordinary temperatures.

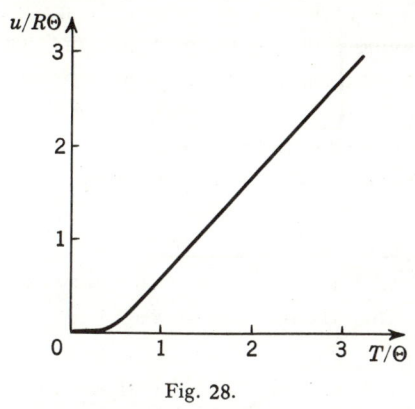

Fig. 28.

Molar energy of quantum-mechanical rotators in terms of temperature (units-$R\Theta$, Θ).

Fig. 28 a.

Molar specific heat of quantum-mechanical rotators in terms of temperature (units: R, Θ).

35. Supplement to the theory of radiation and to that of solid bodies

In Sec. 20, eq. (38), we have deduced the following expression for the mean energy of a linear oscillator:

$$(1) \qquad U = \frac{h\nu}{e^{h\nu/kT} - 1}.$$

This is the energy assumed by the oscillator in a radiation cavity of temperature T when it is in equilibrium with the surrounding black-body radiation. We pointed out at the time that this equation can be derived in a much simpler manner by use of statistical methods.

In order to obtain the proof it suffices to recall the linear oscillator which was considered in Sec. 33 A. It is true that the latter was not placed in a radiation cavity, but it was in equilibrium with its surroundings, it being immaterial to consider the way in which equilibrium had been achieved, i. e. whether equilibrium had been brought about by radiation or by the coupling of the molecules of a gas. According to Planck's quantum hypothesis (33.3), we can use the expression in (33.6a) directly for the energy at equilibrium where, naturally, the factor L (Avogadro-Loschmidt number) should be omitted, because we are now considering a single oscillator and not one mol of substance. The value obtained in this way agrees with (1) precisely.

A. Method of Natural Vibrations

In using statistical methods it is irrelevant whether we are concerned with material objects or with states, with economic data or experimental errors, etc. The application of statistical methods to the natural electromagnetic vibrations of a cavity (parallelepiped or cube) has proved to be particularly fruitful. In Vol. II Sec. 44, we have calculated the number and arrangement of natural vibrations of an elastic slab and we have discussed the simplifications which may be introduced in the case of a cavity containing electromagnetic radiation but entirely devoid of matter. In the latter case it is possible to satisfy the boundary conditions exactly ($\mathsf{E}_{tang} = 0$) with the aid of the elementary sine-law, whereas in the former this is so only when the socalled "mixed" boundary conditions are used.

In addition, the following should be noted: Equation (44.16a) in Vol. II was applicable to the number Z of elastic vibrational modes whose frequency is smaller than ν. In order to adapt this equation to the electromagnetic case it is necessary to put:

$$c_{trans} = c = \text{velocity of light}; \qquad c_{long} = \infty,$$

as already mentioned on p. 325 of Vol. II. In this manner we obtain the equation

$$(2) \qquad \mathsf{Z} = \frac{8\pi}{3} \frac{V \nu^3}{c^3}$$

which was deduced by Lord Rayleigh as early as 1900. Here V denotes the volume of the parallelepiped (or of the cube). The number of vibrational modes per interval $d\nu$ becomes

$$(2\text{ a}) \qquad d\mathsf{Z} = \frac{8\pi V}{c^3} \nu^2 \, d\nu.$$

If we now equip each of these with the energy kT in accordance with the law of equipartition (not $\frac{1}{2} kT$ because it is necessary to take into account the potential energy as well), we find that the energy density per frequency interval becomes

$$(3) \qquad u_\nu = \frac{kT}{V} \frac{d\mathsf{Z}}{d\nu} = \frac{8\pi}{c^3} \nu^2 \, kT.$$

This is the Rayleigh-Jeans law, already quoted in eq. (20.16). As we know, it is grossly inconsistent with experiment because it predicts that $u \to \infty$ as $\nu \to \infty$.

If, however, we adopt Debye's suggestions (Ann. d. Phys. Vol. 33), and regard each frequency mode as a quantized oscillator (*cf.* Sec. 33 C) ascribing to it an energy given by (1), we find from (1) and (2a) that

(4) $$u_\nu = \frac{U}{V}\frac{d\mathsf{Z}}{d\nu} = \frac{8\pi\nu^2}{c^3}\cdot\frac{h\nu}{e^{h\nu/kT}-1}$$

instead of (3), i. e. directly Planck's radiation law, as given in eq. (20.39). Planck adopted this method of derivation in the fourth edition of his "Strahlungstheorie" published in 1921, and described it as an "exceedingly simple derivation" of his law radiation.

B. Debye's theory of the specific heat of a solid

The main difference between the elastic body and the radiation cavity consists in the fact that the degrees of freedom of the former are restricted by its lattice structure, whereas (as far as we know to-day), that of the latter is unlimited. Since the number of normal modes of a body is equal to the number of degress of freedom, and since every atom in the lattice possesses 3 degrees of freedom, the number Z of normal modes is found to be equal to its upper limit Z_g. Thus for a solid body of a mass equal to one mol we have

(5) $\qquad \mathsf{Z}_g = 3L \qquad$ (L = Loschmidt-Avogadro number).

To this upper limit there corresponds an upper limit ν_g of the admissible natural frequencies. As explained in detail in Vol. II, Debye cuts off the spectrum of natural vibrations at $\nu = \nu_g$. The energy of the solid body becomes

(6) $$u = \int_{\nu=0}^{\nu_g} U\, d\mathsf{Z}$$

where U is the quantized value of energy from eq. (1). The limiting frequency ν_g defines the *characteristic temperature* Θ of the solid body,

(7) $$\Theta = \frac{h\nu_g}{k} = \text{Debye temperature.}$$

Performing the integration indicated in (6) we find

(8 a) $\qquad u = \dfrac{3\pi^4 RT^4}{5\,\Theta^3}, \qquad c_v = \dfrac{12\pi^4}{5} R\left(\dfrac{T}{\Theta}\right)^3 \quad$ for $\quad T \ll \Theta$,

(8 b) $\qquad u = 3RT, \qquad c_v = 3R \approx 6\,\dfrac{\text{cal}}{\text{deg mol}} \quad$ for $\quad T \gg \Theta$,

as explained in detail in Vol. II, Sec. 44. The limiting case (8b) is seen to contain the *rule* due to *Dulong and Petit*; in the limiting case (8a) u has the same form as in *Stefan-Boltzmann's law of radiation*; c_v is given by *Debye's third power law* (T^3 – law) *for the molar specific heat of a solid*. It has proved itself as an excellent qualitative rule (disregarding details of the crystal structure) and corrects Einstein's assumption from Sec. 33 B. The T^3-law, just as Einstein's law, satisfies Nernst's Third Law, but it does not converge exponentially. The convergence towards absolute zero is that of a parabola of the third degree. The reason for it is clear: Einstein treated the atoms of a solid body as independent oscillators, whereas Debye collects the molecules which simultaneously perform the natural[1] vibrations into groups. The wavelength of these groups increases as the temperature becomes lower; the correlation between molecules increases in like measure which renders Einstein's assumption of independence quite illusory. The sketch in Fig. 73 of Vol. II gives a qualitative description of the variation of u and c_v between the two limiting cases, (8 a) and (8 b).

This argument dispels the last of Kelvin's clouds mentioned in Sec. 32.

36. Partition function in the Γ-space

We have seen in Sec. 29 that Boltzmann's combinatorial method is not directly applicable to real bodies. It is true that the collecting of numerous elementary cells into a higher unit leads to the right answers and to an approximation which is entirely satisfactory from the practical point of view but, since the thermodynamic relations enjoy general validity, a rigorous justification seems desirable. We shall combine this problem with that of representing the fundamental equations of statistical mechanics in the Γ-space, as we shall require the results in Sec. 37.

A. The Gibbs condition

The succeeding argument which throws new light on Boltzmann's hypotheses of the equality of the probability of equal phase cells is due to J. W. Gibbs.[2] On comparing two phase elements at two different points in the phase space, it is found that the probabilities of finding the phase point

---

[1] The fact that these (as distinct from the atoms in the lattice) may be regarded as being independent of each other can be inferred from the circumstance that the velocity of sound is almost the same for all frequencies.

[2] J. W. Gibbs, "Elementary Principles in Statistical Mechanics", Yale University Press, New Haven, 1902.

of an actual system in one place or in the other can be different and can vary with time. Quite generally, we can write that such probabilities are given by

(1) $$\Delta w = f(p_1 \ldots p_F, \; q_1 \ldots q_F; t) \frac{\Delta \Omega}{h^F}.$$

Here $\Delta \Omega$ denotes the phase element, as before, h^F (with $F = Nf$) is the size of an elementary cell, and the function $f(p, q; t)$ indicates the manner in which this probability changes from point to point and with the course of time. It constitutes the probability per elementary cell h^F. This general form of Δw is important, for example, when we inquire into the manner in which errors in our knowledge of the initial conditions of a mechanical system propagate themselves in time.

In studying thermodynamical statistics we concentrate our attention on states of equilibrium. Consequently $f(p, q)$ need not depend on time explicitly and the question now arises as to what are the consequences of such a statement.

We consider two phase elements $\Delta \Omega$ and $\Delta \Omega'$ which result one from the other because of the motion. The time interval between them will be denoted by $\Delta t = t' - t$. According to eq. (1) the two probabilities can be written as

$$\Delta w = f(p, q, t) \frac{\Delta \Omega}{h^F} \quad \text{and} \quad \Delta w' = f(p', q', t + \Delta t) \frac{\Delta \Omega'}{h^F}.$$

Both are equal because the moving phase element encloses the same phase points at any instant. Thus $\Delta w = \Delta w'$.

Since, according to Liouville's theorem, we also have $\Delta \Omega = \Delta \Omega'$, and since at equilibrium f does not depend on time, we obtain the equation

(2) $$f(p, q) = f(p', q'),$$

on condition that (p', q') results from (p, q) in the course of the motion. Thus we have proved the following proposition: *In the case of thermodynamic equilibrium the probability per elementary cell is an integral of the motion.*

In the case of canonical systems there always exists at least one integral of the motion: the energy $E = H(p, q)$. Every function of energy is also an integral. Hence we may put

(3) $$f = f(H).$$

Apart from energy the argument may contain additional integrals: *partial energies*, if the system is composed of independent sub-systems; *angular momentum* if the system is free to rotate, etc. However, all integrals except

H are constant only under certain restrictive conditions. Small modifications of the system which often do not at all disturb the total energy of the system, but which are important for the establishing of equilibrium, remove accidental integrals. Here one may think of the example of perfect gases in which, for all intents and purposes, the individual molecules exist in isolation from each other; they do, however, exchange energy on collision. Another example is afforded by cosmic matter whose rotation is not confined within any "walls of a vessel" so that on applying statistical considerations it is found that the angular momentum must be taken into account in addition to energy. However, generally speaking, thermodynamical statistics is determined by a single function of energy.

What is the form of the function in eq. (3)? At this point we shall make use of Gibbs' assumption which replaces Boltzmann's hypothesis: *Two coupled mechanical systems in statistical equilibrium remain in equilibrium in the limiting case of vanishingly small coupling even if they are separated.* For example, when two systems equalize their temperatures on being brought in contact, the equilibrium temperature which they both have attained will not change if they are subsequently separated. More precisely, the temperature will not change markedly, if the work performed against the forces of cohesion can be neglected in the overall energy balance; for example, when the bodies are sufficiently large in all directions.

Let H_1 and H_2 denote the Hamiltonians of the two sub-systems and let $H = H_1 + H_2 + \delta H$ ($\delta H \to 0$) be the integral of the coupled system, δH denoting the energy of coupling. According to eq. (3) we have to expect the following distribution functions:

$$f_1(H_1) \cdot f_2(H_2) \quad \text{and} \quad f(H) \approx f(H_1 + H_2).$$

Gibbs' hypothesis leads to the equation

(4) $$f_1(H_1) \cdot f_2(H_2) = f(H_1 + H_2)$$

because the probability of finding each system at a given state must be the same before and after separation.

B. Connection with Boltzmann's method

First, it should be recognized that in the limit when $\delta H \to 0$ the left- and right-hand sides of eq. (4) become integrals of the motion so that Gibbs' hypothesis leads to a statement about the initial conditions only, in the same way as Boltzmann's hypothesis did before. In addition, taking the derivative of (4) with respect to H_1 and H_2 we obtain

$$f_1'(H_1 + H_2) = f_1'(H_1) \cdot f_2(H_2) = f_1(H_1) f_2'(H_2),$$

so that the ratios

$$\frac{f_1'(H_1)}{f_1(H_1)} = \frac{f_2'(H_2)}{f_2(H_2)} = -\beta$$

must be constant, because if this were not so, two functions with different variable arguments could not be identical in form. Integration leads at once to Boltzmann's distribution function

(5) $$f(H) = e^{-\alpha - \beta H}.$$

The constant factor has been written in the form $e^{-\alpha}$ in analogy with eq. (29.10). Its value is obtained from the condition that $\Sigma \Delta W = 1$, so that

(6) $$e^\alpha = \int e^{-\beta H} \frac{d\Omega}{h^F} = Z;$$

Z is now the partition function in the Γ-space and appears still in the form of an integral. The factor $1/h^F$ hints at quantum corrections to be introduced in Sec. C.

All quantities which are characteristic of the state of equilibrium can be deduced from Z. The variational derivate is, in particular,

(7) $$-\frac{1}{\beta} \frac{\delta \log Z}{\delta H(p\,q)} = \frac{e^{-\beta H}}{\int e^{-\beta H} d\Omega} = f(p, q).$$

In analogy with eq. (29.16), we obtain an expression for *mean energy*

(8) $$U = -\frac{\partial \log Z}{\partial \beta}.$$

The *temperature* and *entropy* are given by the Second Law:

$$T\,\delta S = \delta U - \int \delta H \cdot f(p, q) \frac{d\Omega}{h^F}.$$

The second term represents the change in energy as required by the Second Law because on varying H, while keeping the distribution of molecules in the phase space constant, the integral represents the amount of work added to the system. Applying eq. (7), we obtain

$$T\,\delta S = \delta U + \frac{1}{\beta} \int \delta H \frac{\delta \log Z}{\delta H(p, q)} d\Omega = \delta U + \frac{1}{\beta} \delta \log Z - \frac{1}{\beta} \frac{\partial \log Z}{\partial \beta} \delta \beta.$$

The term $\delta \log Z$ includes the variation of β in addition to that of H, and the contribution of the former is subtracted in the third term. Applying eq. (8), we have

$$T \, \delta S = \frac{1}{\beta} \, \delta(\beta \, U + \log Z).$$

Comparing the factors in front of and behind the sign of variation and integrating we can show that

(9) $$T = \frac{1}{k \beta}, \qquad S = k \left(\log Z + \frac{U}{kT} \right).$$

The factor k which is at present undetermined is identical with Boltzmann's constant.

That this is so can be seen on comparing eq. (9) with Boltzmann's equation in (29.1). First, from eqs. (6) and (9) we have

$$S = k \, (\alpha + \beta \, U) = k \, \overline{(\alpha + \beta \, H)} = - k \, \overline{\log f(p, q)}.$$

The bar over a symbol in the above equation denotes a mean value, which is explicitly given as

(10) $$S = - k \int f \cdot \log f \cdot \frac{d\Omega}{h^F}$$

and can be directly compared with eqs. (29.5) and (29.5 a), according to which

$$\log W = - N \sum_i \frac{n_i}{N} \log \frac{n_i}{N}.$$

This gives the equation

(11) $$S = k \log W,$$

which is equivalent to (29.1). It will be realized that eq. (5) gives that distribution function, from among all the possible distribution functions f leading to the same energy U, which renders $\log W$ in eq. (11) a maximum. The relevant calculations are identical with those in Sec. 29. This shows that Gibbs' and Boltzmann's hypotheses are equivalent.

If the system consists of N equal, independent parts of f degrees of freedom each, according to $p_1' \ldots q_f'$, $p_1'' \ldots q_f''$, $\ldots$, $p_1^{(N)} \ldots q_f^{(N)}$, then its energy is given by

$$H = H_0(p', q') + H_0(p'', q'') + \ldots + H_0(p^{(N)}, q^{(N)}).$$

It follows at once that the partition function can be split into N factors:

$$Z = \int e^{-\beta H_0(p', q')} \frac{d\Omega'}{h^f} \times \ldots \times \int e^{-\beta H_0(p^{(N)}, q^{(N)})} \frac{d\Omega^{(N)}}{h^f},$$

which differ only in the symbols employed for the variables under the sign of integration. They are, thus, all equal. Consequently

(12) $$Z = Z_0^N$$

and

(13) $$Z_0 = \int e^{-\beta H_0(p, q)} \frac{d\Omega_0}{h^f}$$

is the partition function in the μ-space. This gives us a direct connection with our preceding arguments.

C. Correction for quantum effects

It must be realized that, generally speaking, it is necessary to base the argument on the partition function in the Γ-space. The splitting of Z into factors Z_0 implies very special assumptions which cease to be valid for the case of a perfect gas with quantum effects included. We shall prove this statement in Sec. 37. In order to introduce the quantum corrections into our statistical considerations, we find it necessary to take several consecutive steps. The first two have already been taken, namely in the μ-space. The third step which makes it necessary to go over to the Γ-space will be described in Sec. 37. The fourth step, the quantum-mechanical derivation of the partition function requires so much knowledge of quantum mechanics that we shall have to be satisfied with a simple transposition of the classical partition function, referring the reader to specialized papers.[1]

The first step consists in replacing the integral form of the partition function in (6) by the sum

(14) $$Z = \sum_{(n)} e^{-\beta E(n)}$$

---

[1] M. Delbrück, G. Molière: Proc. Prussian Ac. of Sci., Phys-Math. Class, 1936, No. 1.

in which the summation is extended over all phase cells of size h^F in the Γ-space. The results are, to all intents and purposes, identical with those which followed from (6), because within one cell the variation suffered by the integrand is insignificant. The size of the phase cell is the only parameter of importance. Its value influences the value of the constant in the expression for entropy (*cf.* Sec. 31, eq. (5 c)). It will be noted, however, that it has already been taken into account in (6) by the assumption of $d\Omega/h^F$ for the phase element. In fact, eq. (6) can be regarded as a continuous approximation to the quantum sum (14), in the sense of Euler's sum equation.

We now have to show that the sum (14) can be split according to eq. (12) in the same way as the partition function (6). Dividing the phase space of a single molecule into cells of size h^f we find that the energy $E(n)$ in eq. (14) is given by

$$(15) \qquad E(n) = \sum_i n_i \varepsilon_i.$$

Here n_i denotes the number of molecules in a given cell in the μ-space and ε_i is the energy of a molecule in such a cell. The sign $\sum_i$ denotes summation over all cells in the μ-space and (n) in eq. (14) denotes that the sum should be taken over all decompositions (partitions) per arrangement of the number of molecules

$$(16) \qquad N = \sum_i n_i.$$

The weighting factor g which appeared in eq. (33.11) has been omitted in (14). It is implied in the convention that every value of energy which occurs several times must be written out as many times. The equality of $E(n)$ in various phase cells in the Γ-space can result from the fact that the values of ε_i in different phase cells of the μ-space are equal (degenerate molecular states). Such degeneration leads to the weighting factor in eq. (33.11).

As long as it is assumed that it is possible to distinguish individual molecules of one kind one from another, as is done in classical statistical mechanics, it is possible to obtain definite values of energy $E(n)$ in many ways because the individual molecules can be distributed over the phase cells in a variety of ways. The distribution is, evidently, such that we always have n_1 molecules in the first cell, n_2 molecules in the second, etc. The number of possibilities is given by the permutability in eq. (29.3), so that the partition function (14) must be provided with a weighting factor:

$$(17) \qquad Z = \sum_{(n)}{}' \frac{N!}{n_1! n_2! \ldots} e^{-\beta \sum_i n_i \varepsilon_i}$$

if every partition of N is written down only once, as indicated by the prime added to the summation sign. Putting

(18) $$e^{-\beta \varepsilon_i} = z_i,$$

we see that eq. (17) becomes

(17 a) $$Z = \sum_{(n)}{}' \frac{N!}{n_1! \, n_2! \ldots} z_1^{n_1} z_2^{n_2} \ldots$$

The permutabilities are seen to be equal to the binomial coefficients. Every power product

$$z_1^{n_1} z_2^{n_2} \ldots$$

occurs in the last sum exactly as often as in the calculation of the N-th power of the sum of all z_i. Hence

$$Z = (z_1 + z_2 + \ldots)^N$$

or

(19) $$Z = Z_0^N, \quad Z_0 = \sum_i z_i = \sum_i e^{-\beta \varepsilon_i}$$

which agrees with eqs. (12) and (13). Z_0 is the partition function in the μ-space; this time, however, it is represented as a sum over all phase cells h^f.

The second step towards quantum mechanical statistics is obtained when we no longer consider that the ε_i are the energy values of the different phase cells, but as the energy values of the different quantum states. Instead of the sum over different phase cells of the μ-space we obtain the sum over all quantum states of a molecule. Boltzmann's hypothesis about phase cells having equal probabilities associated with them is now replaced by the hypothesis that quantum states have equal probabilities ascribed to them. The consequences of this change are studied in Secs. 33 to 35. It was seen that the first step led us to the value of the entropy constant; the second step corrects the principle of equipartition. It will be recalled that we still have to root out a sensitive error which we have encountered when deriving Sackur's formula (cf. Sec. 31, eq. (5 c)). We shall achieve this in Sec. 37, when taking the third step; the same step will lead us to gas degeneration. The results contained in Sec. 37 will turn out to be consistent with Nernst's Third Law.

D. Analysis of Gibbs' Hypothesis

We now revert to eq. (5) once more and inquire into the significance of the dependence on H in the Γ-space. When we consider an actual system which is completely isolated from the surroundings, we find that its energy has a definite value. Thus there is no distribution $f(H)$. It follows that by assuming (5) we are not, at first, considering a completely isolated system, and the derivation of (5) from (4) shows that, in fact, the system under consideration has been separated from a larger system. Thus the *canonical* distribution in (5) is seen at first to apply to a thermodynamic system in a thermal bath.

In order to consider an isolated system it is necessary to refer to eq. (3) directly. Here we have

$$(20) \quad f(H) = \begin{cases} \text{const in a narrow interval } U - \delta U < H < U + \delta U \\ 0 \text{ outside of this interval.} \end{cases}$$

This is the so-called *microcanonical* distribution. It is obtained when Gibbs' assumption of equilibrium is replaced by the requirement that $H = U = \text{const}$. The calculation based on the microcanonical distribution and carried out in the Γ-space is less simple, but the result agrees, to all intents and purposes, with the consequences of the assumption of a canonical distribution because the fluctuations about the mean value of energy (provided that the systems are not too small) are, in most cases, exceedingly small. The disturbances introduced by the bath can normally be only of the order of the energy of interaction, and this we could neglect (see, however, Problem IV. 9).

Thus there is no difference in the μ-space. Since in making the transition to the μ-space it is implied that every molecule interacts only slightly with its surroundings, it may be said that every molecule remains in contact with a thermal bath formed by all the others. On the other hand the mean behavior of a single molecule is determined by the canonical distribution in the μ-space, which is the Boltzmann distribution (29.10), irrespective of whether we begin with the canonical or microcanonical distribution.

37. Fundamentals of quantum statistics.[1]

A. Quantum statistics of identical particles

In the view adopted in quantum mechanics, identical particles cannot be distinguished from each other. This is true not only in relation to electrons in the shell of an atom or in a metal, about which it has long been recognized that they possess no individuality, not only in relation to light quanta and elementary particles in general, but also in relation to the atoms and molecules of our gas. They differ from each other only by special features (ionization, excitation, spin moments). For this reason we must not isolate single *particles* from the total number N and distribute them over the cells $\Delta \Omega_i$ of the phase space, as we have done in Sec. 29. We can only distinguish between the *states of the system*, meaning the states of the gas as a whole, and not the states ε_i of single particles. The former is given by

$$(1) \qquad E(n) = \sum n_i \varepsilon_i.$$

Consequently, from the point of view of quantum statistics, it makes no sense to start with the partition function for a single molecule, eq. (29.15),

$$(2) \qquad Z_\mu = Z_0 = {\sum}' e^{-\beta \varepsilon_i},$$

but it is necessary to use the "partition function for the gas" from eq. (36.14)

$$(2\,\text{a}) \qquad Z_\Gamma = Z = {\sum_{(n)}}' e^{-\beta E(n)}.$$

Since particles cannot be distinguished from each other, an interchange of two particles between two different cells does not lead to a new case. Every distribution given by eq. (36.16) must thus be counted only once. Instead of the permutability in eq. (36.17) we must put a factor of unity. In other words we must use the summation $\underset{n}{\sum}'$ as we have done in eq. (36.7).

The summation becomes more difficult to perform after dropping the factor due to permutability. First of all it can no longer be calculated directly from the partition function for the μ-space. The property of being reducible

---

[1] This Section is based on the short, but significant presentation due to Schrödinger, Statistical Thermodynamics, Cambridge University Press, 1948.

to the μ-space applies only to the classical perfect gas. Before proceeding to compute the sum (2 a) in a systematic way, we propose to consider a limiting case which will allow us to perceive why the use of the partition function (36.17) is almost sufficient in many cases.

We now revert to the remark (*cf.* Sec. 29, p. 216) that in the case of a perfect gas under normal conditions at most one of about 30,000 cells contains just one molecule.[1] Thus in this case we only have to consider occupation numbers 0 and 1, so that the terms $n_i!$ in the denominator of the expression for permutability are, practically, all equal to unity. This signifies that the sums (36.17) and (2 a) differ only by a factor $N!$ which is independent of n_i and which may be taken outside the summation sign $\underset{n}{\Sigma'}$. It follows that for perfect gases under normal conditions we may write:

(3) $$Z_{class} = Z_0^N \approx N!\, Z_{quant} = N!\, Z$$

and that the new partition function is given by

$$Z \approx \frac{Z_0^N}{N!},$$

and

(3 a) $$\log Z \approx N \log Z_0 - N (\log N - 1)$$

in which $\log N!$ has been expanded with the aid of Stirling's formula.

The argument in the preceding Section remains unchanged except for an additive constant. It is, however, seen that this constant suffices to correct Sackur's equation. In fact, if we add the term $-N(\log N - 1)$ to the entropy in eq. (31.5 c), we obtain precisely the formula due to *Sackur and Tetrode*, which is

(4) $$S = k N \log \frac{V}{N} (2\pi m k T)^{3/2}\, e^{5/2}/h^3.$$

As required *a priori*, this expression is proportional to N. Nevertheless the equation does not yet satisfy Nernst's theorem. In order to correct for this deficiency it is necessary to perform a more accurate evaluation of the sum and this becomes possible by:

---

[1] Mean number of molecules per phase cell $e^{-\alpha - \beta \varepsilon_i}$, hence the maximum is $e^{-\alpha}$; mean number of cells per particle according to eqs. (29.14 and 15): $e^\alpha = Z_0/N$ so that for monatomic gases $(V/N)(2\pi m k T)^{3/2}/h^3 \approx 30\,000$ for $m = 1.67 \times 10^{-24}$ g.

B. The method due to Darwin and Fowler[1]

The method allows us to take into account the distribution condition contained in eq. (36.16). In mentioning the names of Darwin and Fowler in the title of this Section we are compelled to make one reservation. The two authors applied their method to classical statistical considerations, as witnessed by the date (1922) of the publication just mentioned; it preceded the formulation of quantum statistics. If they had employed the partition function Z, they could have made reference only to its form (36.17). We shall see in greater detail that such a scheme leads to the classical relation between Z and Z_0, that is to no new discovery. In fact, as seen from eqs. (36.17) to (36.19), it is possible to take into account the distribution condition (36.16) in an elementary way.

The object which the two authors have set themselves consisted in demonstrating the way in which it is possible to take into account the energy condition

(5) $$U = \sum n_i \varepsilon_i$$

in Boltzmann's statistics, when the n_i's are small numbers. Since their method relies on the fact that the terms in the sum (5) are integers, they were forced to measure the energies in "such small units" that all ε_i's, and hence the total energy U, could be approximated by integers. We shall denote this unit by ε_0; (it was not introduced by them explicitly). We are forced to introduce it in order to be able to expand the function in the ζ-plane which corresponds to our $Y(\zeta)$ (see eq. (8) below) in terms of integral powers of ζ so as to be able to use Cauchy's theorem. In using this method we cannot avoid passing to the limit $\varepsilon_0 \to 0$ which implies that the energy scale has been divided into infinitely small elements which is inconsistent with the finite dimensions of phase elements. Pascual Jordan mentions this point in his "Statistical Mechanics,"[2] but "passes over these disturbing circumstances without a more detailed discussion." The energy condition (5) has already been taken into account by the form of the partition function. The latter has been deduced in a different way in Sec. 36, and we no longer need use the Darwin-Fowler method in connection with (5). At this stage, however, we have to introduce the distribution condition (36.16) which is not trivial in the realm of quantum

---

[1] C. G. Darwin and R. H. Fowler, Phil. Mag. Vol. **44**, 450, 823, (1922); see also "Statistical Mechanics" by R. H. Fowler, Cambridge University Press, 1929.

[2] Vol. 87 of the series "Wissenschaft", 2. ed. footnote 2, p. 33, Vieweg, 1944.

statistics. We can apply to it the Darwin-Fowler method (*cf*. Schrödinger, *l. c.*, Chap. VII *ff*). In fact its application is less problematical because the terms in (36.17), unlike those in (5), are, by their nature, integral numbers.

In order to calculate the partition function (2 a) we refer to eq. (36.18). Substituting the energy from eq. (1), we obtain

(6) $$Z = \sum_{(n)}{}' z_1{}^{n_1} z_2{}^{n_2} \ldots .$$

If we now extended the summation over *all* n_i's we would introduce a large number of terms which do not belong in the partition function (6). It is, however, possible to reject them by the use of a stratagem: The quantities z_i from eq. (36.18) are replaced by ζz_i, so that

(7) $$\Pi z_i{}^{n_i} \text{ is replaced by } \zeta^{\Sigma} \Pi_i z_i{}^{n_i}.$$

where

$$\Sigma = \sum_i n_i.$$

Let the sysmbol $Y(\zeta)$ denote the sum resulting from (6) with unrestricted values of n_i. We expand it in powers of ζ and concentrate our attention on that group of terms which is multiplied by ζ^N. According to (7) this is our partition function (5 a). Thus we may write

(8) $$Y(\zeta) = \ldots + \zeta^N Z + \ldots .$$

A second stratagem isolates the terms with ζ^N from all others. Following Darwin and Fowler we make use of Cauchy's theorem on residues, and we obtain

(9) $$Z = \frac{1}{2\pi i} \oint Y(\zeta) \, \zeta^{-N-1} \, d\zeta$$

where ζ is regarded as a complex variable; the integration is to be performed along a closed path in the ζ-plane encircling the origin but no other singularity. In this way all terms in the series in (8) denoted by... are excluded and we retain only that residue (the term with ζ^{-1}) which yields Z directly, in accordance with (9).

C. Bose-Einstein and Fermi-Dirac statistics

We now proceed to analyze the auxiliary function $Y(\zeta)$ a little more closely. Having substituted ζz for z in (6) and having expressly lifted the distribution condition (36.16), we obtain the general expression:

$$Y(\zeta) = \sum_{n_1=0}^{\infty} (\zeta z_1)^{n_1} \cdot \sum_{n_2=0}^{\infty} (\zeta z_2)^{n_2} \cdot \ldots \quad (10)$$

The summations are easy to perform if the n_i's can assume all values

$$n_i = 0, 1, 2, \ldots$$

as already indicated in (9). We then have simply

$$Y(\zeta) = \prod_i \frac{1}{1 - \zeta z_i}. \quad (11)$$

From the point of view of wave mechanics the above result means that the eigen-functions of the system are *symmetrical* functions of the coordinates of its components. This case was developed in 1924 by S. N. Bose for the light-quantum gas and extended shortly afterwards by Einstein to include material gases.[1]

There exists another case which is realized in nature and which corresponds to the antisymmetrical eigen-functions of wave mechanics. In such a case we have

$$n_i = 1 \text{ or } 0.$$

The latter case was introduced into wave mechanics in 1926 by Fermi, who made use of *Pauli's exclusion principle*, and independently by Dirac. The most important application of this statistics occurs in relation to metal electrons.

In this case eq. (10) leads directly to

$$Y(\zeta) = \prod_i (1 + \zeta z_i). \quad (11\text{ a})$$

The two cases can be represented by the single equation

$$Y(\zeta) = \prod_i (1 \mp \zeta z_i)^{\mp 1} \quad (12)$$

(the upper sign gives the Bose-Einstein statistics, and the lower sign leads to the Fermi-Dirac statistics).

---

[1] A detailed discussion of the relation of this result to symmetrical and antisymmetrical wave functions in wave mechanics would exceed the scope of this book.

In both cases we have $Y(0) = 1$ and $Y(\zeta)$ can be expanded into a series of integral powers of ζ in the neighborhood of $\zeta = 0$, as already assumed in (8). In the Fermi-Dirac case $Y(\zeta)$ is a holomorphic function which increases monotonically along the real positive axis. In the Bose-Einstein case $Y(\zeta)$ is a meromorphic function which possesses poles at all points where

$$\zeta = \zeta_i = 1/z_i.$$

In accordance with the definition of z_i in eq. (36.18) all such points lie on the positive real axis of ζ. If we normalize the energy so that all $\varepsilon_i \geqslant 0$, the poles are all on the other side of $\zeta = 1$ and for ζ increasing they cluster at infinity. For values $\zeta < (\zeta_i)_{min}$, $Y(\zeta)$ behaves monotonically in the Bose-Einstein case as well.

We then consider the logarithm of the integrand in (9), denoting

(13) $$F(\zeta) = \log Y(\zeta) - (N+1) \log \zeta.$$

It is equal to $+\infty$ at $\zeta = 0$ (because $\log \zeta = -\infty$) and decreases very fast as long as the second term on the right-hand side is predominant. However, the first term begins to predominate before ζ reaches the value of unity; this term increases monotonically like $Y(\zeta)$. Thus on the positive real axis there exists a point ζ_0 at which

(13 a) $$F'(\zeta_0) = 0.$$

The corresponding value $F''(\zeta_0)$ is very large and positive, because the transition from $F(\zeta)$ decreasing to $F(\zeta)$ increasing takes place rapidly; in fact, the change is faster for larger values of N, as we shall show later.

The last remarks serve to prepare the ground for the following. To conclude this section we wish to make a remark regarding the introduction of the two "new" kinds of statistics: The statistics themselves are not new, but the objects to which we apply them are. The new objects consist of indistinguishable particles and their quantum states which are of a symmetrical or an antisymmetrical nature.

D. The saddle-point method

We now proceed to evaluate the integral (9). Making use of the logarithm of the integrand defined in (13) we may replace (9) by

(14) $$Z = \frac{1}{2\pi i} \oint e^{F(\zeta)} d\zeta.$$

We now expand $F(\zeta)$ into a power series at the point $\zeta = \zeta_0$, its linear term vanishing in accordance with (13 a):

(15) $$F(\zeta) = F(\zeta_0) + \tfrac{1}{2} F''(\zeta_0)(\zeta - \zeta_0)^2 + \cdots.$$

Two-dimensional potential functions cannot possess a real maximum or minimum. Since $\nabla^2 \Phi = 0$, the second derivatives $\partial^2\Phi/\partial x^2$ and $\partial^2\Phi/\partial y^2$ must always be of opposite signs which means that the surfaces $u = $ const are always convex upwards in one direction, and convex downwards at right angles to it. Thus points where $\partial\Phi/\partial x = \partial\Phi/\partial y = 0$ are saddle points. The same is true about the real and imaginary parts u, v, of any complex function $F(\zeta)$ at points $\zeta = \zeta_0$ where $F'(\zeta_0) = 0$. When discussing the behavior of our function $F(\zeta)$ along the positive real axis it was shown that it possesses a sharply marked minimum. Considering the topography of a saddle-like surface of the type of a potential function we see that it must possess a sharply marked maximum at the same point and along a line passing through ζ_0 parallel to the imaginary axis.

In evaluating our integral it is necessary to proceed along a path encircling the origin, e. g. along a circle. Drawing the circle through ζ_0 we see that during the process of integration we pass through a steep path ("steepest descent" on the one side, and "steepest ascent" on the other). The only important contribution to the integral comes from the neighborhood of the saddle point; in its neighborhood the circle can be replaced by a segment of the tangent to it and the remainder of the circle may be neglected, see Fig. 29. Along this segment of the tangent we have

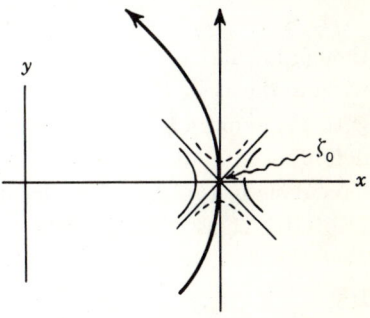

Fig. 29.

The ζ-plane ($\zeta = x + iy$) in the neighborhood of the saddle point ζ_0 with a qualitative representation of lines of constant elevation.

(16) $$\zeta = \zeta_0 + iy; \quad -y_0 < y < +y_0.$$

Neglecting the higher terms we find from (14) and (15) that

(17) $$Z = \frac{1}{2\pi} e^{F(\zeta_0)} \int_{-y_0}^{+y_0} \exp\{-\tfrac{1}{2} F''(\zeta_0) y^2\} dy.$$

Introducing the new variable

$$\eta = y \sqrt{\tfrac{1}{2} F''(\zeta_0)}$$

and assuming that $F''(\zeta_0)$ is sufficiently large, we obtain

$$Z = \frac{1}{2\pi}\sqrt{\frac{2}{F''(\zeta_0)}}\, e^{F(\zeta_0)} \int_{-\infty}^{+\infty} e^{-\eta^2}\, d\eta = \frac{e^{F(\zeta_0)}}{\sqrt{2\pi F''(\zeta_0)}},$$

and hence

(18) $$Z = \frac{Y(\zeta_0)}{\zeta_0^{N+1}\sqrt{2\pi F''(\zeta_0)}}$$

in view of (13).

It is very instructive to prove this result in the classical case first, even though in this case the sum can be calculated exactly without difficulty, and without the use of the stratagem due to Darwin and Fowler. Instead of starting with the definition (6) of the partition function, we would now start with its definition in (36.17 a). If we now drop the partition condition (36.16), as we have done in (6), and if we write the denominators $n_1!$, $n_2!$, ... under the respective signs of summation over n_1, n_2, ..., we obtain

(19) $$Y_{class} = N! \prod_i \left(\sum_{nj} \frac{(\zeta z_i)^{n_i}}{n_i!} \right) = N!\, e^{\zeta(z_1 + z_2 + \cdots)} = N!\, e^{\zeta Z_0}$$

which replaces the preceding function $Y(\zeta)$. This case shows again that the partition function Z_0 has a legitimate meaning in classical mechanics.

From eqs. (13) and (19) we can deduce the relations

(20 a) $$F(\zeta) = \zeta Z_0 + \log N! - (N+1)\log \zeta,$$

(20 b) $$F'(\zeta) = Z_0 - \frac{N+1}{\zeta},$$

(20 c) $$F''(\zeta) = (N+1)/\zeta^2,$$

so that according to (13), we have

(20 d) $$\zeta_0 = \frac{N+1}{Z_0}, \qquad \sqrt{2\pi F''(\zeta_0)} = \left(\frac{2\pi}{N+1}\right)^{\frac{1}{2}} Z_0,$$

and eq. (18) transforms into

(21) $$Z = \frac{N!\, e^{N+1} Z_0^N}{(N+1)^N \sqrt{2\pi(N+1)}} \to Z_0^N.$$

FUNDAMENTALS OF QUANTUM STATISTICS

According to Stirling's formula (29.4 a), the numerical factors are equal to unity; for $N \gg 1$ we have, namely:

(21 a) $$\frac{N!\, e^{N+1}}{(N+1)^N \sqrt{2\pi(N+1)}} = \frac{e}{(1+1/N)^N} \cdot \frac{1}{\sqrt{1+1/N}} \approx 1.$$

Thus our eq. (21) leads to nothing new, but reproduces the already familiar eq. (39.16), as mentioned in Sec. B. The return to the classical case in the preceding argument can thus be regarded as a check on the not very simple analytical methods used in the approximation. Furthermore, we may consider that it offers a substitute for the proof that $F''(\zeta_0)$ is very large for large values of N. We have, in fact, assumed in the transition from (17) to (18) that this was necessary hypothesis for the application of the method of steepest descent (saddle-point method). We can now infer from eq. (20 c) that this assumption is satisfied in the classical case: We know that ζ_0 is finite and that $\zeta_0 < 1$ and conclude from (20 c) that $F''(\zeta_0)$ increases to infinity proportionately with N. We shall assume that the same occurs in quantum statistics.

Reverting to the latter, we form the logarithm of the partition function given in (18), making use of the representation of $Y(\zeta)$ from (12):

$$\log Z = \mp \sum_i \log(1 \mp \zeta_0 z_i) - (N+1)\log \zeta_0 - \frac{1}{2}\log[2\pi F''(\zeta_0)].$$

In this equation we may neglect 1 compared with N. The last term is of order $\log N$ which in the limit $N \to \infty$ is negligible compared with N. It can be seen that in the limit of $N \to \infty$ it can also be neglected in comparison with the others. In this manner we obtain

(22) $$\log Z = \mp \sum_i \log(1 \mp \zeta_0 z_i) - N\log \zeta_0,$$

where ζ_0 is determined by eqs. (13) and (13 a). Thus

(22') $$\frac{d}{d\zeta_0}[\log Y(\zeta_0)] = \frac{N}{\zeta_0}$$

when we neglect 1 against N once more. Consequently

(23) $$\sum_i \frac{\zeta_0 z_i}{1 \mp \zeta_0 z_i} = N,$$

in accordance with eq. (12).

Substituting the expressions (36.18) into eqs. (22) and (23) and putting $\zeta_0 = e^{-\alpha}$, we have

(22 a) $$\log Z = \mp \sum_i \log(1 \mp e^{-\alpha - \beta \varepsilon_i}) + N\alpha$$

and

(23 a) $$\sum_i \frac{1}{e^{\alpha + \beta \varepsilon_i} \mp 1} = N.$$

It follows from (23 a) that for large values of α

$$e^{-\alpha} Z_0 = N, \qquad \alpha = \log Z_0 - \log N$$

and from (22 a) that

$$\log Z = N \log Z_0 - N(\log N - 1)$$

in agreement with eq. (3 a). It is seen that $\alpha \gg 1$ corresponds to the limiting case of ordinary gases. As α becomes small, or even negative, we obtain degenerate gaseous states. The most important example of a degenerate gas is afforded by the conduction electrons in metals which we propose to consider in the following Sections (38 and 39).

38. Degenerate gases

A. Bose-Einstein and Fermi-Dirac distribution

We have based our considerations in Sec. 37 on the partition function in the Γ-space because identical molecules are indistinguishable, but the results, as given in eqs. (37.22 a) and (37.23 a), contain only sums over the μ-space. The sum in eq. (37.22 a)

(1) $$\log Y = \mp \sum_i \log(1 \mp e^{-\alpha - \beta \varepsilon_i}) = \Phi(\alpha, \beta)$$

plays a similar part to that played by the partition function itself. It is a thermodynamic potential. It follows from (37.22′) that:

(2) $$N = -\frac{\partial \Phi}{\partial \alpha}.$$

The internal energy follows from (36.8) and is

(3) $$U = -\frac{\partial \Phi}{\partial \beta}$$

and the number of particles follows from (36.7):

(4) $$\overline{n}_i = -\frac{1}{\beta}\frac{\partial \Phi}{\partial \varepsilon_i} = \frac{1}{e^{\alpha+\beta\varepsilon_i} \mp 1}.$$

Applying eq. (4) we can deduce from (2) and (3) that

(5) $$N = \Sigma\, \overline{n}_i, \qquad U = \Sigma\, \overline{n}_i\, \varepsilon_i.$$

Applying (36.9), we obtain an expression for entropy

(6) $$S = k(\Phi + \alpha N + \beta U).$$

The differential form of Φ has the form:

(7) $$d\Phi = -N\, d\alpha - U\, d\beta - \beta \Sigma\, \overline{n}_i\, d\varepsilon_i.$$

Using the language of thermodynamics we can state it represents the potential in terms of α, β, and ε_i as independent variables.

It is seen from eq. (7) that Φ defined in (1) represents a kind of thermodynamic potential. It differs from previous potentials in that it contains the number of particles as an independent variable, in addition to energy (Problem II. 1). In the case of a completely isolated system the energy remains constant ($U = $ const) and all phase points lie on one surface in the phase space. Every phase element of equal volume between this surface and the energy surface $U + \Delta U$ is of equal probability. We speak of a microcanonical distribution and of a *microcanonical ensemble*.

So far we have carried out our consideration in relation to *canonical* ensembles. The latter occur when the system remains in thermal contact with the surroundings. For every temperature we obtain a canonical distribution function which admits fluctuations of the mean energy; these are due to the energy fluctuations between system and bath. In the case of systems with very many degrees of freedom the differences between the results of calculations for microcanonical systems, as against canonical systems, are, generally speaking, not significant, because the energy fluctuations are exceedingly small.

The function of state, Φ, refers to so-called "*grand canonical ensembles*". When α is constant and is not eliminated, unlike in the following argument, the number of molecules N is also subject to fluctuations. Now there is more than only thermal contact with the surroundings, and an exchange of particles becomes possible. This causes fluctuations in the total number of particles. Again, in the case of large systems, the fluctuations in the number of particles are small, and no appreciable differences result. We shall restrict ourselves to these remarks and will now revert to the canonical distributions, eliminating α with the aid of eq. (2) (see also Sec. 40).

In the limiting case of large values of α, we obtain from (1) that

(1') $$\log Y = \Phi = e^{-\alpha} \Sigma e^{-\beta \varepsilon_i} = e^{-\alpha} Z_0.$$

With the aid of eqs. (2) to (4) we can deduce from it that

$$N = e^{-\alpha} Z_0, \quad U = -e^{-\alpha} \frac{\partial Z_0}{\partial \beta}, \quad \overline{n}_i = -\frac{e^{-\alpha}}{\beta} \frac{\partial Z_0}{\partial \varepsilon_i},$$

or, eliminating α, we have

(8) $$U = -N \frac{\partial \log Z_0}{\partial \beta}, \quad \overline{n}_i = -\frac{N}{\beta} \frac{\partial \log Z_0}{\partial \varepsilon_i}$$

in agreement with (29.6). Here $\overline{n}_i$ represents the ordinary Boltzmann distribution function.

The same calculation can be performed without assuming that α is very large, but the elimination of α ceases to be elementary. In any case, we can now perform our calculations in the μ-space. It will be noticed that in the Γ-space we always have the distribution function $\exp\{-\alpha - \beta E(n)\}$ and that the only change occurs in the number of energy steps in $E(n)$. On the other hand, according to (4), in the μ-space, the Boltzmann distribution function $\exp(-\alpha \beta \varepsilon_i)$ is now replaced by new distribution functions. In the *Bose-Einstein case*:

(4 a) $$\overline{n}_i = \frac{1}{e^{\alpha + \beta \varepsilon_i} - 1},$$

and in the *Fermi-Dirac case*:

(4 b) $$\overline{n}_i = \frac{1}{e^{\alpha + \beta \varepsilon_i} + 1}.$$

Accordingly quantum mechanical gases can also be described in the μ-space if we, as it is said, adopt new statistics. It is usual to derive eqs. (4 a) and (4 b) with the aid of combinatorial methods which take into account the fact that particles cannot be distinguished one from another, as well as Pauli's principle, right in the μ-space. We shall, however, refrain from pursuing this point.

The entropy equation (6) leads to a number of important conclusions. It follows from eq. (4) that

$$e^{\alpha + \beta \varepsilon_i} = \frac{1 \pm \overline{n}_i}{\overline{n}_i}.$$

Substituting this expression into eq. (6), we obtain

$$\frac{1}{k} S = \sum_i \left\{ \mp \log \left(1 \mp \frac{\overline{n}_i}{1 \pm \overline{n}_i} \right) + \overline{n}_i \log \frac{1 \pm \overline{n}_i}{\overline{n}_i} \right\}$$

or, after an elementary rearrangement:

(9) $$\frac{1}{k} S = - \sum_i \{\overline{n}_i \log \overline{n}_i \mp (1 \pm \overline{n}_i) \log (1 \pm \overline{n}_i)\}.$$

The first term is identical with Boltzmann's log W in eq. (36.10). The whole equation represents the quantum mechanical expression for log W, because the distribution functions (4 a) and (4 b) yield, precisely,

(10) $$\log W^\pm = - \sum_i \{f_i \log f_i \mp (1 \pm f_i) \log (1 \pm f_i)\}$$

subject to the additional conditions that $N = \Sigma f_i = $ const, $U = \Sigma f_i \varepsilon_i = $ const for the maximum.

These changes in the thermodynamic probability can be easily interpreted in the Fermi-Dirac case. Taking into account that

(11) $$f_i^0 = 1 - \overline{n}_i, \qquad f_i' = \overline{n}_i$$

represent the probabilities of finding either *no* molecule or *one* molecule in the i-th quantum state, we can infer from (10) that

(12) $$\log W^- = - \sum_i (f_i^0 \log f_i^0 + f_i' \log f_i').$$

This expression differs from that due to Boltzmann by one term which represents Boltzmann's thermodynamic probability of the empty places. Fermi's distribution can now be obtained by determining the maximum of W^- subject to the conditions

(13) $$f_i^0 + f_i' = 1, \qquad \sum_i f_i' = N, \qquad \sum_i \varepsilon_i f_i' = U.$$

The same is true of the Bose-Einstein case. In this case also it is necessary to take into account the Boltzmann thermodynamic probabilities of the empty spaces. Let $f_i^{(n)}$ denote the probability of finding n molecules at the i-th quantum state; the thermodynamic probability is then equal to the sum of expressions of Boltzmann's type:

$$(14) \qquad \log W^+ = - \sum_i (f_i^0 \log f_i^0 + f_i' \log f_i' + f_i'' \log f_i'' + \ldots).$$

The Bose-Einstein distribution (4 a) follows from this expression and $\log W^+$ follows from eq. (9) on calculating the maximum of (14), subject to the additional conditions that

$$(15) \qquad \sum_{n=0}^{\infty} f_i^{(n)} = 1, \quad \sum_{i,n} n f_i^{(n)} = N, \quad \sum_{i,n} n f_i^{(n)} \varepsilon_i = U.$$

A proof is given in Problem IV. 7.

B. Degree of gas degeneration

In the present Section we shall consider monatomic quantum mechanical perfect gases. In this case the sum in eq. (1) can be written as

$$(16) \qquad \Phi = \mp \frac{4\pi V}{h^3} \int_0^{\infty} \log (1 \mp e^{-\alpha - \beta p^2/2m}) \, p^2 \, dp$$

if a continuous approximation with the aid of an integral is used. Alternatively

$$(16\,a) \qquad \Phi = \frac{V}{h^3} \left(\frac{2\pi m}{\beta} \right)^{3/2} \cdot \chi(\alpha),$$

where

$$(16\,b) \qquad \chi(\alpha) = \mp \frac{2}{\sqrt{\pi}} \int \log (1 \mp e^{-\alpha - t}) \sqrt{t} \, dt$$

with $\beta p^2 = 2 m t$. A correction to this equation is given later in (26). The *number of particles* can be calculated from (16 a) with the aid of (2), (3) and (4), when we find

$$(17) \qquad N = - \frac{V}{h^3} (2\pi m k T)^{3/2} \chi'(\alpha).$$

Similarly the *energy* is:

(17 a) $$U = \frac{3}{2} \frac{V}{h^3} (\sqrt{2\pi m k T})^3 \cdot kT \cdot \chi(\alpha) = \frac{3}{2} NkT \cdot [-\chi(\alpha)/\chi'(\alpha)],$$

and the *pressure* is:

(17 b) $$p = \frac{1}{\beta} \frac{\partial \Phi^*}{\partial V} = \frac{1}{h^3} (\sqrt{2\pi m k T})^3 \cdot kT \cdot \chi(\alpha),$$

(taking into account that $\Phi + N\alpha = -\beta F$ and that $dF = -S\,dT - p\,dV$). From the last two equations we can deduce the relation

(18) $$pV = \frac{2}{3} U$$

which is independent of $\chi(\alpha)$ and which thus remains valid in quantum mechanics.

According to eq. (17) α is a function of the ratio

(19) $$\rho = \frac{N}{V} \cdot \frac{h^3}{(\sqrt{2\pi m k T})^3}$$

which is a measure of the deviation of eqs. (17), (17 a), and (17 b) from the ideal gas state. The quantity ρ is defined as the *degree of gas degeneration*. From eqs. (16 b), (17), and (19), we find that

(20) $$\rho = -\chi'(\alpha) = \frac{2}{\sqrt{\pi}} \int_0^\infty \frac{\sqrt{t}\,dt}{e^{\alpha+t} \mp 1}$$

showing that ρ is small for large values of α. A small value of ρ also means that the behavior of the gas is classical. We have estimated in Sec. 37 that for a gas under normal conditions $\rho \approx 1/30,000$. The influence of ρ on the energy of a Bose-Einstein and of a Fermi-Dirac gas is shown graphically in Fig. 30. It serves to eliminate α from eqs. (17) and (17 a).

An example for large values of ρ and large degeneration is afforded by the conduction electrons in a metal. We shall see in Sec. 39 that to a first approximation, the conduction electrons behave like free particles. In other words they move through the metal like gaseous molecules in a vessel, but the electron gas is highly degenerate. On the assumption that one electron per atom in copper is a conduction electron we find $\rho \approx 5000$.

The electron gas is of the Fermi-Dirac type because electrons behave in accordance with Pauli's principle. For small values of ρ both types of gases (i. e. Fermi-Dirac and Bose-Einstein) behave identically, but in the limiting case of large values of ρ the differences are considerable, and it is necessary to discuss them separately.

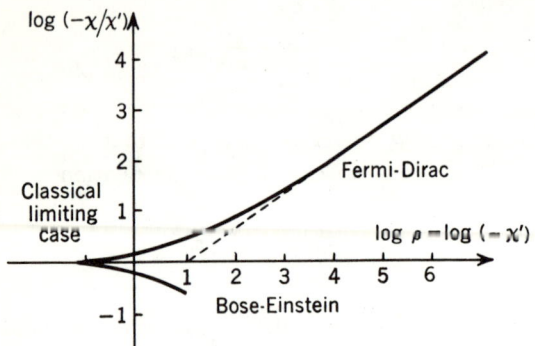

Fig. 30.

The logarithm of the energy factor $U/U_{Boltzm.} = -\chi/\chi'$ from eq. (38.17 a) plotted in terms of the logarithm of the degree of degeneration $\varrho = -\chi'$ as seen from eq. (38.20). (after F. L. Bauer).

C. Highly degenerate Bose-Einstein gas

Until recently, apart from light quanta, no instances of appreciable degeneration in the case of Bose-Einstein statistics were known. It appears, however, that superfluidity[1] in helium at very low temperatures is linked with gas degeneration. In any case it is a fact that superfluidity occurs only in the isotope He^4 which satisfies Bose-Einstein statistics. It is absent in He^3 which behaves in accordance with Fermi-Dirac statistics.

According to eqs. (17) and (20) the number of particles of a Bose-Einstein gas is proportional to

$$(21) \qquad -\chi'(\alpha) = \frac{2}{\sqrt{\pi}} \int_0^\infty \frac{\sqrt{t}\, dt}{e^{\alpha+\beta t} - 1}.$$

This integral reaches its maximum $\alpha = 0$. Values of $\alpha < 0$ must be excluded because $-\chi'(\alpha)$ then diverges. The fact that the above integral just converges for $\alpha = 0$ and assumes the value $\zeta(3/2) = 2.612$ (ζ = Zeta function)

---

[1] F. London, "Superfluids", Vol. I, Introduction, Sec. 4; Structure of matter series, New York, London, 1950.

could be interpreted to mean, in accordance with eq. (17), that only a finite number of Bose-particles can be accommodated in a finite volume, and that their number decreases like $T^{3/2}$ as the temperature decreases and becomes equal to zero, at absolute zero. Such a consequence does not appear plausible, and contradicts the assumption that in the case of Bose-Einstein statistics it is possible to have an arbitrary number of particles in each quantum state.

In fact, the conclusion is false. It is a consequence of the continuous approximation used for the sum (1). At low temperatures when $\beta \varepsilon_i \gg 1$ it is not permissible to replace the sum by an integral. If the energy is so normalized that its lowest value is equal to zero, and if we arrange the values of energy in their order of magnitude

$$0 = \varepsilon_0 < \varepsilon_1 \leqslant \varepsilon_2 \leqslant \ldots ,$$

we find that the mean numbers of molecules for the different energy states are given by eq. (4). They are

(22) $$\bar{n}_0 = \frac{1}{e^\alpha - 1}, \quad \bar{n}_i = \frac{g_i}{e^{\alpha + \beta \varepsilon_i} - 1} \quad (i = 1, 2, \ldots).$$

The factors g_i which appear here indicate the multiplicity of the energy values ε_i, in the same way as in eq. (33.11). The ground state is assumed to be simple. Since in the limit $\alpha \to 0$, $\bar{n}_0$ increases without limit, the number of particles ceases to be bounded.

The fact that (21) remains finite at $\alpha = 0$ means only that the number of particles $\bar{n} = \sum_{i=1}^{\infty} \bar{n}_i$ present in excited states cannot exceed an upper limit so that an increase in the number of particles beyond that limit contributes only to the ground state. The particles condense to a certain extent into the ground state (Einstein condensation). The number $\bar{n}$ is determined by the integral (21). Defining the degree of degeneration in the same way as before, in eq. (19), we see that eq. (17) yields[1]:

(23) $$\bar{n} = -\frac{N}{\rho} \chi'(\alpha) \leqslant \frac{N}{\rho} \zeta\left(\frac{3}{2}\right) = N_0,$$

because the disappearance of the lowest energy level in the definition of $\bar{n}$ makes practically no difference. It is possible to estimate that the error is of the order $\delta \bar{n}/\bar{n} = (\rho/N)^{1/3}$, which is negligible even in cases of appreciable degeneration.

---

[1] It is evident that eq. (20) ceases to apply because it implies $\bar{n} = N$.

We now direct our attention to the consideration of high degrees of degeneration. According to (23) we then have $\bar{n} \ll N$, and practically all particles are at the ground state. Thus eq. (20) shows that α is very small. We have $N_0 \ll N$ and

(24) $$\bar{n} = N_0, \quad \bar{n}_0 = N - N_0, \quad \alpha = \frac{1}{N - N_0}.$$

As a first approximation the numbers of particles in *excited* states is given by

(25) $$\bar{n}_i \approx n_i{}^0 = \frac{g_i}{e^{\beta \varepsilon_i} - 1}$$

where $\alpha \approx 0$ has been substituted into (22). We now inquire into when we may substitute the $n_i{}^0$'s, defined in eq. (25), for the $\bar{n}_i$'s, or, in other words we ask for the conditions under which we have

$$\frac{n_i{}^0 - \bar{n}_i}{\bar{n}_i} = \frac{(e^\alpha - 1) e^{\beta \varepsilon_i}}{e^{\beta \varepsilon_i} - 1} = \frac{e^{\beta \varepsilon_i}/\bar{n}_0}{e^{\beta \varepsilon_i} - 1} \ll 1.$$

First it follows that this expression is small for large values of $\beta \varepsilon_i$. Furthermore, we may have $\beta \varepsilon_i \ll 1$ on condition that for all excited states ε_i we also have:

$$\beta \varepsilon_1 \gg \frac{1}{\bar{n}_0} \approx \frac{1}{N}.$$

The lowest excited quantum state ε_1 is of the order

$$\varepsilon_1 \approx h^2/2\, m\, V^{2/3},$$

which corresponds to a de Broglie wavelength of the order of magnitude of the linear dimensions of the volume. Thus the above inequality is satisfied if

$$\frac{h^2}{2\, m\, k\, T} \cdot V^{-2/3} \gg \frac{1}{\bar{n}_0}.$$

In view of (19), the following condition may be written down

(26) $$\rho \pi^{3/2} \gg \frac{N}{\bar{n}_0{}^{3/2}}$$

which is satisfied in the limiting case of large values of ρ, because we then have $\bar{n}_0 \approx N$.

The numbers of particles of a highly degenerate Bose-Einstein gas occupying excited states are given by eq. (25) and depend on temperature alone. By way of a first approximation they can be calculated from the distribution function which we have already encountered in connection with light quanta (Sec. 20). The more exact values in eq. (22) can be deduced from the modified potential:

$$(16') \quad \Phi = -\log(1-e^{-\alpha}) - \frac{V}{h^3}(2\pi m k T)^{3/2} \cdot \frac{2}{\sqrt{\pi}} \int_0^\infty \log(1-e^{-\alpha-t})\sqrt{t}\, dt.$$

The first term constitutes a correction to eq. (16) for the lowest quantum state. The pressure is independent of the correction term. According to (17 b) we have

$$(27) \quad p = \chi(\alpha)\left(\frac{2\pi m}{h^2}\right)^{3/2}(kT)^{5/2} \approx \chi(0)\left(\frac{2\pi m}{h^2}\right)^{3/2}(kT)^{5/2}.$$

Thus a kind of vapor-pressure curve has been obtained.

In conclusion we wish to show that the degeneration of a Bose-Einstein gas is compatible with Nernst's Third Law. Taking into account (1) to (3) we can deduce from (6) that

$$\frac{1}{k}S = \sum_i \left\{-\log(1-e^{-\alpha-\beta\varepsilon_i}) + \frac{\alpha+\beta\varepsilon_i}{e^{\alpha+\beta\varepsilon_i}-1}\right\}.$$

The value of α in this equation can be derived from (2):

$$N = \sum \frac{1}{e^{\alpha+\beta\varepsilon_i}-1}.$$

However, in the neighborhood of absolute zero all excited states satisfy the condition $\beta\varepsilon_i \gg 1$, so that all terms in the sum for which $i \neq 0$ vanish at least in proportion to $e^{-\beta\varepsilon_i}$. Only the terms which are independent of temperature remain:

$$\frac{1}{k}S_0 = -\log(1-e^{-\alpha}) + \frac{\alpha}{e^\alpha - 1}, \quad N = \frac{1}{e^\alpha - 1}.$$

This equation shows that S tends exponentially to the constant value

$$S_0 = k(\log N + 1).$$

The *specific entropy* at absolute zero is given by

$$(28) \quad s_0 = \frac{S_0}{N} = k\frac{\log N + 1}{N} \approx 0$$

except for terms of an order which has been neglected anyway.

So far we have assumed that the lowest energy state is $\varepsilon_0 = 0$, and it remains to enquire into the changes in the argument which would have to be introduced if we had $\varepsilon_0 \neq 0$. Since, according to eq. (2), α depends on the temperature, we may replace α by $\alpha' - \beta \varepsilon_0$ and thus we are led to eq. (1) in the same form as before. Equations (2) and (4) also remain unaltered. Equation (3) is replaced by

$$U = -\left(\frac{\partial \Phi}{\partial \beta}\right)_{\alpha = const} = -\left(\frac{\partial \Phi}{\partial \beta}\right)_{\alpha' = const} - \left(\frac{\partial \Phi}{\partial \alpha'}\right)\frac{\partial \alpha'}{\partial \beta} = U' + N\varepsilon_0.$$

from which it follows that

$$U' = U - N\varepsilon_0.$$

Consequently if α' is kept constant instead of α in finding the derivative with respect to β, the energy will decrease by its zero-point value $N\varepsilon_0$.

39. Electron gas in metals

A. Introductory remark to Drude's method

Since the discovery of the electron there could be no doubt that an electric current is carried by electrons. The suggestion that the electrons in a metal behave like molecules of a gas and participate in thermal equilibrium is due to P. Drude. The greatest success achieved by Drude's theory consists in the derivation of the Wiedemann-Franz law which states that the ratio of thermal conductivity, $\varkappa$, to electrical conductivity, σ, is given by the equation

(1) $$\frac{\varkappa}{\sigma} = 3 \frac{k^2}{e^2} T.$$

Here $-e$ denotes the elementary charge of an electron. Drude was also able to derive an expression for electrical conductivity which is of some importance even to-day:

(2) $$\sigma = \frac{e^2 l n}{m \bar{v}}$$

(l = "mean free path" of an electron, n = number of *free* electrons per m³, $\bar{v}$ = mean velocity, e = elementary charge, m = mass of an electron). Numerous thermoelectrical and thermomagnetic phenomena, such as thermal emf, voltaic emf, thermal emission of electrons from metals etc., can be explained with the aid of Drude's supposition, at least qualitatively, and their inner connection can be recognized.

However, the fundamental supposition is inconsistent with experimental facts in the matter of specific heats. According to Drude's hypothesis every free electron should possess an energy of $\frac{1}{2} k T$ per degree of freedom at a state of equilibrium, and the molar energy of electrons should have the value of $3 R/2 = 3$ cal/mol deg which is in flat contradiction to the Dulong-Petit rule. When taking into account Maxwell's velocity distribution in a more precise way, it turns out that the numerical factor in eq. (1) should have a value of 2, instead of 3, as shown, which destroys its agreement with the measurements due to Jäger and Diesselhorst. The above and other additional difficulties disappear only if it is assumed *ad hoc* that the number of free electrons is considerably smaller than that of atoms.

The preceding difficulties completely destroyed the faith in Drude's hypothesis of the existence of an "electron gas". The lack of success appeared, essentially, comprehensible, because the electrons in a metal do not move in a *zero field of forces* but in a periodical potential field created by the ions of the metal. In addition the interaction between electrons must also play a part.

However, the existence of metal electrons which are not subjected to any forces can, nevertheless, be justified to a certain extent on the ground of wave mechanics. Moreover, Sommerfeld successfully reverted in 1928 to Drude's assumption of the existence of free electrons and was able to show that the difficulties just discussed disappear if it is taken into account that the electron gas possesses the properties of a highly degenerate Fermi-Dirac gas (*cf.* Sec. 38, eq. (20)). We now proceed to discuss some of the consequences of this theory basing our considerations on a paper by A. Sommerfeld and H. Bethe.[1]

B. The completely degenerate Fermi-Dirac gas

We now wish to recall eq. (38.1) remembering that the lower sign applies to the Fermi-Dirac case. For α we substitute $\alpha = -\beta \zeta$ because it can be shown that in the limiting case of $T \to 0$, ζ and not α remains finite.[1] In this way we obtain

$$(3) \qquad \Phi = \sum_i \log\left(1 - e^{-\beta(\varepsilon_i - \zeta)}\right).$$

---

[1] "Elektronentheorie der Metalle", in "Handbuch der Physik" edited by H. Geiger and K. Scheel, Vol. XIV 2, Chap. 3, I, pp. 333–368.

[2] It is no longer possible to confuse the present ζ with that in Sec. 37.

The quantity ζ in the above equation has a simple thermodynamical interpretation. The free enthalpy in eq. (7.4), namely

$$G = U - TS + pV$$

can now be evaluated with the aid of eqs. (38.6) and (38.17 b). We thus obtain

$$G = U - \frac{1}{\beta}\Phi + \zeta N - U + \frac{V}{\beta}\frac{\partial \Phi}{\partial V}.$$

The terms in this equation which contain U and Φ cancel each other in pairs. In the latter case it should be noted that Φ is proportional to V so that $V\,\partial\Phi/\partial V = \Phi$. Hence we obtain

(4) $$\zeta = -\frac{\alpha}{\beta} = \frac{G}{N},$$

ζ is seen to represent the *free enthalpy per electron*. Corresponding equations can also be derived for the Bose-Einstein case.

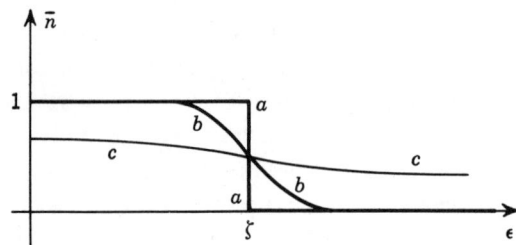

Fig. 31.
Mean number of particles per phase cell
a) at absolute zero   b) for $\beta = 10/\zeta$ c) for $\beta = 1/\zeta$ (ζ = const).

As distinct from the latter, α can assume negative values so that ζ becomes positive. According to eq. (38.4) the particle density is given by

(5) $$\bar{n}_i = \frac{1}{e^{\beta(\varepsilon_i - \zeta)} + 1}.$$

In the limiting case of $T = 0$ ($\beta \to \infty$) we obtain a value of either 0 or 1, depending on whether $\varepsilon_i > \zeta$ or $\varepsilon_i < \zeta$. Thus in the case of *complete degeneracy* we obtain

(6) $$\bar{n}_i = \begin{cases} 1 & \text{for } \varepsilon_i < \zeta_0 \\ 0 & \text{for } \varepsilon_i > \zeta_0 \end{cases}$$

where the quantities at absolute zero have been denoted by the subscript 0 (see curve a in Fig. 31), and where ζ_0 plays the part of a limiting energy. All energy levels below ζ_0 are occupied, all levels above it are empty. Thus it is seen that at absolute zero the lowest levels are occupied; according to Pauli's principle every level is associated with two electrons, one each for each possible orientation of the spin of the electrons.

The total number of electrons determines the *limiting energy*. Defining the limiting momentum P_0 by

(7) $$\zeta_0 = \frac{P_0^2}{2m}, \qquad P_0 = \sqrt{2m\zeta_0},$$

we see that the number of particles is

(7 a) $$N = 2\frac{V}{h^3} \cdot \frac{4\pi}{3} P_0^3,$$

and that the *energy* is

(7 b) $$U = 2\frac{V}{h^3} \cdot \frac{4\pi}{5} \frac{P_0^5}{2m}.$$

The numerical factor 2 is due to the fact that according to quantum mechanics every phase cell includes two quantum states, owing to the two spins and hence also two electrons. The evaluation of P_0 gives

(8) $$P_0 = h\left(\frac{3}{8\pi} \cdot \frac{N}{V}\right)^{1/3}.$$

Thus the *energy* and the *limiting energy* from (7 b) and (7) at absolute zero are given by

(9) $$U_0 = \frac{3Nh^2}{10m}\left(\frac{3}{8\pi} \cdot \frac{N}{V}\right)^{2/3}, \qquad \zeta_0 = \frac{h^2}{2m}\left(\frac{3}{8\pi}\frac{N}{V}\right)^{2/3},$$

so that the pressure, eq. (38.18), becomes

(10) $$p_0 = \frac{8\pi h^2}{15m}\left(\frac{3}{8\pi} \cdot \frac{N}{V}\right)^{5/3}.$$

In the case of copper this pressure has a value of $p_0 \approx 3.8 \times 10^5$ atm. This enormous value corresponds to the electrical attraction between electrons and ions. The limiting energy for copper is $\zeta_{Cu} \approx 11.3 \times 10^{-12}$ erg ≈ 7 e V. It is comparable with the ionization energy of hydrogen atoms (13.54 e V). The total energy per mol is $U_0 = 3/5 \, L \, \zeta_0$ which is roughly equal to the heat of combustion of carbon. Furthermore, the force acting on the cross-section of an ion is equal to the electrostatic attraction: $p_0 \, r^2 \approx e^2/4\pi \, \varepsilon_0 \, r^2$ for $r \approx (V/N)$ which is of the order of several Ångström units. Leaving r provisionally undetermined we can see that, generally speaking, there will be no equilibrium. Furthermore

$$p_0 \, r^2 = \gamma \, e^2/4\pi \, \varepsilon_0 \, r^2,$$

Substituting p_0 from eq. (10), it follows that

$$\gamma = \frac{4\pi \, \varepsilon_0 \, r^4}{e^2} \cdot \frac{1}{5} \left(\frac{3}{8\pi}\right)^{2/3} \frac{h^2}{m \, r^5}$$

or, with $\varLambda = h/m \, c$ (Compton wavelength) and $\alpha = e^2/4\pi \, \varepsilon_0 \, h \, c$ (fine structure constant):

$$\gamma = \frac{2\pi}{5} \left(\frac{3}{8\pi}\right)^{2/3} \frac{\varLambda}{\alpha \, r} \approx \frac{1\text{Å}}{r}.$$

When $\gamma < 1$ (i. e. when $r > 1$ Å) the pressure is too small to balance the forces of attraction. The electrons, and with them the ions, draw closer to each other. When $\gamma > 1$ ($r < 1$ Å) the pressure becomes too large, and the metal will expand. It is seen that the conduction electrons are mainly responsible for the cohesion of the metal.

The preceding numerical values refer to absolute zero. In actual fact, however, the pressure, energy, and the limiting energy depend on temperature. Nevertheless, the degree of degeneration is so high that their dependence on temperature is very weak. It is seen that the degree of degeneration, eq. (38.19), becomes equal to unity at the *characteristic temperature of degeneration*.

$$\Theta = \frac{h^2}{2\pi m k} \left(\frac{N}{V}\right)^{2/3} = \frac{\zeta_0}{\pi k} \left(\frac{8\pi}{3}\right)^{2/3} \approx 100{,}000 \text{ K}.$$

C. Almost Complete Degeneracy

With increasing temperature the step-function from eq. (6) becomes smoothed out (see Fig. 31, curves b and c), but the transition between $\bar{n}_i = 1$ and $\bar{n}_i = 0$ takes place very rapidly at absolute zero, i. e. as long as $T \ll \Theta \approx 100{,}000$ K, or $kT \ll \zeta_0$, i. e. $\beta \, \zeta_0 \gg 1$. This enables us to evaluate the integral (38.16) and the integrals derived from it with a good degree of accuracy.

39. 11a ELECTRON GAS IN METALS

With reference to eq. (38.17) we put $t + \alpha = t - \beta \zeta = x$, and add the spin factor 2. Thus we obtain for the *number of particles*

$$(11) \qquad N = \frac{4\pi V}{h^3} \left(\frac{2m}{\beta}\right)^{3/2} \int_{-\beta\zeta}^{\infty} \frac{(x + \beta \zeta)^{1/2} dx}{e^x + 1}.$$

In a similar way by partial integration we obtain from eq. (38.17 a) the following expression for *energy*:

$$(12) \qquad U = \frac{4\pi V}{2m h^3} \left(\frac{2m}{\beta}\right)^{5/2} \int_{-\beta\zeta}^{\infty} \frac{(x + \beta \zeta)^{3/2} dx}{e^x + 1}.$$

Both integrals can now be further transformed by partial integration. It follows from (11) that

$$N = \frac{8\pi V}{3 h^3} \left(\frac{2m}{\beta}\right)^{3/2} \int_{-\beta\zeta}^{\infty} \frac{e^x}{(e^x + 1)^2} (x + \beta \zeta)^{3/2} dx.$$

The first factor in the integrand is a symmetrical function of x. For large values of $\beta \zeta$ the second factor does not vary considerably in the interval in which $e^x/(e^x + 1)^2$ differs appreciably from zero. This enables us to expand the root in powers of x. Since, moreover, the integral decreases exponentially on both sides we can replace the lower limit of integration by $-\infty$. In this manner we obtain finally

$$(11\,\text{a}) \qquad N = \frac{8\pi V}{3 h^3} (2m\zeta)^{3/2} \int_{-\infty}^{+\infty} \frac{e^x}{(e^x + 1)^2} \left(1 + \frac{3x}{2\beta\zeta} + \frac{3x^2}{8\beta^2\zeta^2} + \ldots\right) dx.$$

The integral in the middle vanishes because the integrand is odd. The first integral can be evaluated without difficulty, its value being unity. The last integral is proportional to

$$\int_{-\infty}^{+\infty} \frac{x^2 e^x}{(e^x + 1)^2} dx = 2 \int_0^{\infty} x^2 \left(e^{-x} - 2 e^{-2x} + 3 e^{-3x} - + \ldots\right) dx =$$

$$= 2 \times 2! \left(1 - \frac{1}{2^2} + \frac{1}{3^2} - + \ldots\right) = \frac{\pi^2}{3}.$$

Substituting the above values of the integrals into eq. (11 a), we have

(11 b) $$N = \frac{8\pi V}{3h^3}(\sqrt{2m\zeta})^3\left(1 + \frac{\pi^2}{8}\frac{k^2 T^2}{\zeta^2} + \cdots\right).$$

The deviation from the limiting case of complete degeneracy is of the order $(T/\Theta)^2$. A corresponding evaluation of (12) yields

(12 a) $$U = \frac{4\pi V}{5mh^3}(\sqrt{2m\zeta})^5\left(1 + \frac{5\pi^2}{8}\frac{k^2 T^2}{\zeta^2} + \cdots\right).$$

The first terms of (11 b) and (11 a) are, naturally, identical with (7 a) and (7 b), respectively.

Denoting the value of ζ at absolute zero by ζ_0, as before, (Fermi's limiting energy), and equating the values of N for $T = 0$ and $T \neq 0$, we obtain from eq. (11 b) that

$$\left(\sqrt{\frac{\zeta}{\zeta_0}}\right)^3 \cdot \left(1 + \frac{\pi^2}{8}\frac{k^2 T^2}{\zeta^2} + \cdots\right) = 1$$

or

(13) $$\zeta = \zeta_0\left(1 - \frac{\pi^2}{12}\frac{k^2 T^2}{\zeta_0^2} + \cdots\right)$$

and by (12 a)

(14) $$U = U_0\left(1 + \frac{5\pi^2}{12}\frac{k^2 T^2}{\zeta_0^2} + \cdots\right)$$

so that according to eq. (38.18), we now have

(15) $$p = p_0\left(1 + \frac{5\pi^2}{12}\frac{k^2 T^2}{\zeta_0^2} + \cdots\right).$$

D. Special problems

The dependence on temperature is due to the fact that some electrons now exceed the Fermi limit, instead of all of them being below it, as at absolute zero. According to (11) and (7 a) their number is given by

$$\delta N/N = \frac{3}{2}\left(\sqrt{\frac{\zeta}{\zeta_0}}\right)^3 \frac{1}{\beta\zeta}\int_{\beta(\zeta_0-\zeta)}^{\infty}\frac{(1 + x/\beta\zeta)^{1/2}\,dx}{e^x + 1}$$

which is equal to the expression for $\zeta = \zeta_0$:

(16) $$\frac{\delta N}{N} = \frac{3}{2} \frac{1}{\beta \zeta_0} \int_0^\infty \frac{dx}{e^x + 1} = \frac{3}{2} \log 2 \cdot \frac{kT}{\zeta_0}$$

except for correction terms of the order T/Θ. If we take into account that only electrons which have exceeded Fermi's limit can make themselves felt thermally as well as the free places given by eq. (38.12), we shall come to the conclusion that only the small fraction δN of electrons is significant for Drude's hypothesis, and not the total number, N, of free electrons. Thus the theory leads us to a decrease in the number of effective electrons in a natural way.

The *specific heat* of electrons is seen to be so small as not to affect the Dulong-Petit rule. The heat capacity can be calculated from eq. (14), and we have

$$C = \frac{dU}{dT} = U_0 \cdot \frac{5\pi^2 k^2 T}{6 \zeta_0^2} = \frac{4\pi^3}{3} \frac{V}{h^3} (\sqrt{2m\zeta_0})^3 \frac{kT}{\zeta_0} \cdot k.$$

In the transformation use is made of equations (7 b) and (9). By (7 a) we have

$$C = \frac{\pi^2}{2} \frac{kT}{\zeta_0} \cdot Nk.$$

Evaluating per mol ($N = L$ = Loschmidt-Avogadro number):

(17) $$c_{electr} = \frac{\pi^2}{2} \frac{kT}{\zeta_0} \cdot R = \frac{\pi^2}{6} \frac{kT}{\zeta_0} \cdot c_{Dulong\text{-}Petit}.$$

The fraction $c_{electr}/c_{Dulong\text{-}Petit}$ is of the order of $\delta N/N$ from eq. (16). In particular, for copper we obtain $T/54{,}000$, so that for $T \approx 300$ K it is equal to $1/180$. The contribution due to the electrons is seen to be less than 1%.

We now propose to apply eqs. (5) or (38.8 b) to the calculation of *Richardson's effect*, i. e. to the emission of electrons from incandescent cathodes. Our assumption that the electrons move as if no forces acted between them implies a constant potential in the interior of the metal, Fig. 32. On the other hand at the boundary strong forces balance the electron pressure, so that the potential must increase rapidly from negative values, reaching the value of zero in the exterior. The quantity $W - \zeta_0$ denotes the gain in energy which occurs when an external electron assumes the Fermi limiting energy in the interior; W denotes the full difference between the potential energy of an electron in the interior and one in the exterior. The real variation in the potential is, naturally, different from that shown in Fig. 32. In the interior there are the periodic variations, already mentioned, and at the boundary the transition does not occur in a step, but continuously, even if it is very rapid. The simplification introduced in Fig. 32 does not, however, destroy the characteristic features of Richardson's effect.

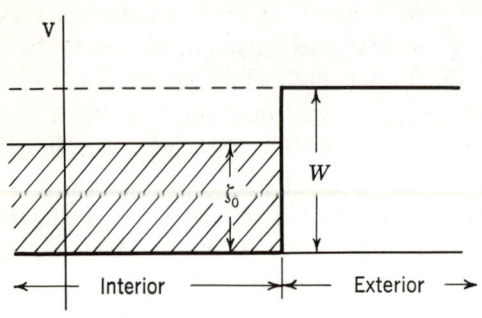

Fig. 32.
Illustrating Richardson's effect.

When the energy of the electrons in the metal becomes sufficiently large, the electrons become capable of overcoming the threshold, and can escape from the metal. The condition for this to happen is that the portion of the kinetic energy of the electron which is due to its motion at right angles to the wall exceeds W. Assuming that the z-axis is normal to the boundary, we can write

$$p_z^2/2m > W.$$

The density of the current due to the electrons leaving the metal is given by the integral

$$I_z = 2\frac{e}{m} \cdot \frac{1}{h^3} \int_0^\infty \frac{p_z\, dp_x\, dp_y\, dp_z}{1 + \exp[(\beta\, \mathbf{p}^2/2m) - \beta\zeta]}.$$

The lower limits of integration are: $p_x, p_y = -\infty$, and $p_z = \sqrt{2mW}$. Putting $\beta/2m\,(p_x^2 + p_y^2) = t$, $(\beta p_z^2/2m) - \beta W = s$, and $\zeta \approx \zeta_0$, we have

$$I_z = \frac{\pi e}{mh^3}\left(\frac{2m}{\beta}\right)^2 \int_0^\infty \frac{ds\, dt}{1 + \exp[\beta(W - \zeta_0) + s + t]}.$$

Since $W > \zeta_0$ and, in general, also $\beta(W - \zeta_0) \gg 1$, we may neglect the unity in the denominator as compared with the exponential function, and we can obtain Richardson's formula:

(18) $$I_z = \frac{4\pi e m}{h^3} k^2 T^2 \exp\left(-\frac{W-\zeta_0}{kT}\right).$$

If we had based our calculation on the Boltzmann distribution instead of that due to Fermi, we would have obtained

(18') $$I_z = \frac{e N_0/V}{\sqrt{2\pi m}} \sqrt{kT} \exp(-W/kT).$$

It is at once clear that the quantum value of the exponent should be $(W-\zeta_0)/kT$ instead of W/kT, because it must be equal to the difference in energy required to overcome the Fermi threshold. In the classical formula N_0 denotes the number of free electrons. Equating the factors in (18) and (18') we have

(19) $$N_0 = \frac{2V}{h^3}(\sqrt{2\pi m k T})^3 = \frac{3}{4}\sqrt{\pi} \cdot N \cdot \left(\sqrt{\frac{kT}{\zeta_0}}\right)^3.$$

This is, again, much smaller than the number of free electrons, but the ratio differs from that given in eqs. (16) and (17). Hence it can be clearly seen that it is not possible to correct Drude's theory simply by introducing a reduced number of free electrons, and thus to obtain the modern electron theory.

We are not yet in a position to discuss conductivity, or to derive the Franz-Wiedemann law, and we propose to defer this topic until we reach Chap. V. In conclusion we shall show that the electron gas satisfies Nernst's Third Law. According to eq. (38.16), the equation for the potential of spin electrons is

$$\Phi = \frac{4\pi V}{h^3}\left(\sqrt{\frac{2m}{\beta}}\right)^3 \int_0^\infty \log(1 + e^{\beta\zeta - t})\sqrt{t}\, dt.$$

Integration by parts gives

$$\Phi = \frac{8\pi V}{3h^3}\left(\frac{2m}{\beta}\right)^{3/2} \int_0^\infty \frac{t^{3/2}\, dt}{e^{t-\beta\zeta_0} + 1}$$

or, in view of (12):

(20) $$\Phi = \frac{2}{3}\beta U.$$

In accordance with eq. (38.6), the entropy is given by

(21) $$S = k\beta\left(\frac{5}{3}U - \zeta N\right).$$

Further, according to eqs. (7 a), (7 b) and (14), we have

(21 a) $$U_0 = \frac{3}{5}\zeta_0 N, \qquad U = \frac{3}{5}\zeta_0 N\left(1 + \frac{5\pi^2}{12}\frac{1}{\beta^2\zeta_0^2}\right).$$

Substituting these values together with the value of ζ into eq. (21), we obtain

$$S = Nk\beta\zeta_0\left(1 + \frac{5\pi^2}{12}\frac{1}{\beta^2\zeta_0^2} - 1 + \frac{\pi^2}{12}\frac{1}{\beta^2\zeta_0^2}\right)$$

or

(22) $$S = \frac{\pi^2}{2} Nk \cdot \frac{kT}{\zeta_0} \quad (= C_{electr} \text{ for } kT \ll \zeta_0).$$

The entropy is seen to vanish in the limiting case of $T \to 0$; it increases in proportion to T. It is seen from (16) that it is of the order $k\,\delta N$.

40. The mean square of fluctuations

So far we have dealt with mean values or even with magnitudes associated with the maximum of probability implying that they are identical with those observed on a macroscopic scale. Such an attitude is justified by the fact that laws involving mean values are identical with the laws of thermodynamics and that the properties of substances required in thermodynamics can be calculated with the aid of suitable molecular models.

It is by no means evident that this must be so. The concept of a mean value includes the possibility of larger or smaller deviations, and single measurements can yield values which *fluctuate* to a larger or lesser extent about this mean value. The good agreement between the statistical mean values and the macrophysical experimental data may be interpreted to signify that the fluctuations encountered in statistical considerations of a thermodynamical nature are, generally speaking, very small; this is a conclusion which is a consequence of the *law of large numbers*.

In order to prove this proposition we shall require a measure for the fluctuations. The mean value of the fluctuations is certainly equal to zero, because the mean value of a quantity is so defined as to render the deviations in both directions equally probable. A possible measure is given by the mean value of the squares of the fluctuations: *the mean square*, for short.

Denoting the mean values by a bar, as we have already done on occasions, we find from eq. (36.14) that the mean value of the energy of a system in the Γ-space is given by:

$$(1) \qquad \overline{E} = \frac{\sum_n E(n)\, e^{-\beta E(n)}}{\sum_n e^{-\beta E(n)}} = -\frac{\partial \log Z}{\partial \beta}.$$

The fluctuations are equal to the differences between special measured values $E(n)$ and the mean value E, or

$$\Delta E(n) = E(n) - \overline{E}.$$

Thus the mean value of the square of the fluctuation in energy is given by

$$(2) \qquad (\Delta E)^2 = \overline{(\Delta E(n))^2} = \overline{(E(n) - \overline{E})^2}.$$

Here ΔE is the root mean square of the fluctuation, and is usually implied when referring to the mean fluctuation for short. Written explicitly eq. (2) becomes

$$(3) \qquad (\Delta E)^2 = \frac{\sum_n [E(n) - \overline{E}]^2 \, e^{-\beta E(n)}}{\sum_n e^{-\beta E(n)}}.$$

Since the formation of a mean is a linear process (and since any mean value is a constant with respect to further operations of taking a mean: $\overline{\overline{E}} = \overline{E}$), we infer from (2) or (3) that

$$(4) \qquad (\Delta E)^2 = \overline{E^2} - 2\,\overline{E\,\overline{E}} + \overline{E}^2 = \overline{E^2} - \overline{E}^2.$$

The mean square of a fluctuating quantity is equal to the difference between the mean value of the square of the quantity and the square of its mean value. It is, incidentally, clear that this difference must always be positive.

We shall now make use of eq. (4) to calculate the mean square of the fluctuation in energy. Since

$$\overline{E^2} = \frac{\sum_n E(n)^2 \, e^{-\beta E(n)}}{\sum_n e^{-\beta E(n)}}$$

we can see that the numerator can be obtained by differentiating the denominator twice with respect to β. Hence

(5) $$\overline{E^2} = \frac{1}{Z} \frac{\partial^2 Z}{\partial \beta^2}.$$

Substituting the mean values (1) and (5) into (4), we have

$$\overline{(\Delta E)^2} = \frac{1}{Z} \frac{\partial^2 Z}{\partial \beta^2} - \frac{1}{Z^2} \left(\frac{\partial Z}{\partial \beta}\right)^2.$$

This is exactly the derivative of the quotient Z'/Z, and we can write finally that:

(6) $$\overline{(\Delta E)^2} = \overline{(E - \overline{E})^2} = \frac{\partial^2 \log Z}{\partial \beta^2}.$$

In view of (1) and of the definition of β we may also write

(7) $$\overline{(\Delta E)^2} = -\frac{\partial \overline{E}}{\partial \beta} = k T^2 \frac{\partial U}{\partial T} = k T^2 C.$$

The mean square of the fluctuation in energy is seen to be determined by thermodynamic quantities only. It is proportional to the heat capacity, C.

For monatomic perfect gases we have $U = 3/2\, N k T$ see eq. (22.6 a), and

(8) $$\overline{(\Delta E)^2} = \frac{3}{2} N k^2 T^2, \quad \frac{\Delta E}{U} = \sqrt{\frac{2}{3N}}.$$

For one mol of gas ($N = L$) the mean fluctuation is equal to the one -10^{12} th part of the mean energy, which is utterly unobservable. For $N = 150$ we should have $\Delta E/U = 6.7\%$.

The preceding examples illustrate the effect of very large numbers, and demonstrate that fluctuations are unimportant in relation to large masses, but may play a significant part in small regions. Fluctuations in energy of the order of 6.7% at room temperature correspond to temperature fluctuations of ± 20 C. The importance of fluctuations in relation to small regions has already been discussed in connection with the study of Brownian motion (*cf.* Sec. 24).

Equation (7) is valid universally. In the case of a system of quantum-mechanical oscillators eq. (33.8) leads to

(9) $$\overline{(\Delta E)^2} = \frac{N (k \Theta)^2}{(e^{\Theta/T} - 1)^2} e^{\Theta/T},$$

or, in analogy with eq. (8), to

(9 a) $$\Delta E/U = 1/\sqrt{N}.$$

In the case of a solid body eqs. (35.8 a) and (35.8 b) lead to

(10) $$\frac{\Delta E}{U} = \begin{cases} \dfrac{2}{\pi^2}\sqrt{\dfrac{5}{3}}\,(\Theta/T)^{3/2}\,\dfrac{1}{\sqrt{N}} & \text{if } T \ll \Theta \\ \dfrac{1}{\sqrt{3N}} & \text{if } T \gg \Theta. \end{cases}$$

It is seen that at absolute zero the fluctuations themselves die out, but that their ratio to the thermal energy U need not do so. The fact that ΔE tends to zero as $T \to 0$ is also a consequence of Nernst's Third Law (cf. Sec. 12) because $C = \partial U/\partial T$ in eq. (7) is proportional to the specific heat. In the case of an electron gas, we have from eq. (39.21 a) that

(11) $$\frac{\Delta E}{U} = \frac{5\pi}{3\sqrt{2}}\left(\frac{kT}{\zeta_0}\right)^{3/2}\frac{1}{\sqrt{N}}$$

which is analogous to the first eq. (10).

The mean squares of fluctuation can be easily calculated for the occupation numbers n_i. In analogy with eq. (38.4), we have

(12) $$\overline{(\Delta n_i)^2} = \overline{(n_i - \overline{n}_i)^2} = \frac{1}{\beta^2}\left(\frac{\partial^2 \log Z}{\partial \varepsilon_i^2}\right)_{\alpha,\beta} = -\frac{1}{\beta}\left(\frac{\partial \overline{n}_i}{\partial \varepsilon_i}\right)_{\alpha,\beta}$$

that is, in the case of a Bose-Einstein, or a Fermi-Dirac gas, respectively

(13) $$\overline{(\Delta n_i)^2} = \frac{e^{\alpha + \beta \varepsilon_i}}{(e^{\alpha + \beta \varepsilon_i} \mp 1)^2}$$

and

(14) $$\Delta n_i = \sqrt{\overline{n}_i(1 \pm \overline{n}_i)}.$$

In the Fermi-Dirac case the above expression markedly differs from zero only at the Fermi threshold. Directly at $\varepsilon_i = \zeta > \zeta_0$ (cf. eq. (39.5)) Δn_i assumes its maximum value $\Delta n_i = \frac{1}{2}$. For the excited states of a highly degenerate Bose-Einstein gas we have $\overline{n}_i \ll 1$, and hence $\Delta n_i = \sqrt{\overline{n}_i}$. The same equation is valid for the classical limiting case. The ground state of a Bose-Einstein gas is of particular interest; since $1 \ll \overline{n}_0 \,(\approx N)$, we have

$$\Delta n_0 \approx \overline{n}_0,$$

and the fluctuation is seen to be of the order of the mean value, and hence numerically very large.

In this connection it should be noted that eq. (14) is valid only for grand canonical ensembles because in the derivation of eq. (12) we have kept α, and not the number of molecules N, constant. Moreover, the sub-systems of the grand canonical ensemble continually exchange particles which explains the manner in which large fluctuations may originate.

From the practical point of view the fluctuations which occur in groups of phase cells are much more important in view of the small size of a single phase cell. Let $\sum_i'$ denote the sum taken over a well-defined group of cells; it is then seen from eq. (38.4) that the mean number of particles within it is given by

$$\bar{n}_j = \sum_i' \bar{n}_i = -\frac{1}{\beta} \sum_i' \frac{\partial \log Z}{\partial \varepsilon_i}$$

and that the mean square of fluctuation becomes

(15) $$(\Delta n_j)^2 = \overline{(n_j - \bar{n}_j)^2} = \frac{1}{\beta^2} \sum_{i_1, i_2}' \frac{\partial^2 \log Z}{\partial \varepsilon_{i_1} \varepsilon_{i_2}}$$

Since in eq. (38.4) the $\bar{n}_i$ depend only on the ε_i, the mixed terms in the double sum disappear. Consequently we have

(16) $$(\Delta n_j)^2 = -\frac{1}{\beta} \sum_i' \frac{\partial \bar{n}_i}{\partial \varepsilon_i} = \sum_i' \bar{n}_j (1 \pm \bar{n}_i).$$

For example, the fluctuations in a volume element ΔV, as given by eq. (16) become

(17) $$(\Delta n)^2 = \frac{4\pi \Delta V}{h^3} \int_0^\infty \frac{e^{\alpha + \beta p^2/2m}}{(e^{\alpha + \beta p^2/2m} \mp 1)^2} p^2 \, dp,$$

(where the spin factor 2 has been omitted). Thus in the limiting case of Boltzmann's statistics, we have

$$(\Delta n)^2 = \frac{4\pi \Delta V}{h^3} e^{-\alpha} \left(\sqrt{\frac{2m}{\beta}}\right)^3 \frac{\sqrt{\pi}}{4}.$$

Substituting the expression for e^α form eq. (31.4), we obtain

(18) $$(\Delta n)^2 = \frac{N \Delta V}{V},$$

or, introducing the mean number of molecules in ΔV from $\bar{n} = N \Delta V/V$:

(18 a)
$$\frac{\Delta n}{\bar{n}} = \sqrt{\frac{V}{N \Delta V}}.$$

Such fluctuations in density can be observed because they cause the scattering of light. The blue color of the sky is due to the scattering of solar light and is caused by the fluctuations in the density of the atmospheric air.

It is, naturally, possible to deduce eq. (18 a) from the fact that the mean density is constant, if direct combinatorial methods are used, and the method would, certainly, be much simpler. On the other hand the preceding argument allows us to show its relation to the fundamental equations of statistical mechanics which would otherwise be lost. The direct method of derivation is discussed in Problem IV. 8.

In the case of an electron gas, eq. (17) yields

$$(\Delta n)^2 = 8 \pi \frac{\Delta V}{h^3} \left(\sqrt{\frac{2m}{\beta}}\right)^3 \int_{-\beta \zeta_0}^{\infty} \frac{e^x}{(e^x+1)^2} \sqrt{x + \beta \zeta} \, \frac{dx}{2}$$

(now with the spin factor 2 included). The evaluation is identical with that in Sec. 39, and in the case of complete degeneration the integration gives:

$$(\Delta n)^2 = 8 \pi \frac{\Delta V}{h^3} (\sqrt{2m \zeta_0})^3 \frac{1}{2 \beta \zeta_0}$$

or, in accordance with eqs. (39.7) and (39.7 a),

$$(\Delta n)^2 = \frac{3}{2} \frac{N \Delta V}{V} \frac{1}{\beta \zeta_0}.$$

Hence, with $\bar{n} = N \Delta V/V$, we have

(19)
$$\frac{\Delta n}{n} = \sqrt{\frac{V}{N \Delta V}} \cdot \sqrt{\frac{3 k T}{2 \zeta_0}}.$$

The density fluctuations of the Fermi gas are seen to disappear at absolute zero.

In conclusion we wish to remark that the higher powers of fluctuation can also be deduced from the partition function. Thus, for example, we have

(20)
$$\overline{(E(n) - \bar{E})^3} = -\frac{\partial^3 \log Z}{\partial \beta^3},$$

since

$$-\frac{\partial^3 \log Z}{\partial \beta^3} = -\left(\frac{Z'}{Z}\right)'' = \left(\frac{Z''}{Z} - \frac{Z'^2}{Z^2}\right)' = -\left(\frac{Z'''}{Z} - \frac{3 Z'' Z'}{Z^2} + \frac{2 Z'^3}{Z^3}\right) =$$

$$= \overline{E^3} - 3\, \overline{E^2}\, \overline{E} + 2\, \overline{E}^3 = \overline{(E - \overline{E})^3}.$$

In the case of a perfect gas, we have $\log Z = -3/2\, N \log \beta + \ldots$ and it follows that

$$\overline{(E - \overline{E})^3}/\overline{E}^3 = \frac{8}{9} \frac{1}{N^2}, \quad \text{or} \quad \left\{\overline{\left(\frac{E - \overline{E}}{\overline{E}}\right)^3}\right\}^{1/3} = \sqrt[3]{\frac{8}{9}}\, N^{-2/3}.$$

CHAPTER V

OUTLINE OF AN EXACT KINETIC THEORY OF GASES.

In the study of statistical mechanics given in Chap. IV we have succeeded in providing an atomistic justification for thermodynamics, i.e. for the science of thermal equilibrium. It has been rendered more complete by demonstrating that the thermodynamic potentials, for example the free energy in eq. (29.12′), and in particular the partition function in eq. (29.15), can be deduced from atomistic data.

An atomistic description of non-equilibrium processes is much less simple. The phenomenological propositions have already been given in Sec. 21. In the present Chapter we shall restrict ourselves to the consideration of the behavior of molecules in perfect gases thus following the course of historical development. In this manner the object of the present chapter may be stated as an attempt to provide an exact formulation of the kinetic theory of gases given in Chap. III. The kinetic theory of condensed matter has been considerably advanced in recent times, but to describe it here we would have to exceed the scope of this text-book. In the present Chapter we shall be forced more often to refer the reader to specialized papers. This is true, in particular, in relation to the methods of solving the collision equation,[1] about to be derived.

The lack of completeness in the present Chapter is not only due to the limitation of its objects. It follows also from the fact that explicit calculations are only possible for very crude molecular models, and this is particularly true when it is desirable to take into account the quantum mechanical properties of molecules. An important step forward in this field was taken with the development of the theory of conduction electrons in metals, particularly when Sommerfeld succeeded in deriving the Wiedemann-Franz law, whose description will be given at the end of this Chapter.

41. The Maxwell-Boltzmann collision equation

A. Description of a state in the kinetic theory of gases

The perfect gas is characterized by the fact that the state of any of its molecules is independent of that of all the others, except for the instant of collision. We can describe it completely by specifying the position and velocity

---
[1] See M. Born, "Cause and Chance", footnote on p. 203.

of every molecule at a given instant. We shall restrict our considerations to monatomic molecules which we shall assume to be rigid spheres. We shall denote position by specifying the *space coordinates* $\mathbf{r} = (x_1, x_2, x_3)$ and we shall also specify the velocity $\mathbf{v} = (\xi_1, \xi_2, \xi_3)$.[1]

The volume elements in the physical and velocity spaces are

$$dx = dx_1 dx_2 dx_3, \qquad d\xi = d\xi_1 d\xi_2 d\xi_3. \tag{1}$$

In the present Chapter, unlike Chap. IV, we shall disregard the quantum nature of these volume elements, except in Sec. 45. We assume them to be so large that they can contain a large number of molecules, and yet small enough to allow us to disregard variations in density within one element. Mathematically this means that dx and $d\xi$ in eq. (1) may be treated like differentials in spite of their large size. This change in definition may serve to justify the difference between the notation in eq. (1) and that employed earlier.

We now proceed to determine the number of molecules, dv, at a place $(\mathbf{r}, \mathbf{v})$ in the phase element of the μ-space, but replace momentum by velocity, so that the latter becomes equal to $dx \, d\xi$, and

$$dv = f(\mathbf{r}, \mathbf{v}, t) \, dx \, d\xi. \tag{2}$$

Thus the total number of particles is given by the integral

$$N = \int f(\mathbf{r}, \mathbf{v}, t) \, dx \, d\xi. \tag{3}$$

The integration is extended over the whole space (or over the volume of a vessel) and over all velocities. Hence the *total* mean value of a function $g(\mathbf{r}, \mathbf{v})$ is given by

$$\overline{g} = \frac{1}{N} \int g(\mathbf{r}, \mathbf{v}) f(\mathbf{r}, \mathbf{v}, t) \, dx \, d\xi. \tag{4}$$

However, in the kinetic theory of gases importance is attached only to *local* mean values. They determine mean values in the velocity space which, generally speaking, vary from point to point. If $\phi(\mathbf{v})$ denotes any function of velocity, then the local mean of ϕ is given by the integral

$$\overline{\phi}(r) = \frac{1}{n} \int \phi(\mathbf{v}) f(\mathbf{r}, \mathbf{v}) \, d\xi. \tag{5}$$

---

[1] The fact that we may regard the molecules as being rigid spheres and that we may restrict ourselves to the consideration of translational motion finds its justification in quantum mechanics. *Cf.* here the footnote on p. 242, Sec. 34, regarding the rotational energy of an electron.

Here n denotes the local particle density

(5 a) $$n = \int f(\mathbf{r}, \mathbf{v}) \, d\xi.$$

Thus $n \, dx$ is the number of particles in a volume element dx irrespective of its velocity. For example, the mean velocity is given by

(6) $$\mathbf{u} = \frac{1}{n} \int \mathbf{v} f(\mathbf{r}, \mathbf{v}) \, d\xi,$$

or, written in terms of components, the triple integral

(6 a) $$u_i(x_1, x_2, x_3) = \frac{1}{n} \int \int \int \xi_i f(x_1, x_2, x_3; \xi_1, \xi_2, \xi_3) \, d\xi_1 \, d\xi_2 \, d\xi_3.$$

We assume that in addition to intermolecular forces, about which we shall have more to say later, there acts an external force, given as a function of space coordinates:

(7) $$\mathbf{F} = \mathbf{F}(\mathbf{r}) = [X_1(\mathbf{r}), X_2(\mathbf{r}), X_3(\mathbf{r})].$$

We shall disregard forces which depend on velocities, such as the forces acting on a charged particle as it moves through a magnetic field.

These assumptions are sufficient to provide a complete justification for the thermodynamics of gases in motion. A beautiful and non-trivial example of the theory under consideration is afforded by the discovery of the effect known as thermal effusion made by Clausius and Waldmann. The effect is obtained in the process of finding a higher-order approximation to the solution of the collision equation for several molecular species. We know from thermodynamics that perfect gases do not change their temperature on mixing. However, the process of mixing itself is accompanied by thermal effects. They are implied in the calculations due to Chapman[1] and Enskog,[2] but their importance in experimental science was first recognized by Clausius and Waldmann, who were also the first ones to observe it. Concerning the relation with thermal diffusion reference should be made to Sec. 21 C (reciprocal relations).

---

[1]Chapman, Phil. Trans. **211** (1911) 433, **216** (1916) 279, **217** (1916) 115.
[2]D. Enskog, Kinetic energy of processes in moderately dense gases, Inaugural dissertation (Uppsala 1917), Ark. for Matem. **16** (1921) No. 16, Kungl. Svenska Akad. **63** (1922) 4.

B. The Variation of f with Time

The Maxwell-Boltzmann collision equation is obtained by inquiring into the variation of f with time. We assume that f is continuous and sufficiently differentiable, this being possible owing to our definition of the volume elements dx and $d\xi$. The phase density $f(\mathbf{r}, \mathbf{v}, t)$ in the μ-space changes owing to the motion of the particles and to their collisions. We now consider a time interval Δt which is, on the one hand, large compared with the duration of a collision τ_s, so that most collisions which have begun during Δt are also completed within it. On the other hand, we shall stipulate that Δt is small compared with the mean collision time τ, i. e. with the interval between two collisions. Thus, generally speaking, *one* molecule will suffer at most *one* collision with another molecule during the interval Δt. This implies that the radius of action of intermolecular forces is sufficiently small compared with the distance between atoms, and, *a fortiori*, small compared with the mean free path (Sec. 27).

If no collisions occurred during Δt we could make the transformations

$$\mathbf{r} \to \mathbf{r}' = \mathbf{r} + \mathbf{v}\Delta t \quad \text{and} \quad \mathbf{v} \to \mathbf{v}' = \mathbf{v} + \frac{1}{m}\mathbf{F}\Delta t$$

so that

(8) $$f(\mathbf{r}, \mathbf{v}, t)\, dx\, d\xi \to f(\mathbf{r} + \mathbf{v}\Delta t, \mathbf{v} + \frac{1}{m}\mathbf{F}\Delta t, t + \Delta t)\, dx'\, d\xi' =$$

$$= \left[f(\mathbf{r}, \mathbf{v}, t) + \Delta t \left\{ \mathbf{v}\cdot\frac{\partial f}{\partial \mathbf{r}} + \frac{1}{m}\mathbf{F}\cdot\frac{\partial f}{\partial \mathbf{v}} + \frac{\partial f}{\partial t} \right\} + \ldots \right] dx'\, d\xi'.$$

The last equation applies in cases when we may neglect higher-order terms, i. e. when f does not change appreciably during the interval Δt. It may be noted that such an assumption is compatible with considerable changes within one mean free path because $\Delta t \ll \tau$. Furthermore, according to Liouville's theorem (Sec. 28), we can write

(8 a) $$dx'\, d\xi' = dx\, d\xi$$

for the preceding motion which is not impeded by collisions. Consequently the differential factors can be cancelled on both sides of eq. (8).

The collisions between molecules cause some molecules to leave $dx\, d\xi$, and some pass from $dx_1\, d\xi_1$ to $dx\, d\xi$. They are equivalent to a loss or gain in particles in $dx\, d\xi$ owing to collisions. Thus the balance equation for the particles consists in stating that the change in the number of particles, according to (8) and due to flow, must be equal to the difference between the numbers gained (J_{gain}) and lost (J_{loss}) as a consequence of collisions. Hence per unit time and phase space element we may write

$$\text{(9)} \qquad \frac{\partial f}{\partial t} + \mathbf{v}\frac{\partial f}{\partial \mathbf{r}} + \frac{1}{m}\mathbf{F}\frac{\partial f}{\partial \mathbf{v}} = J_{gain} - J_{loss}.$$

This is the Maxwell-Boltzmann collision equation. The quantities J_{gain} and J_{loss} must be calculated in accordance with the laws of elastic collision. In order to shorten the equations, i. e. eqs. (8) and (9), we have denoted the spatial gradient of f by $\partial f/\partial \mathbf{r}$ and that in the velocity space by $\partial f/\partial \mathbf{v}$.

C. The laws of elastic collision

These depend on the nature of the forces acting between molecules. In the preceding Sections we only stipulated that the radius of action of the forces was small. Thus, e. g. a force law $\mathbf{F} \sim 1/r^n$ would be compatible with the assumption, provided that n were large enough. This case is the one normally discussed. It leads to particularly simple results for $n = 5$, and the limiting case of $n = \infty$ corresponds to rigid molecules. We shall restrict ourselves here to the latter assumption. The diameter of such a sphere will be denoted by s. It indicates the smallest distance between the centers of two atoms imagined to be spherical in shape. It is evident that real molecules differ greatly from what we have assumed here. Monatomic molecules at moderate temperatures seem to be nearest to this model.[1]

When two molecules collide, the total energy and total momentum must preserve their values. Denoting the velocities of the two particles before collision by $\mathbf{v}, \mathbf{v}_1$ and by $\mathbf{v}', \mathbf{v}_1'$, after collision, we may write

$$\text{(10)} \qquad \mathbf{v} + \mathbf{v}_1 = \mathbf{v}' + \mathbf{v}_1',$$

$$\mathbf{v}^2 + \mathbf{v}_1^2 = \mathbf{v}'^2 + \mathbf{v}_1'^2.$$

If, further, $\mathbf{V}$ denotes the relative velocity before collision, i. e.

$$\text{(11)} \qquad \mathbf{V} = \mathbf{v}_1 - \mathbf{v},$$

we can write down the solution of (10) in the form

$$\text{(12)} \qquad \mathbf{v}' = \mathbf{v} + (\mathbf{V}\,\mathbf{e})\,\mathbf{e}, \quad \mathbf{v}_1' = \mathbf{v}_1 - (\mathbf{V}\,\mathbf{e})\,\mathbf{e}.$$

---

[1]The higher energy levels of atoms do not become excited at normal temperatures so that the law of attraction is determined by the polarization of the atoms.

Here **e** denotes an arbitrary unit vector. Hence the relative velocity after collision is given by

(12 a) $$\mathbf{V}' = \mathbf{v}_1' - \mathbf{v}' = \mathbf{V} - 2(\mathbf{V}\,\mathbf{e})\,\mathbf{e},$$

so that

(12 b) $$(\mathbf{V}'\,\mathbf{e}) = -(\mathbf{V}\,\mathbf{e}).$$

Solving eq. (11) for $\mathbf{v}$ and $\mathbf{v}_1$ yields

(13) $$\mathbf{v} = \mathbf{v}' + (\mathbf{V}'\,\mathbf{e})\,\mathbf{e}, \quad \mathbf{v}_1 = \mathbf{v}_1' - (\mathbf{V}'\,\mathbf{e})\,\mathbf{e}.$$

This equation is identical in form with eq. (12) and it is seen that the transformations for the velocities are mutually reciprocal.

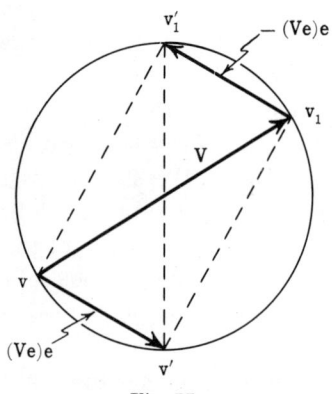

Fig. 33.

Vector diagram for the velocities associated with the elastic collision of two equal spheres.

The collision equation (11) for identical particles can be represented graphically in a simple way; we shall employ the same geometrical construction later. The points $\mathbf{v}$ and $\mathbf{v}_1$ in Fig. 33 represents the end points of the vectors $\mathbf{v}$ and $\mathbf{v}_1$. The vector drawn from $\mathbf{v}$ to $\mathbf{v}_1$ represents the relative velocity $\mathbf{V}$. We now draw a ray through $\mathbf{v}$ in the direction $+\mathbf{e}$ and one through $\mathbf{v}_1$ in the direction $-\mathbf{e}$ and obtain the vectors $\pm(\mathbf{V}\,\mathbf{e})\,\mathbf{e}$ from eq. (11) by projecting $\mathbf{V}$ onto the two parallel rays. These projections determine the points $\mathbf{v}'$ and $\mathbf{v}_1'$, as shown. The four points $(\mathbf{v}, \mathbf{v}_1, \mathbf{v}', \mathbf{v}_1')$ are seen to lie on a rectangle irrespective of the direction of $\mathbf{e}$. They all lie on a sphere whose diameter is $\mathbf{V}$ and whose center is at the mid-point of the vector $|\mathbf{V}|$; furthermore, the pairs $(\mathbf{v}, \mathbf{v}_1)$ and $(\mathbf{v}', \mathbf{v}_1')$ are each diametrically opposed, Fig. 33.

It is now necessary to indicate the meaning of **e** for the collision of rigid spheres. According to eq. (12) we may write

(14) $$\mathbf{V} - \mathbf{V}' = 2(\mathbf{V}\,\mathbf{e})\,\mathbf{e}.$$

During a collision there is a transfer of momentum. *On the one hand* this must be normal to the plane tangent to the two spheres at the point of impact, i. e. in the direction of the line through the two centers, the so-called *central axis*. On the other hand, the amount of momentum transferred is equal to the change in the momentum of either sphere, or to

(14 a) $$\mathbf{v}' - \mathbf{v} = \mathbf{v}_1 - \mathbf{v}_1' = (\mathbf{V}\,\mathbf{e})\,\mathbf{e}.$$

This shows that the vector **e** lies along the central axis. Moreover, since **V** and **V'** are equal in magnitude, as seen from eq. (12 a) and Fig. 33, the central axis bisects the angle formed **V** and **V'**. Fig. 34 represents the directions of the vectors and of the central axis as seen by an observer moving with either of the two spheres, say with that whose velocity is **v** before collision.

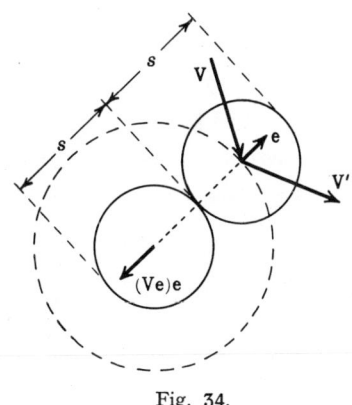

Fig. 34.
The kinematics of an elastic collision.

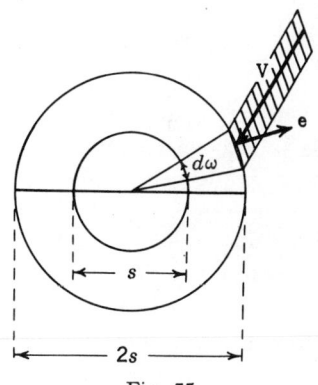

Fig. 35.
Sphere of influence and rate of collision.

D. Boltzmann's collision integral

The preceding geometrical picture provides a starting point for the calculation of the right-hand side of eq. (9). It is perhaps somewhat more convenient to represent the motion of the center of the impinging sphere relative to the sphere of influence of radius s, drawn dotted in Fig. 35. The number of molecules which impinge on an area $s^2\,d\omega$, where $d\omega$ is an elementary solid angle, during an interval of time Δt is

$$s^2\,d\omega\,|\mathbf{V}\,\mathbf{e}|\,\Delta t \cdot f(\mathbf{r},\mathbf{v}_1,t)\,d\xi_1.$$

The first term, $s^2\,d\omega\,|\mathbf{V}\,\mathbf{e}|\,\Delta t$, represents the volume of the oblique cylinder from which particles of a given direction and a given velocity arrive during time Δt, as shown in Fig. 35. The second term denotes the density of these special particles.

Now, in the gas there are $f(\mathbf{r}, \mathbf{v}, t)\, dx\, d\xi$ molecules present whose velocity is $\mathbf{v}$. Thus the total number of collisions between molecules whose velocities are $\mathbf{v}$ and $\mathbf{v}_1$ and whose central axis vector is $\mathbf{e}$ [1] is

$$\text{(15)} \qquad s^2\, d\omega\, |\mathbf{V}\, \mathbf{e}|\, \Delta t \cdot f(\mathbf{r}, \mathbf{v}_1, t) \cdot f(\mathbf{r}, \mathbf{v}, t)\, dx\, d\xi\, d\xi_1$$

Integration over all velocities $\mathbf{v}_1$ and over all directions $\mathbf{e}$ gives the total number of collisions which deflect the paths of the particles whose velocity is $\mathbf{v}$. This number, computed per unit time and per unit $dx\, d\xi$, is the one entering eq. (9). Thus

$$\text{(16)} \qquad J_{loss} = \frac{s^2}{2} \int |\mathbf{V}\, \mathbf{e}|\, f(\mathbf{r}, \mathbf{v}_1, t)\, f(\mathbf{r}, \mathbf{v}, t)\, d\omega\, d\xi_1.$$

The factor $\tfrac{1}{2}$ is due to the fact that the integration is performed over the whole sphere, whereas the physical argument applies only to half of it, as it is easy to see by considering the variation of $\mathbf{e}$ on shifting $\mathbf{v}$ parallel to itself.

The calculation of J_{gain} is quite analogous. We must now arrange it in such a way as to make the velocities after collision equal to $\mathbf{v}$ and $\mathbf{v}_1$, because J_{gain} corresponds to the increase in the number of particles moving with a velocity $\mathbf{v}$ due to collisions. Let the corresponding velocities before the impact be $\mathbf{v}'$ and $\mathbf{v}_1'$. The number of impacts on an element of area $s^2\, d\omega$ of the sphere of influence is analogous to (15):

$$\text{(15 a)} \qquad s^2\, d\omega\, |\mathbf{V}'\, \mathbf{e}|\, \Delta t \cdot f(\mathbf{r}, \mathbf{v}', t)\, f(r, \mathbf{v}'_1 t)\, dx\, d\xi'\, d\xi_1'.$$

According to (12 b) we have $|\mathbf{V}'\, \mathbf{e}| = |\mathbf{V}\, \mathbf{e}|$, and according to (13) (and also to Liouville's theorem given in Sec. 28) we have, furthermore, $d\xi'\, d\xi_1' = d\xi\, d\xi_1$. Moreover the same result follows from the Jacobian of (13)

$$\text{(16 a)} \qquad \frac{\partial(\mathbf{v}, \mathbf{v}_1)}{\partial(\mathbf{v}', \mathbf{v}_1')} = \begin{vmatrix} 1 & . & . & 0 & . & . \\ . & 1 & . & . & 0 & . \\ . & . & 0 & . & . & 1 \\ \hline 0 & . & . & 1 & . & . \\ . & 0 & . & . & 1 & . \\ . & . & 1 & . & . & 0 \end{vmatrix} = -1,$$

---

[1] A short expression, like the one used here, implies that the vectors have the given values within the elements $d\omega$ and $dx\, d\xi$.

where we may assume that $\mathbf{e} = (0, 0, 1)$ without loss of generality.[1] Thus the gain in momentum (again for $\Delta t = 1$ and $dx\,d\xi = 1$) is:

(16 b) $\quad J_{gain} = \dfrac{s^2}{2} \displaystyle\int |\mathbf{V}\,\mathbf{e}|\,f(\mathbf{r},\mathbf{v}+(\mathbf{V}\,\mathbf{e})\,\mathbf{e},t)\,f(\mathbf{r},\mathbf{v}_1-(\mathbf{V}\,\mathbf{e})\,\mathbf{e},t)\cdot d\omega\,d\xi_1.$

Equations (16) and (16 b) are known as *Boltzmann's collision integrals*. Substituting these into eq. (9) and introducing the usual abbreviations

(17) $\quad\begin{aligned}f &= f(\mathbf{r},\mathbf{v},t), \qquad f_1 = f(\mathbf{r},\mathbf{v}_1,t),\\ f' &= f(\mathbf{r},\mathbf{v}',t) = f(\mathbf{r},\mathbf{v}+(\mathbf{V}\,\mathbf{e})\,\mathbf{e},t),\\ f_1' &= f(\mathbf{r},\mathbf{v}_1',t) = f(\mathbf{r},\mathbf{v}_1-(\mathbf{V}\,\mathbf{e})\,\mathbf{e},t),\end{aligned}$

we obtain the *Maxwell-Boltzmann collision equation*:

(18) $\quad \dfrac{\partial f}{\partial t} + \mathbf{v}\dfrac{\partial f}{\partial \mathbf{r}} + \dfrac{1}{m}\mathbf{F}\dfrac{\partial f}{\partial \mathbf{v}} = \dfrac{s^2}{2}\displaystyle\int |\mathbf{V}\,\mathbf{e}|\,(f'f_1' - f f_1)\,d\omega\,d\xi_1.$

The mathematical problem of an exact kinetic theory of gases consists in the solution of the preceding non-linear, integro-differential equation.

E. Boltzmann's hypothesis about molecular chaos

Before proceeding to study the properties of this equation it is necessary to discuss an important assumption implied in the preceding argument. In deriving the expression for the collision integrals we need the probability $W(\mathbf{r},\mathbf{v};\mathbf{r}_1,\mathbf{v}_1)$ of finding a molecule at phase point $(\mathbf{r},\mathbf{v})$ in collision with another at phase point $(\mathbf{r}_1,\mathbf{v}_1)$. However, we only know at first the probabilities $W(\mathbf{r},\mathbf{v})$ and $W(\mathbf{r}_1,\mathbf{v}_1)$ of finding a molecule at $(\mathbf{r},\mathbf{v})$, or $(\mathbf{r}_1,\mathbf{v}_1)$, respectively. The latter can be easily calculated from $f(\mathbf{r},\mathbf{v})$, because according to (3) we can write per unit volume in the phase space:

$$W(\mathbf{r},\mathbf{v}) = \dfrac{1}{N} f(\mathbf{r},\mathbf{v}).$$

The occurrence of products of two functions f in eqs. (15) and (15 a) means that $W(\mathbf{r},\mathbf{v};\mathbf{r}_1,\mathbf{v}_1) = 0$ for $\mathbf{r}\neq\mathbf{r}_1$, and that for $\mathbf{r}=\mathbf{r}_1$ we have assumed that

$$W(\mathbf{r},\mathbf{v};\mathbf{r},\mathbf{v}_1) = W(\mathbf{r},\mathbf{v})\cdot W(\mathbf{r},\mathbf{v}_1)$$

---

[1] The change in sign in the Jacobian is compensated by that in the factor $\mathbf{V}'\mathbf{e}$. It is necessary to remember that $\tfrac{1}{2}|\mathbf{V}''\mathbf{e}|\,d\xi'\,d\xi_1'$ originated from $(\mathbf{V}'\,\mathbf{e})\,d\xi\,d\xi_1'$ with $(\mathbf{V}'\,\mathbf{e}) > 0$.

i. e. that the first function is simply a product of the other two. This implies that the probabilities for the possible *velocities* of one particle are independent of the velocity of the other.

That this constitutes an additional assumption can be seen by considering the function $W(v, v_1)$. The function $W(v, v_1)$ must be symmetrical in v and v_1, since no molecule is preferred compared with another. On integration we find that

$$\int W(v, v_1) \, d\xi_1 = \int W(v_1, v) \, d\xi_1 = W(v).$$

Evidently, owing to symmetry, the same function $W(v)$ is obtained in either case. It is, however, by no means permissible to conclude that $W(v, v_1) = W(v) \times W(v_1)$. In general this will not be the case at all.

The separation of W into a product is a consequence of Boltzmann's hypothesis of *complete molecular chaos*. It corresponds to Maxwell's assumption in Sec. 23, to the assumption of equal probabilities in Sec. 28, or to Gibbs' hypothesis in Sec. 36. This assumption, essentially, justifies the validity of the entropy theorem which we shall derive from eq. (18) in Sec. 42. Born[1] and his co-workers have recently given a more detailed analysis of this hypothesis.

42. The H-theorem and Maxwellian distribution

A. The H-theorem

We now turn our attention to the entropy theorem. The expression for entropy is given by eq. (36.10), namely.[2]

(1) $$\overline{H} = -k \int \log f \cdot f \, dx \, d\xi.$$

It should, however, be noted that this equation refers to the μ-space, whereas originally it has been written for the Γ-space, and that we do not now attribute quantum properties to the phase cells. The expressions in eq. (29.5) for a Boltzmann gas, in eq. (38.12) for a Fermi-Dirac gas, and in eq. (38.14) for an Einstein-Bose gas, referred to the μ-space. In the preceding Chapter, when dealing with thermodynamic equilibrium, we have restricted ourselves to

---

[1] *Cf.* "Cause and Chance", *l. c.* p. 223, footnote 1.
[2] With Boltzmann, we now use the symbol H instead of S.

42.6 THE H-THEOREM AND MAXWELLIAN DISTRIBUTION

the consideration of the conditions under which the integral (1) attained its maximum. In the present Section we shall investigate the variation of H with time. It might be expected that $\bar{H}$ will never decrease under the conditions known from thermodynamics.

When studying non-equilibrium processes it is found that the *local* entropy is more revealing than the total entropy. Hence, instead of using eq. (1) we shall investigate the variation of

$$(2) \qquad H = -k \int \log f \cdot f \, d\xi$$

with time. We obtain

$$\frac{\partial H}{\partial t} = -k \int (1 + \log f) \frac{\partial f}{\partial t} \, d\xi,$$

or according to eq. (41.18)

$$(3) \qquad \frac{\partial H}{\partial t} = k \int (1 + \log f) \left(\mathbf{v} \frac{\partial f}{\partial \mathbf{r}} + \frac{1}{m} \mathbf{F} \frac{\partial f}{\partial \mathbf{v}} - J_f \right) d\xi,$$

where J_f denotes the collision integral appearing on the right-hand side of eq. (41.18), namely

$$(4) \qquad J_f = \frac{s^2}{2} \int |\mathbf{V} \mathbf{e}| \, (f' f_1' - f f_1) \, d\omega \, d\xi_1.$$

The first term on the right-hand side of eq. (3) can be transformed as follows:

$$\int (1 + \log f) \, \mathbf{v} \frac{\partial f}{\partial \mathbf{r}} \, d\xi = \operatorname{div} \int \mathbf{v} \log f \cdot f \, d\xi.$$

The integral

$$(5) \qquad \mathbf{S} = -k \int \mathbf{v} \log f \cdot f \, d\xi$$

denotes the flux vector associated with H defined in eq. (2). Thus eq. (3) assumes the following form

$$(6) \qquad \frac{\partial H}{\partial t} + \operatorname{div} \mathbf{S} = -k \int (1 + \log f) J_f \, d\xi.$$

This has taken into account that the second term on the right-hand side of (3) vanishes. In fact, since **F** depends only on the space coordinates, the second term can be written

$$\frac{k}{m} \mathbf{F} \int (1 + \log f) \frac{\partial f}{\partial \mathbf{v}} d\xi = \frac{k}{m} \mathbf{F} \int \frac{\partial}{\partial \mathbf{v}} (f \log f) \, d\xi.$$

This integral can be represented in the form of a surface integral over the sphere at infinity in the velocity space. Since the energy is finite, f vanishes, and we have

(7) $$\frac{k}{m} \mathbf{F} \int (1 + \log f) \frac{\partial f}{\partial \mathbf{v}} d\xi = 0.$$

In this manner only the integral on the right-hand side of eq. (6) remains and we can substitute J_f from eq. (4), when we obtain the form

(8) $$-\frac{k s^2}{2} \int |\mathbf{V} \mathbf{e}| (1 + \log f) (f' f_1' - f f_1) \, d\omega \, d\xi \, d\xi_1.$$

Interchanging the two triples of variables of integration, **v** and $\mathbf{v}_1$, does not affect the value of the integral, so that we may also write

$$-\frac{k s^2}{2} \int |\mathbf{V} \mathbf{e}| (1 + \log f_1) (f' f_1' - f f_1) \, d\omega \, d\xi \, d\xi_1.$$

It should be noted that the following transformation may be performed. First $\mathbf{V} \to -\mathbf{V}$ and

$$\mathbf{v}' = \mathbf{v} + (\mathbf{V} \mathbf{e}) \mathbf{e} \to \mathbf{v}_1 - (\mathbf{V} \mathbf{e}) \mathbf{e} = \mathbf{v}_1'$$

$$\mathbf{v}_1' = \mathbf{v}_1 - (\mathbf{V} \mathbf{e}) \mathbf{e} \to \mathbf{v} + (\mathbf{V} \mathbf{e}) \mathbf{e} = \mathbf{v}'.$$

Consequently, we can write the right-hand side of eq. (6) in the following, more symmetrical form

(9) $$-\frac{k s^2}{4} \int |\mathbf{V} \mathbf{e}| (2 + \log f + \log f_1) (f' f_1' - f f_1) \, d\omega \, d\xi \, d\xi_1.$$

Instead of integrating with respect to **v** and $\mathbf{v}_1$, we may also integrate with respect to $\mathbf{v}'$ and $\mathbf{v}_1'$. Thus, according to eq. (41.8 a), eq. (9) becomes

$$-\frac{k s^2}{4} \int |\mathbf{V} \mathbf{e}| (2 + \log f + \log f_1) (f' f_1' - f f_1) \, d\omega \, d\xi' \, d\xi_1'.$$

Now it is necessary to assume that $\mathbf{v}$ and $\mathbf{v}_1$ have been eliminated with the aid of eq. (41.13) rather than eliminating $\mathbf{v}'$ and $\mathbf{v}_1'$ with the aid of eq. (41.11). Having done this we can change our notation and write $\mathbf{v}$ and $\mathbf{v}_1$ respectively for $\mathbf{v}'$ and $\mathbf{v}_1'$. This will cause no confusion because $\mathbf{v}$ and $\mathbf{v}_1$ do not appear in the equation. Having performed the change of variables it will, nevertheless, be found convenient to define new variables $\mathbf{v}'$ and $\mathbf{v}_1'$ with the aid of eq. (41.11), so that the integral on the right-hand side of (8), now denoted by G, becomes

$$(10) \qquad G(\mathbf{r}, t) = -\frac{k\,s^2}{4} \int |\mathbf{V}\,\mathbf{e}|\,(2 + \log f' + \log f_1')\,(f f_1 - f' f_1')\,d\omega\,d\xi\,d\xi_1.$$

It is easy to see that the factor $|\mathbf{V}\,\mathbf{e}|$ in the integrand remains unchanged. It will turn out that $G \equiv 0$, cf. Sec. 21, eq. (8).

The integral can be made even more symmetrical if it is replaced by half the sum of the two equal expressions in eqs. (8) and (10). In doing so it is necessary to note the change in the sign of the term in the last bracket in the integrand. Thus we obtain

$$(11) \qquad G(\mathbf{r}, t) = -k \int (1 + \log f) \cdot J_f\,d\xi =$$
$$= -\frac{k\,s^2}{8} \int |\mathbf{V}\,\mathbf{e}|\,(\log f + \log f_1 - \log f' - \log f_1')\,(f' f_1' - f f_1)\,d\omega\,d\xi\,d\xi_1.$$

or, after a simple rearrangement:

$$(12) \qquad G(\mathbf{r}, t) = \frac{k\,s^2}{8} \int |\mathbf{V}\,\mathbf{e}|\left(\log\frac{f' f_1'}{f f_1}\right)(f' f_1' - f f_1)\,d\omega\,d\xi\,d\xi_1.$$

At this stage it might be remarked that we shall encounter an identical transformation of an integral of the type (8), except that an arbitrary function $\psi(\mathbf{v})$ will occur instead of $1 + \log f$. We would then obtain

$$(13) \qquad \int |\mathbf{V}\,\mathbf{e}|\,\psi(\mathbf{v})\,(f' f_1' - f f_1)\,d\omega\,d\xi\,d\xi_1 =$$
$$= \frac{1}{4} \int |\mathbf{V}\,\mathbf{e}|\,(\psi + \psi_1 - \psi' - \psi_1')\,(f' f_1' - f f_1)\,d\omega\,d\xi\,d\xi_1$$

in complete analogy with the preceding case. The different ψ-functions in eq. (13), are defined in the same way as the f's in eq. (41.17).

First we note that the integrand in eq. (12) cannot be negative, because $\log(f' f_1'/f f_1)$ and $f' f_1' - f f_1$ always have the same signs. Hence

$$(14) \qquad \dot{H} + \mathrm{div}\,\mathbf{S} = G \geqslant 0.$$

The relation between this equation and eq. (21.10) will be discussed later. Integration over a finite volume yields

$$\text{(15)} \qquad \frac{d}{dt}\int H\,dx + \int S_n\,d\sigma = \int G\,dx \geqslant 0,$$

where the volume integral in the second term has been transformed with the aid of Gauss' theorem into a surface integral.[1] The integral $\int H\,dx$ is seen to change owing to two causes; first, there is a flow of entropy through the surface and, secondly, there exists within the volume a distribution of sources which are either zero or positive. When the system is isolated from the surroundings there is no flow of entropy across the surface and we must have

$$\text{(16)} \qquad \frac{d}{dt}\int H\,dx = \int G\,dx \geqslant 0.$$

The entropy of an isolated system cannot decrease. It should be realized that the scope of eq. (14) exceeds that of the entropy principle in thermodynamics. It determines the magnitude of the irreversible change in H. Furthermore, eq. (5) defines the entropy flux.

B. Maxwellian distribution

When $G = 0$ the change in entropy is determined solely by the flow of entropy. Since the integrand in eq. (12) cannot be negative, this can occur only if

$$\text{(17)} \qquad f'f_1' = f f_1.$$

Putting

$$\text{(17 a)} \qquad \log f = \psi$$

we find that (17) is equivalent to

$$\text{(17 b)} \qquad \psi' + \psi_1' = \psi + \psi_1.$$

The sum $\psi + \psi_1$ is seen to remain constant during a collision; it is an *additive invariant of the collision.*

---

[1] $d\sigma$ denotes a surface element on the surface and S_n is the component of the vector **S** in the direction of the normal outwards.

We can at once write down five functions which satisfy eq. (17 b), namely a constant, and the expressions for momentum and energy:

(18) $\quad \psi_0 = 1, \quad \psi_1 = \xi_1, \quad \psi_2 = \xi_2, \quad \psi_3 = \xi_3, \quad \psi_4 = \dfrac{1}{2} \mathbf{v}^2.$

In fact, these are the only additive invariants for a collision.

In order to prove this proposition we revert to the representation in Fig. 33, Sec. 41. We shall call $\psi(\mathbf{v})$ an antipodal function if for antipodal points $\mathbf{v}$ and $\mathbf{v}_1$ on an *arbitrary* sphere in the velocity space we have

$$\psi(\mathbf{v}) + \psi(\mathbf{v}_1) = \text{const.}$$

The constant may, evidently, vary from sphere to sphere. Since on collision the points $\mathbf{v}$ and $\mathbf{v}_1$ change to antipodal points on the same sphere, eq. (17 b) is seen to be satisfied. Antipodal functions are thus equivalent to the additive invariants for a collision.

It is easy to show[1] that a continuous antipodal function vanishes identically if it vanishes at the following five points:

(19) $\quad \mathbf{v} = (0, 0, 0);\ (1, 0, 0);\ (0, 1, 0);\ (0, 0, 1);\ (-1, 0, 0).$

Using the five functions (18) it is always possible to construct a function, by the use of linear superposition, which would assume arbitrarily prescribed values at the characteristic points (19), i. e. one that would assume the same values at those points as an arbitrary antipodal function. Since the difference between the prescribed and the so constructed antipodal function is also an antipodal function, namely one which vanishes at the five points (19), it must vanish identically. In other words the only antipodal functions, i. e. the only additive collision invariants are

(20) $\quad \psi = a_0 + \mathbf{a}\, \mathbf{v} + a_4\, \mathbf{v}^2.$

According to eq. (17 a) we may also write

(20 a) $\quad \log f = \alpha - \gamma(\mathbf{v} - \mathbf{u})^2,$

with a different set of constants. Putting $a = e^\alpha$ we obtain Maxwell's distribution law

(21) $\quad f = f_0(\mathbf{v}) = a\, e^{-\gamma(\mathbf{v}-\mathbf{u})^2}$

with the difference that $a, \gamma,$ and $\mathbf{u}$ may still be functions of $\mathbf{r}$ and t. We refer to it as to the *local* Maxwellian distribution.

---

[1] The proof was given by Harold Grad, Comm. pure appl. Maths., **2** (1949) 311.

It now remains to prove Grad's lemma. We begin by considering at first only the ξ_x, ξ_y-plane, as shown in Fig. 36a. Of the first points in (19) the ones denoted by ⊠ lie in this plane; they are A, B, C, and D. At those points we have $\psi = 0$ by definition. The same can be said about all nodal points of the quadratic lattice in Fig. 36 a, because we can always find pairs of antipodal points of which we know that $\psi = 0$ for three points and hence must be so at the fourth. For example $(A, D; B\ 1)$, $(C, D; B\ 2)$ $(A, C; D, 7)$ etc.

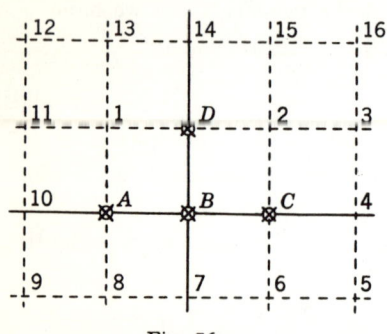 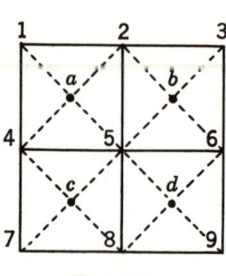

Fig. 36 a,   Fig. 36 b.

Illustrating the derivation of local Maxwellian distribution.

Using the same construction we can find additional points at which ψ vanishes. The nodal points from Fig. 36a have been denoted by + in Fig. 36b. It is seen from the diagram that ψ must vanish also at the midpoints denoted by •, as it is easy to infer from the following antipodal pairs: (a, b; 2,5), (b, d; 5,6), (c, d; 5,8), (a, c; 4,5); (a, d; b, c). Since

$$\psi(a) + \psi(b) = \psi(2) + \psi(5) = 0, \quad \psi(b) + \psi(d) = \psi(5) + \psi(6) = 0,$$
$$\psi(c) + \psi(d) = \psi(5) + \psi(8) = 0, \quad \psi(a) + \psi(c) = \psi(4) + \psi(5) = 0,$$
$$\psi(a) + \psi(d) = \psi(b) + \psi(c)$$

we also have

$$\psi(a) = \psi(b) = \psi(c) = \psi(d) = 0.$$

This procedure leads to another quadratic lattice which is now smaller and oblique, and can, therefore, be continued. In this manner we can obtain a lattice of points which is as dense as we please and at whose nodes ψ vanishes. Assuming continuity we have $\psi(\mathbf{v}) = 0$, q.e.d.

Making use of all points in eq. (19) we can easily extend the construction and the proof to three dimensions.

C. Equilibrium distributions

Equation (21) contains all distributions which are compatible with an entropy whose value changes only owing to a flow. It must, however, be realized that a, γ and $\mathbf{u}$ considered as functions of $\mathbf{r}$ and t in eq. (21) cannot be arbitrary if they are to be compatible with the Maxwell-Boltzmann equation. Moreover, it follows from eq. (41.18) that

$$(22) \qquad \frac{\partial f}{\partial t} + \mathbf{v}\,\frac{\partial f}{\partial \mathbf{r}} + \frac{1}{m}\,\mathbf{F}\,\frac{\partial f}{\partial \mathbf{v}} = 0.$$

Introducing eq. (20 a) we can deduce the following equations from (22):

$$(23) \qquad \begin{aligned} &\operatorname{grad}\gamma = 0, \qquad \frac{\partial u_i}{\partial k} + \frac{\partial u_k}{\partial i} = \frac{\dot{\gamma}}{\gamma}\,\delta_{ik}; \\ &\dot{\alpha} = -(\mathbf{u}\,V)\,(\alpha - \gamma\,\mathbf{u}^2) - \dot{\gamma}\,\mathbf{u}^2 \\ &\mathbf{F} = m\left\{\frac{\partial \mathbf{u}}{\partial t} + \frac{1}{2\gamma}\,\operatorname{grad}(\alpha - \gamma\,\mathbf{u}^2) + \frac{\dot{\gamma}}{\gamma}\,\mathbf{u}\right\} \end{aligned}$$

in which i and k represent x, y, or z.

The first two lead to the following form:

$$(24) \qquad \gamma = \gamma(t), \qquad \mathbf{u} = \frac{\dot{\gamma}}{2\gamma}\,\mathbf{r} + \mathbf{a}(t) \times \mathbf{r} + \mathbf{b}(t)$$

Since $\mathbf{u}$ denotes the mean local velocity, eq. (24) is seen to represent a special superposition of a *translation*, a *rotation*, and a radial *expansion*. The whole motion remains *isotropic*, because during time δt, $\mathbf{r}$ transforms to $\mathbf{r} + \mathbf{u}\,\delta t$ and $d\mathbf{r} \to d\mathbf{r} + d\mathbf{u}\,\delta t$, so that

$$d\mathbf{r}^2 \to d\mathbf{r}^2 + 2(d\mathbf{r}\,d\mathbf{u})\,\delta t,$$

if the term with δt^2 is dropped. According to eq. (24), we have

$$d\mathbf{u} = \frac{\dot{\gamma}}{2\gamma}\,d\mathbf{r} + \mathbf{a} \times d\mathbf{r},$$

and

$$(25) \qquad d\mathbf{r}^2 \to \left(1 + \frac{\dot{\gamma}}{\gamma}\,\delta t\right) d\mathbf{r}^2,$$

so that all the distances are seen to vary in the same ratio.

The last eq. (23) determines the field of forces in which a local Maxwellian distribution may occur. In particular, when $\mathbf{u} = 0$, eqs. (23) and (24) lead to

(26) $$\alpha = \alpha(\mathbf{r}), \quad \gamma = \text{const}, \quad \mathbf{b} = 0, \quad \mathbf{a} = 0,$$

and in particular

(27) $$\mathbf{F} = +\frac{m}{2\gamma} \operatorname{grad} \alpha.$$

Thus, in the absence of a local velocity of flow, thermal equilibrium can exist only in potential fields which do not vary with time (*cf.* the barometric formula).

43. Fundamental equations of fluid dynamics

A. Series expansion for the distribution function

In order to evaluate the collision integral in eq. (41.18) it is necessary to know the distribution function f which results from the solution of the collision equation. It will differ from the local Maxwellian distribution (42.21), because we are not dealing with a state of equilibrium. However, the deviations from the equilibrium distribution are, generally speaking, small. For this reason it is useful to begin the construction of f with the local Maxwellian distribution. Without any essential loss of generality we can put

(1) $$f = \left(1 + a_k \frac{\partial}{\partial \xi_k} + a_{kl} \frac{\partial^2}{\partial \xi_k \partial \xi_l} + a_{klm} \frac{\partial^3}{\partial \xi_k \partial \xi_l \partial \xi_m} + \cdots \right) f_0$$

because, in essence, f/f_0 is an expansion in Hermite polynomials in three variables, i. e. a complete system (*cf.* Vol. VI), and we may expect that the coefficients of the expansion decrease rapidly, if the deviations are small.

The subscripts $k, l, m \ldots$ in eq. (1) represent the co-ordinates x, y, z. A summation is implied whenever identical indices occur, so that, for example, the second term denotes the sum

$$a_x \frac{\partial f_0}{\partial \xi_x} + a_y \frac{\partial f_0}{\partial \xi_y} + a_z \frac{\partial f_0}{\partial \xi_z}.$$

The same convention applies to the higher-order terms. Generally speaking, the coefficients $a_k, a_{kl}, a_{klm}, \ldots$ depend on $\mathbf{r}$ and t, but they are independent of $\mathbf{v}$ by definition. They form tensors of the first, second, and higher orders, and may be assumed to be symmetrical in all subscripts.

The expansion of f in eq. (1) becomes unique only if the coefficients a, α, γ, and $\mathbf{u}$, which are independent of the velocity, are defined with the aid of the local Maxwellian distribution

(2) $$f_0 = e^{\alpha - \gamma(\mathbf{v}-\mathbf{u})^2} = a\, e^{-\gamma(\mathbf{v}-\mathbf{u})^2}.$$

This follows from the requirement that it should be possible to evaluate particular integrals exactly with the aid of f_0 alone. Thus the *particle density* is:

(3 a) $$n = \int f\, d\xi = \int f_0\, d\xi = a\, (\pi/\gamma)^{3/2},$$

the *mean velocity* is:

(3 b) $$\overline{\mathbf{v}} = \frac{1}{n} \int \mathbf{v} f\, d\xi = \frac{1}{n} \int \mathbf{v} f_0\, d\xi = \mathbf{u},$$

and the *mean, isotropic, thermal-kinetic pressure* from eq. (22.3 a) becomes:

(3 c) $$p = \frac{m}{3} \int (\mathbf{v}-\overline{\mathbf{v}})^2 f\, d\xi = \frac{m}{3} \int (\mathbf{v}-\overline{\mathbf{v}})^2 f_0\, d\xi = \frac{n\, m}{2\, \gamma}.$$

It represents the pressure which would be exerted on the walls of a small volume moving with the mean velocity.

Equation (3 a) determines the factor a in eq. (2). If we define the temperature T by putting

(4) $$\gamma = m/2\, k\, T$$

and if we use $\rho = n\, m$ as the *mass density*, we obtain

(4 a) $$a = \frac{\rho}{m} \left(\frac{m}{2\pi\, k\, T} \right)^{3/2}.$$

The two remaining equations yield conditions for the coefficients in the expansion (1). The i-th component from eq. (3 b) gives

$$\int \xi_i \left(1 + a_k \frac{\partial}{\partial \xi_k} + \ldots \right) f_0\, d\xi = \int \xi_i f_0\, d\xi.$$

Since f_0 vanishes together with all its derivatives on the sphere at infinity in the velocity space, we may perform as many partial integrations as we like, without having to retain surface integrals. Thus we have

$$\int \xi_i f_0\, d\xi - a_i \int f_0\, d\xi = \int \xi_i f_0\, d\xi.$$

The left-hand side contains only two terms. Since already the first term is identical with that on the right-hand side, we have

(5) $$a_i = 0.$$

In a similar way, inserting f from eq. (1) and integrating by parts, we obtain

$$\int (\xi_i - \bar{\xi}_i)^2 f_0 \, d\xi - 2 a_i \int (\xi - \bar{\xi}_i) f_0 \, d\xi + a_{ik} \delta_{ik} \int f_0 \, d\xi = \int (\xi - \bar{\xi}_i)^2 f_0 \, d\xi.$$

The first term is cancelled by the first term on the right-hand side, and according to eqs. (3 b) and (5) the second term is identically equal to zero. The third term gives $a_{ik} \delta_{ik} = 0$, where

$$\delta_{ik} = \begin{cases} 1 \text{ for } i = k \\ 0 \text{ for } i \neq k \end{cases}$$

denotes the Kronecker symbol. Thus the sum of the diagonal elements of a_{ik} must vanish, i. e.

(6) $$a_{jj} = 0.$$

In the following, eq. (1) is replaced by

(7) $$f = \left(1 + \frac{1}{2\rho} \sigma_{kl} \frac{\partial^2}{\partial \xi_k \partial \xi_l} - \frac{1}{6\rho} Q_{klm} \frac{\partial^3}{\partial \xi_k \partial \xi_l \partial \xi_m} + \right. \\ \left. + \frac{1}{24\rho} R_{klmn} \frac{\partial^4}{\partial \xi_k \partial \xi_l \partial \xi_m \partial \xi_n} + \ldots \right) f_0.$$

The coefficients in the expansion have now been denoted by symbols which will prove convenient later. In conclusion we can deduce an additional condition from (6): The trace of the tensor σ_{kl} vanishes, or

(7') $$\sigma_{jj} = 0.$$

B. Maxwell's transport equation

A moment is defined here as the local mean value of a power of velocity calculated in accordance with eq. (41.5). For example

$$\overline{\xi_i \xi_k} = \frac{1}{n} \int \xi_i \xi_k f \, d\xi.$$

In the following the moments of local relative velocity are more important than the former:

(8) $$\mathbf{c} = \mathbf{v} - \bar{\mathbf{v}} = \mathbf{v} - \mathbf{u}, \qquad c_i = \xi_i - \bar{\xi}_i = \bar{\xi}_i - u_i.$$

Putting $dc = d\xi$, we can write the second-order moments as

$$\overline{c_i c_k} = \frac{1}{n} \int c_i c_k f \, dc.$$

According to (8), the first-order moment, the mean values, can be written:

(9) $$\bar{c}_i = \int (\xi_i - \bar{\xi}_i) f \, dc = 0.$$

The advantage of the expansion in eq. (1) or (7) consists in the fact that a finite number of coefficients in the expansions is always sufficient to calculate a moment, the number being n for a moment of n-th order.

All moments can be reduced to those calculated with the aid of the Maxwellian distribution. For an arbitrary velocity function $\phi(\mathbf{c})$, we shall make use of eq. (41.5):

(10) $$\bar{\phi} = \frac{1}{n} \int f \phi \, d\xi \quad \text{together with} \quad \bar{\phi}^0 = \frac{1}{n} f_0 \phi \, d\xi.$$

On integrating by parts, we have[1]

(9 a) $$\overline{\phi(\mathbf{c})} = \bar{\phi}^0 + \frac{1}{2\rho} \sigma_{kl} \frac{\partial^2 \bar{\phi}^0}{\partial c_k \, \partial c_l} + \frac{1}{6\rho} Q_{klm} \frac{\partial^3 \bar{\phi}^0}{\partial c_k \, \partial c_l \, \partial c_m} + \cdots$$

All odd moments calculated with the aid of the Maxwellian distribution vanish since the former is symmetrical with respect to the origin ($\mathbf{c} \to -\mathbf{c}$). Furthermore, since the Maxwellian distribution is invariant with respect to mirror reflections and rotations we have

(10 a) $$\overline{c_i c_k}^0 = \frac{p}{\rho} \delta_{ik}, \qquad \overline{c_i c_j c_k c_l}^0 = \frac{p^2}{\rho^2} (\delta_{il} \delta_{jk} + \delta_{jl} \delta_{ki} + \delta_{kl} \delta_{ij}),$$

---

[1] This transformation can lead to difficulties in the higher-order terms if factors $c = \sqrt{\overline{c_i^2}}$ occur, because the integrals may diverge, in spite of the fact that the integral (10) is convergent. In such cases it is necessary to abandon the integration by parts or to take the "finite parts" of the integrals (cf. Laurent Schwartz, "Théorie des distributions", Hermann & Cie., Paris, 1950.)

except for numerical factors. These can be calculated from the special integrals

$$\overline{c_z^{2^0}} = \sqrt{\left(\frac{\gamma}{\pi}\right)^3} \int c^2\, \zeta^2\, e^{-\gamma c^2} c^2\, dc\, d\xi\, d\phi = \frac{1}{2\gamma} = \frac{p}{\rho},$$

$$\overline{c_x^2 c_y^{2^0}} = \overline{c_x^{2^0}} \cdot \overline{c_y^{2^0}} = (\overline{c_z^{2^0}})^2 = \frac{p^2}{\rho^2}.$$

In accordance with eq. (9 a), the mean moments are:

(9 b) $\qquad \overline{c_i c_k} = \dfrac{p}{\rho}\, \delta_{ik} + \dfrac{1}{\rho}\, \sigma_{ik}, \qquad \overline{c_i c_k c_l} = \dfrac{1}{\rho}\, Q_{ikl},$

$$\overline{c_i c_j c_k c_l} = \frac{p^2}{\rho^2}(\delta_{il}\delta_{jk} + \delta_{jl}\delta_{ki} + \delta_{kl}\delta_{ij}) + \frac{p}{\rho^2}[(\sigma_{il}\delta_{jk} + +) + (\sigma_{jk}\delta_{il} + +)] + \frac{1}{\rho} R_{ijkl},$$

the remaining following from these by cyclic transposition.

Generally speaking the moments vary with time and position. They satisfy characteristic equations which are consequences of the collision equation (41.18). Multiplying this by $\phi(\mathbf{v})$ and integrating over the velocity space we obtain directly the *transport* (or *transfer*) equation for the quantity $\phi(\mathbf{v})$:

(11) $\qquad \dfrac{\partial}{\partial t}(\rho\, \overline{\phi}) + \operatorname{div}(\rho\, \overline{\phi\, \mathbf{v}}) - \mathbf{f}\left(\overline{\dfrac{\partial \phi}{\partial \mathbf{v}}}\right) = J(\phi),$

where

(12) $\qquad \rho = m n, \quad \mathbf{f} = \dfrac{\rho}{m}\, \mathbf{F} = n\, \mathbf{F}$

denote the mass and force density respectively; $\rho\, \overline{\phi}$ and $\rho\, \overline{\phi\, \mathbf{v}}$ represent the density and the flux of the quantity ϕ; $J(\phi)$ denotes the integral

(12 a) $\qquad J(\phi) = \dfrac{m s^2}{2} \int \phi(\mathbf{v})\, (f' f_1' - f f_1) \cdot |\mathbf{V}\, \mathbf{e}|\, d\omega\, d\xi\, d\xi_1$

which can be transformed to

(13) $\qquad J(\phi) = \dfrac{m s^2}{8} \int (\phi + \phi_1 - \phi' - \phi_1')\, (f' f_1' - f f_1)\, |\mathbf{V}\, \mathbf{e}|\, d\omega\, d\xi\, d\xi_1,$

in accordance with (42.13). We shall call it the *collision* moment of the quantity $\phi(\mathbf{v})$.

Substituting the additive collision variables from eq. (42.18) into eq. (11), we find that the right-hand side vanishes. All additive collision invariants have been given in eq. (42.18). Accordingly, we obtain five equations which correspond to the five conservation laws: those for mass, energy and the three components of momentum. They are:

$$\frac{\partial}{\partial t}(\rho\,\overline{\psi}) + \operatorname{div}(\rho\,\overline{\psi\,\mathbf{v}}) = n\,\mathbf{F}\left(\overline{\frac{\partial \psi}{\partial \mathbf{v}}}\right). \tag{14}$$

$J(\psi) = 0$ signifies that the mass, energy and momentum do not change *on collision*. The momentum need not be absolutely constant, since we admit an external field of forces.

C. Conservation of mass

This follows from $\psi = \psi_0 = 1$. Substituting this expression into eq. (14), we obtain

$$\frac{\partial \rho}{\partial t} + \operatorname{div}(\rho\,\mathbf{u}) = 0 \tag{15}$$

in view of (3 b). This is the familiar (Vol. II, eq. (5.4′)) equation of continuity of fluid dynamics. Its validity is more general than would appear from the assumptions required in the derivation of Boltzmann's equation.

The physical significance of this equation becomes clear on taking an integral over an arbitrary, finite volume V. The expression

$$\int \rho\,dx = M$$

gives the total mass enclosed by the volume V, so that the first term in eq. (15) leads to dM/dt. The integral over the second term can be simplified with the aid of Stokes' theorem:

$$\int \operatorname{div}(\rho\,\mathbf{u})\,dx = \int \rho\,u_n\,d\sigma,$$

where $d\sigma$ denotes an element on the surface area O of volume V; u_n is the component of the mean velocity in the direction of the outward normal to the surface with its positive direction outwards, and $\rho\,u_n\,d\sigma$ gives the mass flow through the element $d\sigma$. It is positive when the flow is outwards, and negative when the flow is in the opposite direction. The integral must be extended over the whole surface O of V.

Hence, eq. (15) yields

$$\text{(15 a)} \qquad -\frac{dM}{dt} = -\frac{d}{dt}\int \rho\, dx = \int \rho\, u_n\, d\sigma.$$

The decrease in mass in V is equal to the flow of mass outwards. The density of mass flow (flux) is

$$\text{(15 b)} \qquad \mathbf{s} = \rho\, \mathbf{u};$$

it is determined by the transport of mass with the mean velocity **u**.

D. CONSERVATION OF MOMENTUM

In order to derive the momentum equation it is necessary to write eq. (14) in component form:

$$\text{(14 a)} \qquad \frac{\partial}{\partial t}(\rho\, \overline{\psi}) + \frac{\partial}{\partial \xi_k}(\rho\, \overline{\psi\, \xi_k}) = f_k \left(\overline{\frac{\partial \psi}{\partial \xi_k}} \right)$$

Substituting $\psi = \psi_i = \xi_i$ in accordance with eq. (42.18), we obtain

$$\text{(16)} \qquad \frac{\partial}{\partial t}(\rho\, u_i) + \frac{\partial}{\partial \xi_k}(\rho\, \overline{\xi_i\, \xi_k}) = f_i.$$

The first term can be transformed with the aid of the equation of continuity, giving

$$\frac{\partial}{\partial t}(\rho\, u_i) = \rho\, \frac{\partial u_i}{\partial t} + u_i\, \frac{\partial \rho}{\partial t} = \rho\, \dot{u}_i - u_i\, \frac{\partial}{\partial \xi_k}(\rho\, u_k)$$

$$= \rho \left(\frac{\partial}{\partial t} + u_k\, \frac{\partial}{\partial \xi_k} \right) u_i - \frac{\partial}{\partial \xi_k}(\rho\, u_i\, u_k).$$

Thus eq. (16) now assumes the following form:

$$\text{(17)} \qquad \rho\, \frac{du_i}{dt} \equiv \rho \left(\frac{\partial}{\partial t} + u_k\, \frac{\partial}{\partial \xi_k} \right) u_i = -\frac{\partial}{\partial \xi_k}\left[\rho\, (\overline{\xi_i\, \xi_k} - u_i\, u_k) \right] + f_i,$$

where $d/dt = (\partial/\partial t + u_k\, \partial/\partial \xi_k)$ denotes the substantive derivative referred to a volume moving with the mean velocity (Vol. II, eq. (11.3)).

According to eq. (9 b) the expression in the square bracket becomes

$$\rho\, (\overline{\xi_i\, \xi_k} - u_i\, u_k) = \rho\, \overline{(\xi_i - u_i)(\xi_k - u_k)} = \rho\, \overline{c_i\, c_k} = p\, \delta_{ik} + \sigma_{ik}.$$

Reverting to vector notation we see that eq. (17) leads rigorously to

(18) $$\rho \frac{d\mathbf{u}}{dt} \equiv \rho \left(\frac{\partial}{\partial t} + \mathbf{u}\, \nabla\right) \mathbf{u} = -\operatorname{grad} p - \operatorname{Div} \sigma + \mathbf{f}.$$

The tensorial divergence $\operatorname{Div} \sigma$ is a vector whose components are

(18 a) $$(\operatorname{Div} \sigma)_i = \partial \sigma_{ik}/\partial k.$$

Equation (18) is identical with the equations of motion of fluid dynamics. Since $\sigma_{jj} = 0$, σ_{ik} represents a stress tensor which leads to shearing stresses only (Vol. II, Sec. 10). The assumption

(19 a) $$\sigma_{ik} = 0$$

leads to Euler's equation. Putting

(19 b) $$\sigma_{ik} = -\eta \left(\frac{\partial u_i}{\partial k} + \frac{\partial u_k}{\partial i} - \frac{2}{3} \frac{\partial u_j}{\partial j} \delta_{ik}\right),$$

we obtain the Navier-Stokes equations. We shall deduce approximately this form of σ_{ik} from the collision equation (see Sec. 44).

In the case of pure shear flow (Couette flow, Sec. 27) for which $\mathbf{u} = (u(y), 0, 0)$, we obtain eq. (27.4), namely

$$\sigma_{xy} = -\eta \frac{\partial u}{\partial y} = \sigma_{yx}.$$

Reciprocally, eq. (19 b) follows from (27.4) when we take into account the transformation properties for the rotation of the system of coordinates.

On integrating over a small (!) volume flowing with the mean velocity, we obtain the following equation which is analogous to eq. (15 a):

(20) $$\frac{d}{dt} \int \rho\, u_i\, dx = -\int (p\, n_i + \sigma_{in})\, d\sigma + \int f_i\, dx.$$

Here n_i denotes the i-th component of the unit vector $\mathbf{n}$ in the direction of the external normal, and $\sigma_{in} = \sigma_{ik} n_k$. The increase in momentum per unit time is composed of the flow of momentum through the surface (from outside inwards because of the negative sign) and the total force acting on the volume which results from the force density $\mathbf{f}$.

E. Conservation of Energy

The energy equation is obtained by putting $\psi = \tfrac{1}{2}\psi_4 = \tfrac{1}{2}v^2$. In this case eq. (14) becomes

$$(21) \qquad \frac{\partial}{\partial t}\left(\frac{1}{2}\rho \overline{v^2}\right) + \operatorname{div}\left(\frac{1}{2}\rho \overline{v^2 \cdot v}\right) = \overline{f \cdot v}.$$

According to eqs. (8) and (10)

$$\overline{v^2} = \overline{(c+u)^2} = \overline{c^2} + u^2 = \frac{3p}{\rho} + u^2$$

so that the first term of (21) becomes:

$$\frac{\partial}{\partial t}\left(\frac{3}{2}p + \frac{\rho}{2}u^2\right) = \frac{3}{2}\dot p + \frac{\dot \rho}{2}u^2 + \rho u \frac{\partial u}{\partial t}.$$

Transforming with the aid of the continuity equation (15) and the momentum equation (18), we obtain:

$$\frac{3}{2}\dot p - \frac{1}{2}u^2 \operatorname{div}(\rho u) - u\left[\rho(u\nabla)u + \operatorname{grad} p - \operatorname{Div}\sigma - f\right]$$

or

$$(22) \qquad \frac{\partial}{\partial t}\left(\frac{\rho}{2}\overline{v^2}\right) = \frac{3}{2}\dot p - (u\nabla)p - \operatorname{div}\left(\frac{\rho}{2}u^2 \cdot u + \sigma \times u\right) + \sigma\varepsilon + u f.$$

Here the product $\sigma \times u$ of the tensor σ with the vector u is a vector whose components are:

$$(22\text{ a}) \qquad (\sigma \times u)_i = \sigma_{ik} u_k;$$

ε denotes that part of the strain tensor of the flow (Vol. II, Sec. 1) which relates to shear:

$$(22\text{ b}) \qquad \varepsilon = \varepsilon_{kl} = \frac{1}{2}\left(\frac{\partial u_l}{\partial k} + \frac{\partial u_k}{\partial l} - \frac{2}{3}\frac{\partial u_j}{\partial j}\delta_{kl}\right)$$

and $\sigma\varepsilon$ is the scalar product of the two tensors σ and ε, defined as

$$(22\text{ c}) \qquad \sigma\varepsilon = \sigma_{kl}\varepsilon_{kl} = \sigma_{kl}\frac{\partial u_l}{\partial k}.$$

The equivalence of the last two expressions is a consequence of the symmetry $\sigma_{lk} = \sigma_{kl}$ subject to the trace condition that $\sigma_{jj} = 0$.

Correspondingly we can transform the mean value in the second term of (21), and we obtain

(23) $\quad \overline{\mathbf{v}^2 \cdot \mathbf{v}} = \overline{(\mathbf{c}+\mathbf{u})^2 \cdot (\mathbf{c}+\mathbf{u})} = \overline{\mathbf{c}^2 \cdot \mathbf{c}} + \overline{\mathbf{c}^2} \cdot \mathbf{u} + 2\,\overline{(\mathbf{c}\,|\,\mathbf{c})} \times \mathbf{u} + \mathbf{u}^2 \cdot \mathbf{u}.$

Writing the components of this term in accordance with eq. (9 b), we find

$$\overline{c_j{}^2\, c_i} = \frac{1}{\rho} Q_{jji}.$$

Introducing the vector $\mathbf{Q}$ with the components

(24) $\quad\quad\quad\quad\quad\quad\quad\quad Q_i = \frac{1}{2} Q_{jji}$

we have

(24 a) $\quad\quad\quad\quad\quad\quad\quad \overline{\mathbf{c}^2 \cdot \mathbf{c}} = \frac{2}{\rho} \mathbf{Q}.$

By eq. (9 b), the second term becomes

(24 b) $\quad\quad\quad\quad\quad\quad\quad \overline{\mathbf{c}^2} \cdot \mathbf{u} = \frac{3\,p}{\rho}\, \mathbf{u}.$

In the third term $\overline{(\mathbf{c}\,|\,\mathbf{c})}$ denotes the tensor $\overline{c_i\,c_k}$, so that eq. (9 b) gives

(24 c) $\quad\quad\quad\quad\quad 2\overline{(\mathbf{c}\,|\,\mathbf{c})} \times \mathbf{u} = \frac{2\,p}{\rho}\mathbf{u} + \frac{2}{\rho}\sigma \times \mathbf{u}.$

The fourth term need not be transformed.

Substituting eqs. (22) and (23) into (21) and making use of eqs. (24) to (24 c), we obtain

$$\frac{3}{2} \dot p - (\mathbf{u}\,\nabla)\, p + \operatorname{div}\left[\mathbf{Q} + \frac{5}{2} p\,\mathbf{u}\right] + \sigma\varepsilon + \mathbf{f}\mathbf{u} = \mathbf{f}\mathbf{u}$$

An elementary transformation gives

(25) $\quad\quad \dfrac{3}{2}\dfrac{dp}{dt} = \dfrac{3}{2}\left(\dfrac{\partial}{\partial t} + \mathbf{u}\,\nabla\right) p = -\operatorname{div}\mathbf{Q} - \sigma\varepsilon - \dfrac{5}{2} p\,\operatorname{div}\mathbf{u}.$

Interpreting $\mathbf{Q}$ defined in eq. (24 a) as the *heat flux* (flow of energy in the moving element), and introducing the *density of internal energy* (kinetic energy in the moving system of co-ordinates)

(24 d) $\quad\quad\quad\quad\quad\quad Q = \dfrac{1}{2}\rho\, \overline{\mathbf{c}^2} = \dfrac{3}{2} p,$

we can rearrange eq. (25) to read

(26) $$\frac{dQ}{dt} + \text{div } \mathbf{Q} = \left(\frac{\partial}{\partial t} + u\,V\right)Q + \text{div }\mathbf{Q} = -Q\,\text{div }\mathbf{u} - p\,\text{div }\mathbf{u} - \sigma\,\varepsilon.$$

This equation can be transformed in two ways. Transposing the first term on the left-hand side to the right-hand side, we have

(27) $$\frac{\partial Q}{\partial t} + \text{div }(\mathbf{Q} + Q\,\mathbf{u}) = -p\,\text{div }\mathbf{u} - \sigma\,\varepsilon.$$

The right-hand side now contains the work of compression and that due to shearing forces (energy dissipation due to friction, see Sec. 44). The divergence on the left-hand side operates on the term representing the local change in Q due to heat conduction $\mathbf{Q}$ and convection $Q\,\mathbf{u}$.

If, instead of the energy density, we now introduce the energy per unit mass q, we have $Q = \rho\,q$, and hence

$$\frac{\partial Q}{\partial t} = \rho\frac{\partial q}{\partial t} + q\frac{\partial \rho}{\partial t} = \rho\frac{\partial q}{\partial t} - q\,\text{div }(\rho\,\mathbf{u}) = \rho\frac{dq}{dt} - \text{div }(\rho\,q\,\mathbf{u}),$$

and it follows from (27) that

(28) $$\rho\frac{dq}{dt} + \text{div }\mathbf{Q} = \rho\left(\frac{\partial}{\partial t} + \mathbf{u}\,V\right)q + \text{div }\mathbf{Q} = -p\,\text{div }\mathbf{u} - \sigma\,\varepsilon.$$

The right-hand side has the same meaning as before. The left-hand side now contains the flow of heat due to conduction, because from the macroscopic standpoint we are now concerned with a definite element of mass and observe it as it moves.

F. Entropy theorem

We now recall the definition of entropy and entropy flux in eqs. (42.1) and (42.5), as well as eq. (42.14) in which the distribution of entropy sources, G, defined in eq. (42.12) is essentially positive. Let η denote the entropy per unit mass,[1] so that

$$H = \rho\,\eta.$$

---

[1] The symbol η in this Section should not be confused with the viscosity, η, used elsewhere in the present Chapter (*Transl.*).

43. 30a — FUNDAMENTAL EQUATIONS OF FLUID DYNAMICS

Taking into account the equation of continuity (15) we conclude from eq. (42.14) that

$$\dot{H} = \dot{\rho}\,\eta + \rho\,\dot{\eta} = \rho\,\dot{\eta} - \eta\,\mathrm{div}\,(\rho\,\mathbf{u})$$

$$= \rho\left(\frac{\partial}{\partial t} + \mathbf{u}\,\nabla\right)\eta - \mathrm{div}\,(\rho\,\eta\,\mathbf{u}).$$

Thus eq. (42.14) can be written in the form

(29) $$\rho\,\frac{d\eta}{dt} = \rho\left(\frac{\partial}{\partial t} + u\,\nabla\right)\eta = -\mathrm{div}\,(\mathbf{S} - \rho\,\eta\,\mathbf{u}) + G.$$

In this case the quantity of entropy $H \cdot \mathbf{u}$ transferred by convection must be subtracted from $\mathbf{S}$. According to (42.1) and (42.5), we have

(29 a) $$\mathbf{S} - H\,\mathbf{u} = -k\int \mathbf{c}\,\log f \cdot f\,d\xi.$$

We now proceed to calculate approximations for $\mathbf{S} - H\,\mathbf{u}$ and H. We assume that f differs only little from f_0, so that we may put

$$\log f \approx \log f_0.$$

Thus in view of eq. (42.21), we have

$$\log f \approx \log a - \gamma\,c^2.$$

Hence, by eq. (42.1) we find the following expression for H

$$H = -k\,n\,\log a + k\gamma\int \mathbf{c}^2 f\,d\xi\ .$$

or, according to (3 c) and (4 a)

$$H = -\frac{k}{m}\rho\,\log\rho + \frac{3}{2}\frac{k}{m}\rho\,\log T + \mathrm{const} \times \rho.$$

It follows that the entropy per unit mass (see eq. (5.10)) can be written:

(30) $$\eta = -\frac{k}{m}\left(\log\rho - \frac{3}{2}\log T\right) + \mathrm{const}$$

or, taking the substantive derivative

(30 a) $$\frac{d\eta}{dt} = -\frac{k}{m}\left(\frac{1}{\rho}\frac{d\rho}{dt} - \frac{3}{2}\frac{1}{T}\frac{dT}{dt}\right).$$

Multiplying by T, we find

$$T \frac{d\eta}{dt} = \frac{dq}{dt} - \frac{n k T}{\rho^2} \frac{d\rho}{dt},$$

because $q = (3/2) n k T/\rho = (3/2) (k/m) T$. Written in differential form with $p = n k T$, we see that the last equation becomes

(31) $$T \, d\eta = dq + p \, d\left(\frac{1}{\rho}\right),$$

which is the Second Law with dq denoting the differential of internal energy. It is directly linked with the assumption that $\log f \approx \log f_0$ or, in other words, that the flow is not too far removed from the equilibrium distribution.

The last requirement sounds like that in the definition of reversible processes. But, in fact, it is less stringent and allows G in eq. (29) to be different from zero. This follows from the fact that we have only postulated the equality of $\log f$ and $\log f_0$ which both vary very slowly for large arguments. The leeway left by this requirement has already been discussed in Sec. 21 F. It could be justified on the basis of a more exact solution of the Maxwell-Boltzmann collision equation. We shall refrain from doing this here, referring the reader to published papers.[1]

The value of G can be calculated from eq. (29). Substituting Maxwell's expression for $\log f$ into (29 a), we have

$$\mathsf{S} - H \mathbf{u} = k \gamma \int \mathbf{c}^2 \cdot \mathbf{c} f \, d\xi,$$

or, in view of eq. (24 a)

$$n k \gamma \cdot \frac{2}{\rho} \mathbf{Q} = \frac{\mathbf{Q}}{T}.$$

We thus obtain the thermodynamically plausible result:

(32) $$\mathsf{S} - H \mathbf{u} = \frac{\mathbf{Q}}{T}.$$

Substituting this expression, as well as eq. (31), into (21), we find that

$$G = -\frac{k}{m} \frac{d\rho}{dt} + \frac{1}{T}\left(\rho \frac{dq}{dt} + \mathrm{div}\, \mathbf{Q}\right) - \frac{1}{T^2} (\mathbf{Q}\, V) \, T.$$

---

[1] D. Enskog, Phys. Z. 12 (1911) 533; J. Meixner, Z. Phys. Chemie 53 (1943) 235.

The middle term can now be transformed with the aid of the energy equation (28). Since $p/T = nk = (k/m)\rho$, we have

$$G = -\frac{1}{T^2}(\mathbf{Q}\cdot\text{grad } T) - \frac{1}{T}(\Pi\,\sigma) - \frac{k}{m}\left(\frac{d\rho}{dt} + \rho\,\text{div } \mathbf{u}\right).$$

The last term vanishes (equation of continuity). Thus

(33) $$G = -\frac{1}{T^2}(\mathbf{Q}\cdot\text{grad } T) - \frac{1}{T}(\sigma\,\varepsilon).$$

This is the fundamental relation on which irreversible thermodynamics is based and in this connection reference may be made to Sec. 21. Making use of eq. (19 b) and of Fourier's hypothesis (21.3) for heat conduction

(33 a) $$\mathbf{Q} = -\varkappa\,\text{grad } T,$$

which, incidentally, can be justified on the ground of kinetic theory in the same way as eq. (19 c), Sec. 44, we obtain, on inserting into eq. (33), that

(33 b) $$G = +\frac{\varkappa}{T^2}(\text{grad } T)^2 + \frac{\eta}{T}\varepsilon^2 \geqslant 0$$

where G is seen to be essentially positive.

44. On the integration of the collision equation

A. Integration with the aid of moment equations

Numerous approximate methods have been developed for the integration of the Maxwell-Boltzmann collision equation (41.18). Concerning the details of the theory of integration reference may be made to the comprehensive review by K. F. Herzfeld[1] and to the paper by H. Grad[2], already quoted. From among the various approximation we shall make use of only those which are consistent with the expansion (43.1) in terms of the derivatives of the Maxwellian distribution. Moreover, we shall carry the development only far enough to exhibit the systematic character of the method and to justify the relations in eqs. (43.19 b) and (43.33 a) which lead to the Navier-Stokes equations and to the heat conduction equation. In this way we are led to the moment method (see H. Grad[2]).

---

[1] "Freie Weglänge und Transporterscheinungen in Gasen", Hand- u. Jahrbuch d. Chem. Physik, Vol. III 2, Sec. IV, Leipzig, 1939.
[2] *l. c.* p. 288.

We can develop a large class of functions $g(\mathbf{v})$ in terms of the derivatives of Maxwell's distribution. If we put

$$\text{(1)} \qquad g(\mathbf{v}) = \left(A_0 + A_k \frac{\partial}{\partial \xi_k} + A_{kl} \frac{\partial^2}{\partial \xi_k \, \partial \xi_l} + \ldots \right) f_0,$$

we can find the coefficients by calculating the moments

$$n G_0 = \int g(\mathbf{v}) \, d\xi = n A_0;$$

$$\text{(2)} \qquad n G_k = \int \xi_k g(\mathbf{v}) \, d\xi = n(A_0 \overline{\xi_k}^{\,0} - A_k);$$

$$n G_{kl} = \int \xi_k \xi_l g(\mathbf{v}) \, d\xi = n(A_0 \overline{\xi_k \xi_l}^{\,0} - A_k \overline{\xi_l}^{\,0} - A_l \overline{\xi_k}^{\,0} + A_{kl})$$

etc. The mean values $\overline{\xi_k}^{\,0}, \overline{\xi_k \xi_l}^{\,0}, \ldots$ have been defined in eq. (43.10). The equations (2) constitute recurrence formulae for the coefficients of the expansion. They give the following relations:

$$A_0 = G_0,$$

$$\text{(3)} \qquad A_k = G_0 \overline{\xi_k}^{\,0} - G_k,$$

$$A_{kl} = G_0 \overline{\xi_k \xi_l}^{\,0} - G_k \overline{\xi_l}^{\,0} - G_{el} \overline{\xi_k}^{\,0} + G_{kl}, \quad \text{etc.}$$

According to these equations the coefficients in the expansions of two functions are identical if their moments are identical. We now apply this proposition to the collision equation which is valid when all equations for moments (43.11) are satisfied. Instead of solving the collision equation we can integrate all moment equations. The equations for moments constitute a suitable starting point for approximations.

We have already considered the first equations for moments in Sec. 43, that is all for which the contribution from the collision integral vanishes. In addition we now proceed to consider the moment equations for $\phi = \xi_i \xi_k$ and $\phi = \xi_i \xi_j \xi_k$. From eq. (43.11) we obtain:

$$\text{(4)} \qquad \frac{\partial}{\partial t}(\rho \, \overline{\xi_i \xi_k}) + \frac{\partial}{\partial t}(\rho \, \overline{\xi_i \xi_k \xi_l}) - (\overline{\xi_i} f_k + \overline{\xi_k} f_i) = J_{ik},$$

and

$$\text{(5)} \qquad \frac{\partial}{\partial t}(\rho \, \overline{\xi_i \xi_j \xi_k}) + \frac{\partial}{\partial t}(\rho \, \overline{\xi_i \xi_j \xi_k \xi_l}) - (\overline{\xi_j \xi_k} f_i + \overline{\xi_k \xi_i} f_j + \overline{\xi_i \xi_j} f_k) = J_{ijk}.$$

The right-hand sides contain the *collision moments*

(4 a) $$J_{ik} = \frac{1}{2} m s^2 \int \xi_i \xi_k (f' f_1' - f f_1) |\mathbf{V} \mathbf{e}| d\omega d\xi d\xi_1$$

and

(5 a) $$J_{ijk} = \frac{1}{2} m s^2 \int \xi_i \xi_j \xi_k (f' f_1' - f f_1) |\mathbf{V} \mathbf{e}| d\omega d\xi d\xi_1.$$

B. Transformation of the equations for moments

When calculating the mean values and collision moments in eqs. (4) and (5) we must insert a suitable approximation to the distribution function f. The simplest non-trivial approximation is obtained when we consider on both sides of the equation only the highest non-vanishing term. This means that in our approximation it is sufficient to use the Maxwell distribution on the left-hand side.

Referring partly to previous calculations, we replace the mean values of the powers of ξ by the following expressions:

(6)
$$\rho \overline{\xi_i} = \rho u_i, \qquad \rho \overline{\xi_i \xi_k} \approx \rho \overline{\xi_i \xi_k}^0 = p \delta_{ik} + \rho u_i u_k,$$

$$\rho \overline{\xi_i \xi_j \xi_k} \approx \rho \overline{\xi_i \xi_j \xi_k}^0 = p(u_i \delta_{jk} + u_j \delta_{ki} + u_k \delta_{ij}) + \rho u_i u_j u_k,$$

$$\rho \overline{\xi_i \xi_j \xi_k \xi_l} \approx \rho \overline{\xi_i \xi_j \xi_k \xi_l}^0 = \frac{p^2}{\rho} (\delta_{jk} \delta_{il} + +) + p[(\delta_{jk} u_i u_k + +) + (\delta_{il} u_j u_l + +)] + \rho u_i u_j u_k u_l.$$

Hence eq. (4) becomes:

$$\frac{\partial}{\partial t}(p \delta_{ik} + \rho u_i u_k) + \frac{\partial}{\partial l}[p(u_i \delta_{kl} + +) + \rho u_i u_k u_l] - [u_i f_k + u_k f_i] = J_{ik},$$

or, after a simple rearrangement:

$$J_{ik} = \left(\frac{dp}{dt} + \frac{5}{3} p \operatorname{div} \mathbf{u}\right) \delta_{ik} + p \left(\frac{\partial u_i}{\partial k} + \frac{\partial u_k}{\partial i} - \frac{2}{3} \frac{\partial u_l}{\partial l} \delta_{ik}\right)$$
$$+ [\dot{\rho} + \operatorname{div}(\rho \mathbf{u})] u_i u_k + u_i \left(\rho \frac{du_k}{dt} + \frac{\partial p}{\partial k} - f_k\right) + u_k \left(\rho \frac{dv_i}{dt} + \frac{\partial p}{\partial i} - f_i\right).$$

According to eqs. (43.15), (43.18) and (43.25) all terms on the right-hand side vanish, except the second, if it is taken into account that in the present approximation $\sigma_{ik} = 0$ and $Q_l = 0$. Introducing the strain tensor ε_{ik} from eq. (43.22 b), we have

(7) $$J_{ik} = 2 p\, \varepsilon_{ik}.$$

Equation (5) can be transformed in the same way. Introducing the mean values (6) we obtain first in shorthand notation

$$J_{ijk} = \frac{\partial}{\partial t}[p(u_i\,\delta_{jk} + +) + \rho\, u_i u_j u_k] + \frac{\partial}{\partial l}\left[\frac{p^2}{\rho}(\delta_{il}\,\delta_{jk} + +) + p(\delta_{jk} u_i u_l + +)\right.$$
$$\left. + p(\delta_{il} u_j u_k + +) + \rho\, u_i u_j u_k u_l\right] - \frac{p}{\rho}(f_i\,\delta_{jk} + +) - (f_i u_j u_k + +).$$

It is convenient to rearrange terms to obtain the following more lucid form

$$J_{ijk} = \left(\frac{dp}{dt} + p\,\mathrm{div}\,\mathbf{u}\right)(u_i\,\delta_{jk} + +) + \frac{p}{\rho}\left[\delta_{jk}\left(\rho\frac{du_i}{dt} - f_i + \frac{\partial p}{\partial i}\right) + +\right]$$
$$+ p\left[\delta_{jk}\frac{\partial}{\partial i}\left(\frac{p}{\rho}\right) + +\right] + [\dot{\rho} + \mathrm{div}\,(\rho\,\mathbf{u})]\, u_i u_j u_k$$
$$+ \left[u_j u_k\left(\rho\frac{du_i}{dt} - f_i + \frac{\partial p}{\partial i}\right) + +\right] + p\left[\left(\frac{\partial u_j}{\partial i} u_k + \frac{\partial u_k}{\partial i} u_j\right) + +\right].$$

Taking into account eqs. (43.15), (43.18) and (43.25) with $\sigma_{ik} = 0$, $Q_l = 0$ and $p/\rho = (k/m)\,T$, we find that

$$J_{ijk} = \frac{k\,p}{m}\left(\frac{\partial T}{\partial i}\,\delta_{jk} + +\right) + p\left[u_i\left(\frac{\partial u_k}{\partial j} + \frac{\partial u_j}{\partial k} - \frac{2}{3}\frac{\partial u_l}{\partial l}\,\delta_{jk}\right) + +\right],$$

or, according to eq. (4):

(8) $$J_{ijk} = \frac{k\,p}{m}\left(\frac{\partial T}{\partial i}\,\delta_{jk} + +\right) + (u_i J_{jk} + +).$$

We shall see later that J_{ik} and σ_{ik} are proportional. Since on the left-hand side of eqs. (4) and (5) we have assumed $\sigma_{ik} = 0$ we must drop J_{jk} in eq. (8) for reasons of consistency. Thus we obtain finally

(9) $$J_{ijk} = \frac{k\,p}{m}\left(\frac{\partial T}{\partial i}\,\delta_{jk} + \frac{\partial T}{\partial j}\,\delta_{ki} + \frac{\partial T}{\partial k}\,\delta_{ij}\right).$$

It follows for the trace that

(10) $$J_{jjk} = \frac{5\,k\,p}{m}\frac{\partial T}{\partial k}.$$

C. Evaluation of collision moments

These are given by eqs. (43.12 a) or (43.13) and vanish when f is replaced by the Maxwell distribution. Thus in this case it is necessary to take into account the correction terms. The lowest-order term in the expansion of the product $f f_1$ can be written

$$(11) \quad f f_1 = f_0 f_{01} + \frac{1}{2\rho} \sigma_{rs} \left(f_0 \frac{\partial^2 f_{01}}{\partial \xi_{1r} \partial \xi_{1s}} + f_{01} \frac{\partial f_0}{\partial \xi_r \partial \xi_s} \right) -$$

$$- \frac{1}{6\rho} Q_{mrs} \left(f_0 \frac{\partial^3 f_{01}}{\partial \xi_{1m} \partial \xi_{1r} \partial \xi_{1s}} + f_{01} \frac{\partial^3 f_0}{\partial \xi_m \partial \xi_r \partial \xi_s} \right) + \cdots$$

as seen from eq. (43.7). The collision moments are homogeneous and linear in σ_{rs} and Q_{mrs}, because the first term does not contribute to the collision moments, and the quadratic terms in σ_{sr} and Q_{mrs} occur only from the fourth order onwards, which we do not consider.

The moments in eqs. (4 a) and (5 a) are symmetric tensors in the same way as the coefficients σ_{ik} and Q_{ijk}. We can write down the form of the collision moments because no other tensors than σ_{ik}, Q_{ijk} and the unit tensor δ_{ik} play any part. Thus we must have

$$(12) \quad J_{ik} = a\, \sigma_{ik}$$

$$J_{ijk} = b\, Q_{ijk} + c(Q_{rrj}\, \delta_{jk} + Q_{rrj}\, \delta_{ki} + Q_{rrk}\, \delta_{ij}).$$

In the first eq. (12) the term proportional to δ_{jk} has been omitted because the factor which follows from homogeneity, $\sigma_{jj} = 0$.

Making use of eqs. (7) and (9) together with (12), we can now calculate the coefficients σ_{ik} and Q_{ijk} in the expansion. They are proportional to ε_{ik} and $\left(\frac{\partial T}{\partial i} \delta_{jk} + \frac{\partial T}{\partial j} \delta_{ki} + \frac{\partial T}{\partial k} \delta_{ij} \right)$. To be consistent with eqs. (43.19 b) and (43.33 a) we denote the coefficients of proportionality by

$$(13) \quad \sigma_{ik} = -2\eta\, \varepsilon_{ik}, \qquad Q_l = -\varkappa\, \frac{\partial T}{\partial l}.$$

The last equation, as seen from (43.24), is derived from the tensor

$$(13\text{ a}) \quad Q_{ijk} = -\frac{2\varkappa}{5} \left(\frac{\partial T}{\partial i} \delta_{jk} + \frac{\partial T}{\partial j} \delta_{ki} + \frac{\partial T}{\partial k} \delta_{ij} \right).$$

The viscosity, η, and the thermal conductivity, $\varkappa$, can be calculated from eqs. (12). Substituting eqs. (13) and (13 a) into it, we obtain two equations which must be identical with eqs. (7) and (9) respectively:

$$J_{ik} = -2\eta\, a\, \varepsilon_{ik} = 2\, p\, \varepsilon_{ik},$$

$$J_{ijk} = -\frac{2\varkappa\,(b+5\,c)}{5}\left(\frac{\partial T}{\partial i}\delta_{jk} + + \right) = \frac{k\,p}{m}\left(\frac{\partial T}{\partial i}\delta_{jk} + + \right).$$

It follows that

(14) $\quad \eta = -\dfrac{p}{a}, \qquad \varkappa = -\dfrac{5\,k\,p}{2\,m\,(b+5\,c)} = +\dfrac{5}{2}\dfrac{a}{b+5\,c}\dfrac{k}{m}\eta.$

Substituting the expressions (13) and (13 a) into (43.7), we find the distribution function

(15) $\quad f = f_0 - \dfrac{\eta}{\rho}\varepsilon_{ik}\dfrac{\partial^2 f_0}{\partial \xi_i\,\partial \xi_k} + \dfrac{\varkappa}{5\,\rho}\left(\dfrac{\partial T}{\partial i}\dfrac{\partial}{\partial \xi_i}\right)\dfrac{\partial^2 f_0}{\partial \xi_j{}^2}.$

Equations (13) and (13 a) agree with eqs. (43.19 b) and (43.33 a), respectively, and provide a justification for the Navier-Stokes equations as well as for the heat conduction equation. The source density of entropy (43.33) becomes, as it should, essentially positive on condition that $\varkappa$ and η are positive (cf. Sec. D), as already demonstrated in eq. (43.33 b).

D. Viscosity and thermal conductivity

It now remains to evaluate the integrals (4 a) and (5 a) starting with the form (43.13) in which the term with $f'f_1'$ can be transformed. We replace the variables $\mathbf{v}'$, $\mathbf{v}_1'$ with $\mathbf{v}$, $\mathbf{v}_1$, and notice that in accordance with eqs. (41.12 b) and (41.16 a) $d\xi\, d\xi_1$ and $|\mathbf{V}\,\mathbf{e}|$ remain unchanged; the first factor changes sign. In this way (43.13) can be replaced by

(17) $\quad J_\phi = +\dfrac{1}{4}m\,s^2\displaystyle\int (\phi' + \phi_1' - \phi - \phi_1)\,f f_1\,|\mathbf{V}\,\mathbf{e}|\,d\omega\,d\xi\,d\xi_1$

and the integration can be carried out in two steps.

Since the distribution functions f and f_1 are independent of the unit vector $\mathbf{e}$ we may split off the integral over $\mathbf{e}$:

(18) $\quad I_\phi = I_\phi(\mathbf{v},\mathbf{v}_1) = \dfrac{1}{4}m\,s^2\displaystyle\int (\phi' + \phi_1' - \phi - \phi_1)\cdot|\mathbf{V}\,\mathbf{e}|\,d\omega.$

We now consider the special value of I_ϕ for $\phi = \xi_i \xi_k$ and $\phi = \xi_i \xi_j \xi_k$, so that

(18 a) $\qquad I_{ik} = + \dfrac{1}{4} m s^2 \int (\xi_i' \xi_k' + \xi_{1i}' \xi_{1k}' - \xi_i \xi_k - \xi_{1i} \xi_{1k}) |\mathsf{V e}| d\omega$

and

(18 b) $\qquad I_{ijk} = + \dfrac{1}{4} m s^2 \int \xi_i' \xi_j' \xi_k' + \xi_{1i}' \xi_{1j}' \xi_{1k}' - \xi_i \xi_j \xi_k - \xi_{1i} \xi_{1j} \xi_{1k}) |\mathsf{V e}| d\omega.$

In the above equations V denotes the relative velocity. Introducing the mean velocity $\mathsf{U} = \tfrac{1}{2}(\mathsf{v} + \mathsf{v}_1)$, we have

(19 a) $\qquad\qquad \mathsf{v} = \mathsf{U} - \tfrac{1}{2} \mathsf{V}, \qquad v_1 = \mathsf{U} + \tfrac{1}{2} \mathsf{V},$

and from eq. (41.12) we have

(19 b) $\qquad\qquad \mathsf{v}' = \mathsf{U} - \tfrac{1}{2} \mathsf{V}', \qquad \mathsf{v}_1' = \mathsf{U} + \tfrac{1}{2} \mathsf{V}'$

with

(19) $\qquad\qquad\qquad \mathsf{V}' = \mathsf{V} - 2(\mathsf{V e})\,\mathsf{e}.$

After an elementary rearrangement, we find

(20 a) $\quad \xi_i' \xi_k' + \xi_{1i}' \xi_{1k}' - \xi_i \xi_k - \xi_{1i} \xi_{1k}) = 2 (\mathsf{V e})^2 e_i e_k - (\mathsf{V e})(V_i e_k + V_k e_i) = V_{ik}$

and

(20 b) $\quad \xi_i' \xi_j' \xi_k' + \xi_{1i}' \zeta_{1j}' \xi_{1k}' - \xi_i \xi_j \xi_k - \xi_{1i} \xi_{1j} \xi_{1k}) = (U_i V_{jk} + U_j V_{ki} + U_k V_{ij}).$

It is seen from eq. (20 a) that the integral (18 a) depends only on the vector V. Since the result must be a symmetrical tensor whose trace is zero, we must have

(20) $\qquad\qquad I_{ik} = \dfrac{\gamma\, m\, s^2}{4} V \left(V_i V_k - \dfrac{1}{3} V^2 \delta_{jk} \right).$

Furthermore it follows from (18 b) that

(21) $\qquad\qquad I_{ijk} = (U_i I_{jk} + U_j I_{ki} + U_k I_{ij}).$

The constant γ can be calculated from the special integral

$$I_{ik} V_i V_k = \dfrac{\gamma\, m\, s^2}{6} V^5 = \dfrac{m\, s^2}{2} \int [(\mathsf{V e})^4 - V^2 (\mathsf{V e})^2] |\mathsf{V e}| d\omega$$

$$= 2\pi\, m\, s^2 \cdot V^5 \int_0^1 (\zeta^5 - \zeta^3)\, d\zeta = - \dfrac{\pi\, m\, s^2}{6} V^5,$$

so that $\gamma = -\pi$ and

(22) $$I_{ik} = -\frac{\pi m s^2}{4} V\left(V_i V_k - \frac{1}{3} V^2 \delta_{ik}\right).$$

According to (17) and (18) the equation which determines the collision moments is

$$J_\phi = \int I_\phi f f_1 \, d\xi \, d\xi_1,$$

where $f f_1$ is given by eq. (11). The first term gives no contribution, and the two following pairs can be contracted, because eqs. (21) and (22) are symmetrical in $\mathbf{v}$ and $\mathbf{v}_1$. Thus

(23) $$J_\phi = \frac{1}{\rho} \int I_\phi \left(\sigma_{rs} \frac{\partial^2 f_0}{\partial \xi_r \, \partial \xi_s} - \frac{1}{3} Q_{mrs} \frac{\partial^3 f_0}{\partial \xi_m \, \partial \xi_r \, \partial \xi_s}\right) f_{01} \, d\xi \, d\xi_1.$$

Applying partial integration it is possible to transfer the derivatives to the I_ϕ. The product $f f_1$ is replaced by the function

(24) $$f_0 f_{01} = n^2 \frac{\gamma^3}{\pi^3} e^{-2\gamma(\mathbf{U}-u)^2 - \frac{1}{2}\gamma \mathbf{V}^2}.$$

It follows that integrals extending over odd polynomials of $\mathbf{V}/|\mathbf{V}|$ vanish so that in view of (22) and (21), respectively, eq. (23) leads to

(25 a) $$J_{ik} = \frac{\sigma_{rs}}{\rho} \int \frac{\partial^2 I_{ik}}{\partial V_r \, \partial V_s} f_0 f_{01} \, d\xi \, d\xi_1$$

and

(25 b) $$J_{ijk} = \frac{1}{2\rho} \left(Q_{irs} \int \frac{\partial^2 I_{jk}}{\partial V_r \, \partial V_s} f_0 f_{01} \, d\xi \, d\xi_1 + +\right).$$

The sum is cyclic in ijk.

The integral in (25 a) is symmetrical in ik and rs and the trace over $i = k$ vanishes. It then follows that

(25) $$\int \frac{\partial^2 I_{ik}}{\partial V_r \, \partial V_s} f_0 f_{01} \, d\xi \, d\xi_1 = I \left(\delta_{ir} \delta_{ks} + \delta_{is} \delta_{kr} - \frac{2}{3} \delta_{ik} \delta_{rs}\right).$$

The numerical factor is calculated for special values of the subscripts, e. g. for $i = r = x$, $k = s = z$. Thus

$$I = -\frac{\pi m s^2}{4} \int \frac{\partial^2 V\, V_x V_z}{\partial V_x\, \partial V_z} f_0 f_{01}\, d\xi\, d\xi_1$$

$$= -\frac{\pi m s^2}{4} \int V \left(1 + \frac{V_x^2 + V_z^2}{V^2} - \frac{V_x^2 V_z^2}{V^4}\right) f_0 f_{01}\, d\xi\, d\xi_1.$$

It is seen that the integration over U and over the direction of V can be carried out at once on substituting the expression from eq. (24). Thus

$$I = -\frac{\pi m s^2}{4} n^2 \frac{\gamma^3}{\pi^3} \left(\frac{\pi}{2\gamma}\right)^{3/2} \cdot 2\pi \cdot \frac{16}{5} \int_0^\infty V^3 e^{-\frac{1}{2}\gamma V^2}\, dV,$$

or

(26) $$I = -\frac{4}{5} m n^2 s^2 \left(\frac{2\pi}{\gamma}\right)^{1/2} = -\frac{8}{5} n^2 s^2 (\pi m k T)^{1/2}.$$

Substituting eq. (25) into (25 a) and (25 b) we obtain

$$J_{ik} = \frac{2}{\rho} I \sigma_{ik}$$

(27) $$J_{ijk} = \frac{3}{\rho} I Q_{ijk} - \frac{1}{3\rho} I(Q_{irr}\, \delta_{jk} + +),$$

in agreement with eq. (12). A comparison with the latter yields

(27') $$a = 2\frac{I}{\rho}, \qquad b = 3\frac{I}{\rho}, \qquad c = -\frac{1}{3}\frac{I}{\rho}.$$

Taking into account (26), we can deduce from eq. (14) that

(28) $$\eta = \frac{p\rho}{2I} = \frac{5/16}{\pi s^2} (\pi m k T)^{1/2}$$

and

(28 a) $$\frac{\varkappa}{\eta} = \frac{5}{2} \frac{a}{b + 5c} \frac{k}{m} = \frac{15}{4} \frac{k}{m}.$$

Measurements on monatomic gases give the following results

	He	Ne	A	Kr	X
$\dfrac{4 m \varkappa}{15 k \eta}$	0.98	1.00	0.98	1.02	1.03

and the agreement is seen to be astonishingly good because the assumption of rigid molecules must at first appear to be questionable and useful only in a qualitative way.

A remarkable comparison between the two eqs. (13) is obtained in applying the left-hand equation to shear flow (Couette flow, Vol. II) when we assume that $\mathbf{u} = (0, u_y(x), 0)$. In this case

$$\sigma_{xy} = -\eta \frac{\partial u_y}{\partial x} = -\frac{\eta}{m} \frac{\partial p_y}{\partial x}. \tag{29}$$

Here p_y denotes the mean momentum of a molecule in the y-direction and σ_{xy} is the flux of momentum across an element of area of size 1 at right angles to the x-axis. Inserting eq. (28 a) into the right-hand side eq. (13) for the heat flux, and introducing the local heat energy $Q = 3/2\, k\, T$, we find

$$Q_x = -\frac{5}{2} \frac{\eta}{m} \frac{\partial Q}{\partial x}. \tag{29 a}$$

This equation shows that the transfer of heat proceeds more efficiently than the transfer of momentum. This result can be understood in a qualitative way. Large molecular velocities in a given direction enhance transfer in that direction. A change in these velocities exerts no influence on the momentum being transferred, because in the case of friction we are concerned with the transfer of that component of momentum which is normal to the direction of flow of momentum. In the case of energy transfer the conditions are different, because every component of velocity contributes to the energy. It follows that large energies are favored in the process of transfer and that, on the whole, thermal contact is more intimate than momentum contact.

The latter remark does not apply to rotational energy. For this reason, in the case of polyatomic rigid molecules, Eucken replaces the expression $5/2\, Q$ in eq. (29 a) by

$$\frac{5}{2} Q_{transl} + Q_{rot} = \left(\frac{5}{2} \cdot \frac{3}{2} + \frac{f-3}{2}\right) k\, T = \left(\frac{f}{2} + \frac{9}{4}\right) k\, T.$$

This leads to the equation

$$Q_x = -\left(1 + \frac{9}{2f}\right) \frac{\eta}{m} \frac{\partial Q}{\partial x} \tag{29 b}$$

where $Q = \frac{1}{2} f k\, T$ denotes the mean energy of a molecule, and f is the number of degrees of freedom. We thus obtain for

$$f = 3 \quad 5 \quad 6$$

$$1 + \frac{9}{2f} = 2.5 \quad 1.9 \quad 1.75.$$

The results of measurements are as follows:

$$H_2 \quad O_2 \quad CO \quad Air$$

$$1 + \frac{9}{2f} = 2.00 \quad 1.92 \quad 1.81 \quad 1.96 \quad \text{instead of } 1.9$$

and for

$$CH_4 \quad CO_2 \quad C_3H_8$$

$$1 + \frac{9}{2f} = 1.74 \quad 1.64 \quad 1.66 \quad \text{instead of } 1.75.$$

The term $\sqrt{T}$ in eq. (28) is proportional to the mean velocity. It is easy to verify that

$$\overline{c} \approx \overline{c}^0 = \frac{2\sqrt{2}}{\pi m} (\pi k T m)^{1/2}$$

so that

(30) $$\eta = \frac{5\pi}{32\sqrt{2}} \frac{m \overline{c}}{\pi s^2}$$

Substituting $l = 1/n \pi s^2 \sqrt{2}$ from eq. (27.11) for the order of magnitude of the mean free path we obtain

(30 a) $$\eta = \frac{5}{32} \rho l \overline{c} = \frac{\rho l \overline{c}}{2.04}.$$

45. Conductivity and the Wiedemann-Franz law

A. THE COLLISION AND TRANSFER EQUATIONS FOR ELECTRONS IN METALS

The collision equation for metal electrons differs from Boltzmann's collision equation (41.18) in that it is necessary to take into account only collisions between conduction electrons and the ions of the lattice and that the lattice ions are very heavy compared with the electrons. Consequently, on impact there is an exchange of momentum, but, practically speaking, no exchange of energy. Evidently such a statement cannot be strictly true because, as we have already seen in Sec. 39, the electrons participate in thermal equilibrium. Nevertheless, by way of a first approximation, we can neglect the transfer of energy as compared with the transfer of momentum. Let **v** denote the velocity of an electron before it collides with a lattice ion, at first considered to be a rigid sphere. Thus the velocity **v**′ after collision is

(1) $$\mathbf{v}' = \mathbf{v} - 2(\mathbf{v}\,\mathbf{e})\,\mathbf{e}.$$

Here the symbol **e** denotes, as before, the unit vector in the direction of the central axis. The change in momentum is

(2) $$\varDelta \mathbf{p} = m(\mathbf{v}' - \mathbf{v}) = -2\,m(\mathbf{v}\,\mathbf{e})\,\mathbf{e},$$

whereas the change in energy is negligible:

(3) $$E = \frac{m}{2}(\mathbf{v}'^2 - \mathbf{v}^2) = 0.$$

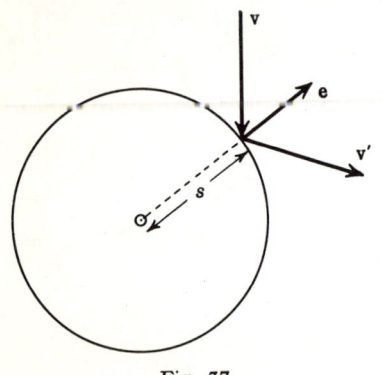

Fig. 37.
Illustrating collision between conduction electrons and lattice ions.

If f denotes the distribution function for the electrons, we can see that the left-hand side of the collision equation (41.18) remains unchanged, but the collision integral is now different. The integrand will contain only one factor f and the second should be replaced by the probability of impinging on a spherical ion. It is determined by the density, n_0, of the lattice ions. Denoting the radius of an ion by s, we obtain

(4) $$J_{loss} = \frac{1}{2} n_0 s^2 \int |\mathbf{v}\,\mathbf{e}|\,f(\mathbf{r}, \mathbf{v}, t)\,d\omega,$$

$$J_{gain} = \frac{1}{2} n_0 s^2 \int |\mathbf{v}\,\mathbf{e}|\,f(\mathbf{r}, \mathbf{v} - 2\,(\mathbf{v}\,\mathbf{e})\,\mathbf{e}, t)\,d\omega,$$

in a way similar to Sec. 41 D.

Introducing the mean free path $l = 1/n_0 \pi s^2$, we can write the collision equation in the form

(5) $$\frac{\partial f}{\partial t} + \mathbf{v}\frac{\partial f}{\partial \mathbf{r}} + \frac{1}{m}\mathbf{F}\frac{\partial f}{\partial \mathbf{v}} = \frac{1}{2\pi l}\int |\mathbf{v}\,\mathbf{e}|\,(f' - f)\,d\omega.$$

The function f' depends on the argument **v**′ from eq. (1) in the same way as f depends on **v**. In accordance with our model, the mean free path l should be a constant, but the model is certainly very crude. We shall adapt it better to reality by assuming that l is a function of the velocity of the electron and of the special properties of the lattice, in particular of its temperature. In order to find a more exact expression it would be necessary to resort to wave mechanics, but we shall refrain from doing so here.

The transfer equation for a function $\phi(\mathbf{v})$ can be obtained in a way similar to eq. (43.11). Multiplying eq. (5) by $\phi(\mathbf{v})$, and integrating with respect to $\mathbf{v}$ gives

(6) $$\frac{\partial}{\partial t}(\rho\,\overline{\phi}) + \operatorname{div}(\rho\,\overline{\phi\,\mathbf{v}}) - n\,\mathbf{F}\left(\overline{\frac{\partial \phi}{\partial \mathbf{v}}}\right) = J(\phi).$$

Here $J(\phi)$ denotes the collision moment of the function ϕ:

(5 a) $$J(\phi) = \frac{m}{2\pi}\int \frac{\phi}{l}(f'-f)\,|\mathbf{v}\,\mathbf{e}|\,d\omega\,d\xi = \frac{m}{2\pi}\int \frac{f}{l}\,d\xi \int (\phi'-\phi)|\mathbf{V}\,\mathbf{e}|\,d\omega.$$

Since $\phi' - \phi = 0$ for $\phi = 1$ and for $\phi = \tfrac{1}{2}\mathbf{v}^2$, it is seen that the mass and energy equations remain valid, as before. The momentum equation, however, is now replaced by the equation

(7) $$\frac{\partial}{\partial t}(\rho\,\overline{\mathbf{v}}) + \operatorname{Div}[\rho(\overline{\mathbf{v}|\mathbf{v}})] - n\,\mathbf{F} = \frac{m}{2\pi}\int \frac{f}{l}\,d\xi \int (\mathbf{v}'-\mathbf{v})\,|\mathbf{v}\,\mathbf{e}|\,d\omega.$$

According to (1) the integral over $d\omega$ becomes

(8) $$\int (\mathbf{v}'-\mathbf{v})\,|\mathbf{v}\,\mathbf{e}|\,d\omega = -2\int |\mathbf{v}\,\mathbf{e}|\,(\mathbf{v}\,\mathbf{e})\,\mathbf{e}\,d\omega = -2\pi v\,\mathbf{v}.$$

This form is due to the fact that the integral is a vector and a homogeneous function of degree 2 of $\mathbf{v}$. The numerical factor is obtained by taking the scalar product of (8) with $\mathbf{v}$ and by dividing by v^3. Thus

$$\frac{\mathbf{v}}{v^3}\int (\mathbf{v}'-\mathbf{v})\,|\mathbf{v}\,\mathbf{e}|\,d\omega = -2\int |\zeta^3|\,d\zeta\,d\phi = -2\pi.$$

Consequently, eq. (7) becomes:

(9) $$\frac{\partial}{\partial t}(\rho\,\overline{\mathbf{v}}) + \operatorname{Div}[\rho(\overline{\mathbf{v}|\mathbf{v}})] - n\,\mathbf{F} = -\rho\left(\overline{\frac{v\,\mathbf{v}}{l}}\right).$$

B. Approximate Solution of the Collision Equation

We now propose to consider the solutions of (5) in the neighborhood of the equilibrium solution f_0, beginning with the entropy principle, as in Sec. 42. Differentiating the entropy density equation (38.9), namely

$$H = -k \int [n \log n + (1-n) \log (1-n)] \cdot \frac{2 m^3}{h^3} d\xi$$

with respect to time, we have

$$\frac{\partial H}{\partial t} = -k \int \log \frac{n}{1-n} \cdot \dot{n} \cdot \frac{2 m^3}{h^3} d\xi$$

or, according to eq. (5) (with $F = 0$), and in view of the fact that $f = \frac{2 m^0}{h^3} n$, that it is equal to

$$+ k \operatorname{div} \int \mathbf{v} [n \log n + (1-n) \log (1-n)] \frac{2 m^3}{h^3} d\xi$$

$$+ \frac{k}{4\pi} \int \frac{|\mathbf{v}\,\mathbf{e}|}{l} \cdot \log \frac{n'/(1-n')}{n/(1-n)} \cdot (n' - n) \, d\omega \, \frac{2 m^3}{h^3} d\xi.$$

Thus we obtain

$$\tag{10} \frac{\partial H}{\partial t} + \operatorname{div} \mathbf{S} = G,$$

where the flux of entropy is:

$$\tag{11} \mathbf{S} = -k \int \mathbf{v} [n \log n + (1-n) \log (1-n)] \frac{2 m^3}{h^3} d\xi$$

and the *source density* is

$$\tag{11 a} G = \frac{k}{4\pi} \int \frac{|\mathbf{v}\,\mathbf{e}|}{l} \log \frac{n'/(1-n')}{n/(1-n)} \cdot (n' - n) \, d\omega \, \frac{2 m^3}{h^3} d\xi.$$

In the case of equilibrium we must have $G = 0$. Since $n/(1-n)$ increases monotonically with n, the integrand cannot be negative, and $G = 0$ can only occur when $n' = n$. Since the only collision invariants are 1 and $\mathbf{v}^2$, f must be a function of energy alone:

$$\tag{12} n = n_0(E), \qquad E = \frac{m}{2} \mathbf{v}^2.$$

It is remarkable that the result is indefinite. This is a reflection of the fact that the interactions between electrons have been neglected. We have to imagine that eq. (11 a) contains an additional term for G which is due to collisions between the electrons, in the same way as in eq. (42.12). Normally it would be small but if we substitute (12), this term alone remains. It is then that we establish a special energy function. If G were given by eq. (42.12), f_0 would be identical with Maxwell's distribution, but we know that we have to assume the Fermi distribution in relation to electrons. Leaving aside the problem of correcting the expression for the entropy source density G, we assume with reference to eq. (39.5) that

$$(12\text{ a}) \qquad f_0 = \frac{2\,m^3}{h^3}\,n_0 = \frac{2\,m^3/h^3}{e^{\beta(E-\zeta)}+1}.$$

Here β and ζ represent parameters which can still depend on time and on space coordinates: $\beta = \dfrac{1}{k\,T}$ and ζ is the free enthalpy per electron (cf. eq. (39.4)). Sommerfeld's theory of conductivity differs from that due to Drude and Lorentz only in the assumption (12 a).

Equations (12) and (12 a) differ from previous results in one more respect, namely in that the equilibrium distribution no longer depends upon the local mean velocity. Mathematically this is a consequence of the fact that the momentum of an electron is not preserved after a collision. It is also understandable on physical grounds that the electron distribution changes if we cause the electrons to move collectively while keeping the lattice fixed. It follows that in the expansion of the distribution function, f, in terms of the equilibrium distribution, f_0, we shall find first-order derivatives as well:

$$(13) \qquad f = f_0 - u_k \frac{\partial f_0}{\partial \xi_k} + \frac{1}{2\rho}\,\sigma_{kl}\,\frac{\partial^2 f_0}{\partial \xi_k\,\partial \xi_l} - \frac{1}{6\rho}\,Q_{klm}\,\frac{\partial^3 f_0}{\partial \xi_k\,\partial \xi_l\,\partial \xi_m} + \ldots.$$

Following Sommerfeld (and Lorentz) we shall use here a slightly different approximation. Since f_0 depends only on the energy E we can write the first-order term as

$$-u_k\,\frac{\partial f_0}{\partial \xi_k} = -m(u_k\,\xi_k)\,f_0',$$

where primes denote differentiation with respect to E. The higher-order terms of the expansion in (13) also contain terms of the same type. If

$$Q_{klm} = \frac{2}{5}\,(Q_k\,\delta_{lm} + Q_l\,\delta_{mk} + Q_m\,\delta_{kl})$$

(or when we separate the term of this form from Q_{klm}), we can see that the contribution from the term of the third order will be

$$-\frac{1}{5\rho} Q_k \frac{\partial^3 f_0}{\partial \xi_k \partial \xi_l^2} = -\frac{m^2}{\rho} (Q_k \xi_k) \left(f_0'' + \frac{2}{5} E f_0''' \right).$$

Characteristically these terms are of the form

$$U_k(E) \xi_k$$

and the coefficients U_k depend on energy, apart from their dependence on time and space coordinates; they are, however, independent of the direction of the velocity. If we perform the same transformation with regard to all terms in eq. (13), we obtain a series of the following type:

(14) $$f = f_0(E) + U_k(E) \xi_k + \tfrac{1}{2} U_{kl}(E) \xi_k \xi_l + \ldots .$$

The higher coefficients of the expansion can be assumed to be symmetrical tensors whose traces all vanish. We have $U_{kk} = 0$, $U_{kkl} = 0$, etc.

If we had $U_{kkl} \neq 0$, we could represent it in the form

$$U_{klm} = U^*_{klm} + (V_k \delta_{lm} + V_l \delta_{mk} + V_m \delta_{kl})$$

where we could have $U^*_{kkm} = 0$. It would only be necessary to put $U_{kkm} = 5 V_m$. Thus the first term would have the desired form and the second would give a contribution to the series (14) of the form

$$\frac{3}{2} (V_k \xi_k) \xi_l^2 = \frac{3E}{m} V_k(E) \xi_k$$

which could, obviously, be included in the second term of the expansion in (14).

We now substitute eq. (14) into the collision equation (5) and take into account only the first non-vanishing terms, restricting ourselves to steady flows only $(\partial f/\partial t) = 0$. Thus we obtain:

(15) $$\left(\mathbf{v}, \frac{\partial f_0}{\partial \mathbf{r}} + \mathbf{F} f_0' \right) = \frac{\mathsf{U}}{2\pi l} \int |\mathbf{v} \, \mathbf{e}| \, (\mathbf{v}' - \mathbf{v}) \, d\omega.$$

By eq. (8) the right-hand side is equal to

$$-\frac{1}{l} v (\mathsf{U} \, \mathbf{v}).$$

Since this equation must apply for arbitrary directions of **v**, we may also write

(16) $$U = -\frac{l}{v}\left(F f_0' + \frac{\partial f_0}{\partial r}\right).$$

Hence the first approximation to the distribution function (14) is:

(17) $$f = f_0 - l\left(\frac{\mathbf{v}}{v}, f_0' \, \mathbf{F} + \frac{\partial f_0}{\partial r}\right).$$

C. Flux of current and energy

In terms of the following integrals:

(18) $$W_n = \int E^n \, \mathbf{v} \, f \, d\xi$$

the *current* and *energy fluxes* are

(18 a) $$\mathbf{I} = -e \mathbf{W}_0.$$
(18 b) $$\mathbf{W} = \mathbf{W}_1.$$

For reasons of symmetry the first term in (17) gives no contribution to (18). Performing the angular integration we obtain from the second term that

$$W_n = -\frac{4\pi}{3} \int_0^\infty \left(f_0' \, \mathbf{F} + \frac{\partial f_0}{\partial r}\right) l \, E^n \, v^3 \, dv.$$

Substituting into eq. (12) the function

(19) $$g(\varepsilon) = -\frac{1}{e^\varepsilon + 1}, \quad \varepsilon = \beta(E - \zeta)$$

and replacing v by $E = \tfrac{1}{2} m v^2$, we have

(20) $$W_n = -\frac{16\pi \, m \, \beta}{3 \, h^3} \int_0^\infty g'(\varepsilon) \left(\mathbf{F} - \frac{1}{\beta}\frac{\partial \beta}{\partial r}\zeta + \frac{E}{\beta}\frac{\partial \beta}{\partial r}\right) l \, E^{n+1} \, dE.$$

Introducing the abbreviations

(21) $$\mathbf{F}_1 = \mathbf{F} - \frac{1}{\beta}\frac{\partial \beta}{\partial r}\zeta, \quad \mathbf{F}_2 = \frac{1}{\beta}\frac{\partial \beta}{\partial r}$$

and putting

(21 a) $$-\frac{16\pi m \beta}{3 h^3} \int_0^\infty g'(\varepsilon)\, l\, E^n\, dE = K_n$$

we can deduce quite generally from eq. (20) that

(22) $$\mathsf{W}_n = K_{n+1}\, \mathsf{F}_1 + K_{n+2}\, \mathsf{F}_2,$$

and in particular that

(22 a) $$\mathsf{I} = -e\,(K_1\, \mathsf{F}_1 + K_2\, \mathsf{F}_2),$$
$$\mathsf{W} = K_2\, \mathsf{F}_1 + K_3\, \mathsf{F}_2.$$

Eliminating

(21 b) $$\mathsf{E}' = -\frac{1}{e}\mathsf{F}_1 - \frac{\zeta}{e}\frac{1}{\beta}\operatorname{grad}\beta = \mathsf{E} + \frac{1}{e}\operatorname{grad}\zeta,$$

we have

(23) $$\mathsf{E}' = \frac{1}{e^2 K_1}\mathsf{I} + \frac{K_2 - \zeta K_1}{e K_1}\frac{1}{\beta}\operatorname{grad}\beta,$$
$$\mathsf{W} = -\frac{K_2}{e K_1}\mathsf{I} + \frac{K_1 K_3 - K_2^2}{K_1}\frac{1}{\beta}\operatorname{grad}\beta$$

which agrees with eqs. (21.18 a, b). The *electrical conductivity* becomes

(24) $$\sigma = e^2 K_1$$

and the *Peltier coefficient* and the absolute thermal emf become, respectively,

(24 a) $$\Pi = \frac{K_2}{e K_1},$$

(24 b) $$\varepsilon = \frac{K_2 - \zeta K_1}{e K_1 T},$$

whereas the thermal conductivity is:

(24 c) $$\varkappa = \frac{K_1 K_3 - K_2^2}{K_1 T}.$$

D. Ohm's law

Ohm's law is obtained on the assumption that ζ and β are constant throughout space. Hence from (21 b) we obtain: $E' = E$, and from (23):

(25) $$I = \sigma E$$

which is Ohm's law. The electrical conductivity can be calculated from (24) with K_1 from eq. (21 a) assuming complete degeneration (*cf.* 39 B):

$$\sigma = \frac{16\pi m \zeta_0 l_0}{3 h^3} e^2.$$

Here ζ_0 and l_0 denote the values of ζ and l, respectively, at Fermi's threshold. The particle density can be calculated with the aid of eqs. (39.7 and 39.7 a) and is

$$n = \frac{8\pi m^3}{3 h^3} v_0^3, \qquad \zeta_0 = \frac{1}{2} m v_0^2.$$

Consequently σ can be written in the form

(25 a) $$\sigma = \frac{n e^2}{m} \cdot \frac{l_0}{v_0}$$

which is Drude's equation (39.2).

In the above equation v_0 denotes Fermi's limiting velocity and l_0 is the value of the mean free path for electrons moving with that velocity. Evidently l_0 can vary with the temperature, since it depends on the lattice properties. On the other hand v_0 is independent of temperature. However, when we evaluate the integrals in accordance with Sec. 39 C for the case of almost complete degeneracy, v shows a weak dependence on temperature. It is of the order (kT/mv_0^2) and cannot be observed.

The mean free path l_0 can also be calculated from measured values of conductivity and so, for example, for silver at room temperature we obtain $l_0 \approx 5 \times 10^{-6}$ cm on the assumption of one conducting electron per atom. This would mean that l_0 is much larger than the distance between lattice ions, a sure sign that the mean free path must be calculated with the aid of wave mechanics.

When evaluating the *Peltier coefficient* it is also sufficient, at least here, to evaluate the integrals K_n for the limiting case of complete degeneration. It follows from eq. (21 a) that in this case

(21 c) $$K_n = \frac{16\pi m}{3 h^3} l_0 \zeta_0^n.$$

Consequently

(26) $$\Pi = \frac{m v_0^2}{2 e}.$$

It is noticed that the mean free path cancels and does not appear in the final expression. This is an elementary value of the Peltier coefficient Π. When grad $\beta = 0$, the first term in eq. (23) gives the energy flux

$$\mathbf{W} = n \cdot \frac{m v_0^2}{2} \cdot \bar{\mathbf{v}}.$$

It is evident that this is the kinetic energy transferred macrophysically, because the electrons which contribute to the mean value $\mathbf{v}$ are near the Fermi threshold. Additional remarks concerning the Peltier effect have been given in Sec. 21. In order to determine its numerical value it is necessary to perform a more accurate calculation.

When the current density $\mathbf{I} = 0$, but the temperature distribution is not uniform, there is a flow of heat, and an electric field is formed. The flow of heat and the strength of the electric field are determined by the thermal conductivity, $\varkappa$, and the absolute emf, ε. According to (21 c) both vanish in the case of complete degeneration.

E. THERMAL CONDUCTIVITY AND ABSOLUTE THERMAL ELECTROMOTIVE FORCE

In order to calculate $\varkappa$ and ε it is necessary to obtain a more accurate expression for the numerator. According to Sec. 39 D the integrals in eq. (21 a) can be written

$$K_n = -\frac{16 \pi m}{3 h^3} \int_{-\beta \zeta}^{\infty} g'(\varepsilon) F_n\left(\zeta + \frac{\varepsilon}{\beta}\right) d\varepsilon$$

where $F_n = l E^n$ so that

(27) $$K_n = +\frac{16 \pi m}{3 h^3}\left[F_n(\zeta) + \frac{\pi^2}{6 \beta^2 \zeta^2} \cdot \zeta^2 F_n''(\zeta)\right].$$

The particle density is given by eq. (39.11 b)

$$n = \frac{8 \pi}{3 h^3} (2 m \zeta)^{3/2} \left(1 + \frac{\pi^2}{8 \beta^2 \zeta^2}\right).$$

Consequently

(27 a) $$K_n = \frac{n}{\sqrt{2m}} l \, \zeta^{n-3/2} \left[1 + \frac{\pi^2}{6\beta^2 \zeta^2}\left(\frac{\zeta^2 F_n''}{F\,n} - \frac{3}{4}\right)\right].$$

In the correction term we can substitute ζ_0 for ζ, i. e. the value at Fermi's threshold. For the first factor it is actually necessary to take into account eq. (39.13), but it is sufficient to substitute the threshold value here as well because by eq. (27 a) we have

$$K_1 K_3 - K_2^2 = \frac{n^2}{2m} l_0^2 \, \zeta_0 \cdot \frac{\pi^2}{6\beta^2 \zeta_0^2} (1 \times 0 + 3 \times 2 - 2 \times 2 \times 1),$$

so that the factors in front of the brackets in (27 a) are seen to be multiplied by higher-order terms only. Thus we have

$$\frac{K_1 K_3 - K_2^2}{K_1} = \frac{n}{\sqrt{2m}} l_0 \zeta_0^{3/2} \cdot \frac{\pi^2}{3\beta^2 \zeta_0^2}.$$

Putting $\zeta_0 = \tfrac{1}{2} m v_0^2$, we obtain the *thermal conductivity* from eq. (24 c):

(28) $$\varkappa = \frac{\pi^2}{3} \frac{k}{m} \frac{l_0}{v_0} \cdot n k T.$$

It is remarkable that the derivatives of $l(\zeta)$ which occur in eq. (27 a) disappear in the expression for $\varkappa$.

The absolute thermal emf can be calculated from

$$K_2 - \zeta K_1 = \frac{n}{\sqrt{2m}} l \, \zeta^{1/2} \frac{\pi^2}{6\beta^2 \zeta^2} \cdot 2 \left(1 + \zeta \frac{l'}{l}\right).$$

Accordingly eq. (24 b) can be written

(29) $$\varepsilon = \frac{\pi^2}{3} \frac{k}{e} \cdot \frac{kT}{\zeta_0} \left(1 + \frac{d \log l_0}{d \log \zeta_0}\right).$$

which contains the first derivative of $l(\zeta)$.

F. The Wiedemann-Franz law

The expressions in Drude's eq. (25 a) and in eq. (28) for thermal conductivity contain only the threshold value l_0/v_0. On forming the ratio $\varkappa/\sigma$ the term l_0/v_0 cancels, and we are led to the Wiedemann-Franz law:

(30) $$\frac{\varkappa}{\sigma} = \frac{\pi^2}{3} \frac{k^2}{e^2} T.$$

Following Lorenz' method, experimental physicists indicate the relation

(31) $$\Lambda = \frac{\varkappa}{\sigma T} = \frac{\pi^2}{3}\frac{k^2}{e^3} \quad \text{(Lorenz number)}.$$

We prefer to compare the dimensionless ratio

(32) $$\Lambda_0 = \frac{e^2 \varkappa}{\sigma k^2 T} = \frac{\Lambda e^2}{k^2} = \frac{\pi^2}{3} = 3.29,$$

with experimental results. Thus we have to multiply the Lorenz number by $e^2/k^2 = 1.344 \times 10^8$ deg² volt⁻². Experiments show that Λ_0 is not constant, but decreases with decreasing temperature; however, at high temperatures the curve tends asymptotically to a constant value. At a temperature of 100 C we obtain the following values:

Cu	Au	Pb	Pt
3.15	3.19	3.46	3.51

The curves $\Lambda_0(T)$ for copper and gold could tend asymptotically to the value $\pi^2/3$, but in the case of lead and platinum it is distinctly higher. Drude has given the value of $\Lambda_0 = 3$ on the basis of a crude estimate. A more accurate calculation on a classical basis performed by H. A. Lorentz gives $\Lambda_0 = 2$. Compared with this, the quantum mechanical value appears to constitute a considerable improvement. In actual fact the values for most substances remain below this value which is consistent with the supposition that the temperature is not yet high enough.

However, the values for many substances exceed this value markedly, as seen from the following examples:

W (polycrystalline)	$T = 273$ K,	$\Lambda_0 = 4.11$,
Bi (fine crystals)	$T = 90$ K,	$\Lambda_0 = 5.56$,
	$T = 273$ K,	$\Lambda_0 = 3.62$.

The case of bismuth shows even a temperature anomaly in that Λ_0 at first increases with decreasing temperature. In this connection it must be remembered that the Lorenz number deals with the electrical current and the heat flux transferred by the electrons. If the lattice itself contributes to the conduction of heat the value of $\varkappa/\sigma T$ must be expected to increase. The value $\Lambda_0 = 5.56$ would indicate that about 2/3 of the heat flux is conducted by the lattice and the thermal conductivity of the lattice (evaluated from the total value of $\varkappa = 0.06$ cal cm^{-1} sec^{-1} deg^{-1}) would be about 0.024 cal cm^{-1} deg^{-1}.[1] This is not entirely unlikely. For example NaCl at 20 C is a good electrical insulator (spec. resistance $\rho = 1 \times 10^{17}$ ohm cm), but at 25 C its thermal conductivity is $\varkappa = 0.02$ cal cm^{-1} sec^{-1} deg^{-1}.

The preceding deviations could be due to purely experimental conditions (insufficiently high temperature, large contribution due to the conductivity of the ion lattice), but the dependence of Λ_0 on temperature certainly cannot be explained by such secondary influences. In this connection the fact that the collision integral in eq. (5) has been calculated with the aid of classical mechanics makes itself felt, and we are thus led to the consideration of the difficulties which are encountered in an improved theory:

We have assumed originally that the metallic ion is a hard elastic sphere and has a definite radius s, so that the mean free path becomes $1/n\pi s^2$. In fact such an assumption has no sense, and the mean free path must be calculated with the aid of quantum mechanics;[2] we have assumed that l was an unknown function of the lattice temperature T_g and of the energy of the electron, E. We must, however, stress here that the concept of the mean free path can be justified on the grounds of quantum mechanics only for extremely high, and also, but with certain reservations, for extremely low temperatures.

---

[1] Data from J. D'Ans and E. Lax, Taschenbuch für Chemiker und Physiker, 2 ed., Springer 1949, p. 1126.
[2] H. A. Bethe and A. Sommerfeld: *l. c.* p. 277, footnote 1; Sec. 31—38.

PROBLEMS

Chapter I

I.1. It is assumed that the three variables x, y, z satisfy a functional relationship $f(x, y, z) = 0$, or, solved for z, $z = f(x, y)$. Prove the identity:

$$\left(\frac{\partial z}{\partial x}\right)_y \left(\frac{\partial x}{\partial y}\right)_z \left(\frac{\partial y}{\partial z}\right)_x = -1.$$

Replacing x, y, z by p, V, T in that order, deduce a relation between the coefficient of thermal expansion, the coefficient of tension and the compressibility (for definitions see (1.5) and (1.6)).

I.2. *On the heating problem.*

Calculate the quantity of heat required to increase the room temperature from 0 C to 20 C.

I.3. *Absolute temperature and perfect gas thermometer.*

Prove that the absolute temperature T defined in eq. (6.7 a) is identical with that defined by the perfect-gas law.

I.4. *Application of the Second Law to the proof of an algebraic inequality.*

a) Two bodies whose heat capacities are C_1, C_2, and whose temperatures are T_1, T_2 exchange heat, both volumes being kept constant. What is the common final temperature of both?

b) Comparing the values of entropy before and after equilibrium has set in ($\Delta S > 0$) for the special case of perfect gases deduce an inequality which is a generalization of that between the arithmetical and geometrical means.

I.5. One mol of a perfect gas expands reversibly until its volume is doubled: a) under constant pressure, b) isothermally, c) isentropically. Calculate the work of expansion, the heat added and the change in entropy for each case.

I.6. Imagine a Carnot cycle with water as the working fluid operating between 2 C and 6 C, so that at 6 C there is isothermal expansion and isothermal compression at 2 C. It is seen that heat is added during both processes, if the pressure is low enough (*cf.* (7.10)), and so heat is converted completely into work in violation of the Second Law. How is it possible to resolve this contradiction? Make a qualitative sketch of the isentropes and isotherms in a T, v-diagram in the neighborhood of 4 C.

I.7. Show that the ratio of isothermal to isentropic compressibility is always equal to the ratio of specific heats at constant volume and at constant pressure as it is for a perfect gas (*cf.* vol. II). In other words, show that

$$\frac{\varkappa_T}{\varkappa_S} = \frac{c_p}{c_v} \quad \text{where} \quad \varkappa_S = -\frac{1}{v}\left(\frac{\partial v}{\partial p}\right)_S, \quad \varkappa_T = -\frac{1}{v}\left(\frac{\partial v}{\partial p}\right)_T$$

In order to do this express the differential dq in terms of dv and dp and prove that

$$T\,ds = dq = c_p\left(\frac{\partial T}{\partial v}\right)_p dv + c_v\left(\frac{\partial T}{\partial p}\right)_v dp.$$

I.8. One kilogram of water is compressed isothermally at 20 C from 1 at to 20 at. Calculate the amount of work required, the quantity of heat rejected and the increase in internal energy. Mean compressibility $\varkappa = 0.5 \times 10^{-4}$/at, mean coefficient of thermal expansion $\alpha = 2 \times 10^{-4}$/deg. Use eq. (7.7) and the relation (2) given in the solution to Problem I.1.

I.9. *Adiabatic equilibrium of the atmosphere.*

During so-called convective (adiabatic) equilibrium of the atmosphere which is particularly well established in the presence of "sirocco" winds the value of $p\,v^\gamma$ is independent of altitude; v denotes here the molar or the specific volume. Making use of the relation between density and pressure which follows from the conditions of equilibrium in the gravitational field it is possible to show that there is a linear temperature decrease with altitude. Measurements give its value as 1 C/100 m; what is the theoretical value? — Calculate the height of the general polytropic atmosphere (defined by $p\,v^n$ = const.; n is known as the polytropic exponent). In particular calculate the height of the adiabatic, and of the isothermal atmosphere (for which $n = 1$), for a ground temperature of 0 C.

I.10. *The flow of gases.*

Calculate the final temperature and the maximum value of the flow velocity for superheated steam of 300 C and 5 at pressure which expands isentropically through a suitably shaped nozzle to a back-pressure of 1 at.

In order to perform the calculation make use of the fact that the kinetic energy can at most be equal to the difference in the enthalpies of the compressed and of the expanded gas (*cf.* Sec. 4 B). The same fact can be proved with the aid of Bernoulli's equation for a compressible fluid (*cf.* Vol. II, Sec. 11), assuming the flow to be irrotational and steady.

I.11. *Isothermal equilibrium of the atmosphere.*

A gas is contained in a closed box placed in the gravitational terrestrial field. The internal energy is then augmented by the potential energy; the latter depends on the altitude above ground.

a) Establish the condition of thermodynamic equilibrium by subdividing the gas into i cells of volume V_i at an elevation z_i above ground. Assume a definite value of specific volume v_i and temperature T_i for each cell and solve the problem by calculating the maximum in entropy for a given total energy, mass and volume.

b) Show that the temperature is independent of altitude.

c) Show that Gibbs' potential is independent of altitude provided that its definition includes the potential energy (compare the electrochemical potential in Sec. 18).

d) Calculate the density in terms of altitude, assuming the validity of the perfect gas law.

e) Calculate the difference in elevation for which the difference in potential energies per mol is equal to RT.

I.12. (After H. Einbinder, Phys. Rev. **74**, 805 (1948)).

Assuming an equation of state of the form $pv = \alpha u(T, v)$, where $u(T, v)$ is the specific internal energy and α is a constant show:

1. That the specific energy u and the specific entropy s can be expressed in the form

$$u = v^{-\alpha} \Phi(T v^\alpha); \qquad s = \psi(T v^\alpha)$$

where Φ is an arbitrary function of the argument, and $\Phi'(x) = x \psi'(x)$.

2. That $u/v = \sigma T^{\frac{\alpha+1}{\alpha}}$ when the energy density u/v depends only on the temperature T. This, for example, is true for black-body radiation with $\alpha = 1/3$ (cf. Sec. 20 B) and for a Bose gas composed of N particles of mass m below the temperature $T_0 \sim \dfrac{h^2}{m\,k} (N/v)^{2/3}$ with $\alpha = 2/3$ (Einstein condensation, cf. Sec. 38).

3. Assume that in the neighborhood of absolute zero $\Phi(T v^\alpha)$ can be represented by the power law

$$\Phi(T v^\alpha) = c\, T^m v^{\alpha m} \quad (m > 0).$$

a) Find the relation between α and m required by the condition of dynamic stability $(\partial p/\partial v)_T \leqslant 0$.

b) Assume that $u \to 0$ for $T \to 0$. According to the uncertainty principle we must also have $v = v(T, p) > 0$ also for $T \to 0$. Determine the relation between α and m implied by the relation $v(T, p) > 0$.

c) Taking into account the results of questions 3 a and 3 b find an expression for the internal energy and for the equation of state.

4. What result can be deduced for small values of T with the same power law but assuming that $u(T, v)$ tends to a finite limit when $T \to 0$?

Chapter II

II.1. Prove with the aid of (14.11 b) that pV is the thermodynamic potential for the variables T, p, μ_i. (It occurs in the theory of grand canonical ensembles in statistical mechanics; cf. Sec. 40 for a special application).

II.2. *The rate of change of latent heat along the vapor-pressure curve.*

In connection with eq. (16.14 a) we have used an experimental value for dr/dT; calculate the same quantity theoretically making use of the definition $r = \Delta h$, forming its differential, and utilizing the Clausius-Clapeyron equation together with some of the relations in the Table of Sec. 7.

II.3. A perfectly insulated vessel having a volume of 20 liters contains 1 kg of H_2O at 10 C partly in the liquid and partly in the vapor phase. Calculate the amount of energy required to bring the system to a temperature of 200 C. It is convenient to imagine that the final state is reached as follows: a) isothermal compression until complete liquefaction has been achieved, b) isothermal compression from the saturation pressure of $p = 0.0125$ kg/cm^2 at 10 C to $p = 15.86$ kg/cm^2 (= saturation pressure at 200 C), c) heating at constant pressure without evaporation ($c_p \approx 1$ cal g^{-1} deg^{-1}), d) isothermal expansion (evaporation) until the initial volume has been reached. The energy input for process b) and the work of expansion for process c) may be neglected (Compare Prob. I.8).

At 10 C: latent heat of evaporation $r = 591.6$ cal/g, specific volume of liquid $v_1 = 1.00$ dm^3/kg, of steam $v_2 = 106.4$ m^3/kg.

At 200 C: $r = 463.5$ cal/g, $v_1 = 1.16$ dm^3/kg, $v_2 = 0.127$ m^3/kg.

II.4. *Realization of thermodynamic temperature scale.*

Prove that the absolute temperature T can be calculated from the Clausius-Clapeyron equation $T\, dp/dT = r/(v_{vap} - v_{liq})$ if r, v_{vap} and v_{liq} are known as functions of pressure p.

II.5. The vapor pressure of mercury is:

        a) 0.0127 torr at 50 C, and 0.0253 torr at 60 C,

        b) 247 torr at 300 C, and 505 torr at 310 C.

Calculate the latent heat of evaporation in each of these intervals assuming r to be constant. Neglect v_{liq} and assume that the vapor behaves like a perfect gas. — Interpolate r linearly between these intervals and determine the vapor-pressure curve for Hg. Extrapolate the curve to temperatures > 300 C and calculate the boiling point at 1 atm (= 760 torr) pressure (accurate value 356.7 C).

II.6. Consider a gas whose molecules are capable of achieving three energy levels ε_0, ε_1, ε_2, ($\varepsilon_1 - \varepsilon_0$ and $\varepsilon_2 - \varepsilon_0$ are thus excitation energies for the first and second excited energy levels respectively) and suppose that it is a mixture of three gases each consisting of molecules of one internal energy level ε_0, ε_1, or ε_2. Deduce the conditions of equilibrium on the assumption that all gases have the same entropy constant and transition is possible between 0 and 1 as well as between 0 and 2. Show that the same conditions apply when transitions from 1 to 2 are also permitted.

II.7. Principle of detailed equilibrium.

a) Calculate the differences in the chemical potentials $\mu_1 - \mu_0$ and $\mu_2 - \mu_0$ for the preceding problem for the case when the molar masses n_i differ only slightly from those at equilibrium, $\overline{n}_i$.

b) Regarding the variation in n_0, n_1, n_2 make the following assumption: The change per second in, say, n_1 is due to the fact that during that time a fraction $k_{11} n_1$ proportional to the amount present changes into the states 0 and 2. At the same time fractions proportional to n_0 and n_2, i. e. $k_{10} n_0$ and $k_{12} n_2$ undergo the transition into state 1. Write down the equations for dn_i/dt. The preceding assumption implies that the number of transitions from one state to another depends only on the number of molecules at that state and that it is proportional to it.

c) Show that the law of mass action (see eq. (3) in the solution to the previous problem) follows from this assumption insofar as $\overline{n}_1/\overline{n}_0$ and $\overline{n}_2/\overline{n}_0$ no longer depend on $\overline{n}_1$ and $\overline{n}_2$. A comparison with the already mentioned eq. (3) yields two relations which must be satisfied by the coefficients k_{ik}.

d) Making use of the relations deduced in a) substitute $\mu_1 - \mu_0$ and $\mu_2 - \mu_0$ for n_j in the equation for dn_i/dt found in b). What is the meaning of the validity of Onsager's reciprocal relations with respect to the mechanism of the reaction?

Chapter III

III.1. A vertical cylinder is fitted with a piston of mass M which can follow the influence of gravity without friction. The cylinder contains a sphere (of mass $m \ll M$) moving up and down in a vertical direction with a velocity c; it is elastically reflected by the piston and by the cylinder head. Neglect the influence of gravity on the motion of the sphere.

a) Establish the condition of equilibrium for the piston and compare it with the perfect-gas equation, ignoring the dimensions of the sphere.

b) Repeat the calculation taking into account that the radius of the sphere is r and compare the result with the van der Waals equation.

c) Imagine that the piston is being withdrawn slowly with a velocity $V_p \ll c$ and compare the loss in energy suffered by the sphere with the work $dW = P\,dV$ of a gas.

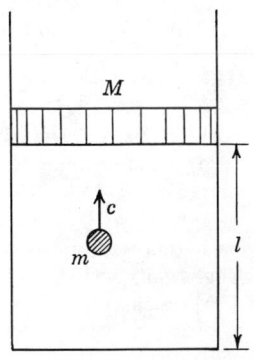

Fig. 38. One-dimensional gas consisting of a single molecule.

III.2. For the Maxwellian velocity distribution (23.9) calculate a) the most probable velocity, b) the mean velocity, and c) the root-mean-square of the velocity.

III.3. Compute the number of H_2 molecules which impinge on an area $\sigma = 1$ cm^2 of a wall in a second with a velocity which exceeds 12000 m/sec, assuming that the temperature is 0 C and that the total number of molecules in 1 cm^3 is 2×10^{19}.

III.4. Calculate the mean value of the number of throws, k, after which the 6 of a dice may be expected to show up. Calculate, further, the mean variance defined as $\overline{(k - \bar{k})^2}$.

III.5. Calculate the pressure exerted by a perfect gas on a wall (assumed placed in the plane $x = 0$ of a rectangular system of coordinates) if the wall attracts the molecules at large distances and repels them at small distances with a force whose potential is $U = - A\,e^{-x} + B\,e^{-2ax}$. a) assuming that the influence of the force extends over a distance which is small compared with the mean free path; b) assuming that both are comparable.

Calculate the distance over which the influence of wall forces would have to extend in helium under normal conditions in order to affect the pressure on the wall.

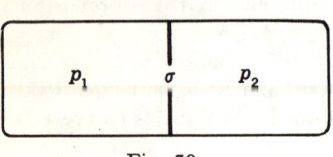

Fig. 39.

The transfer of mass and energy through narrow openings.

III.6. A perfect gas fills two compartments of a vessel which are connected through a very small opening of area σ. The initial temperature in both compartments is equal to, say, T, and the pressures are p_1 and p_2 respectively.

a) Calculate the mass of gas which flows in a unit of time from the compartment with the higher pressure to that with the lower pressure under steady-state conditions ($p_1 = $ const, $p_2 = $ const);

b) Calculate the corresponding rate of transfer of energy;

c) Calculate the mean quantity of energy transferred per particle;

d) Why is it larger than $3/2\,kT$?

e) What measures is it necessary to take to maintain a steady state?

III.7. A gas at temperature T contains a movable plate B placed between two fixed plates A_1 and A_2 at a distance which is small compared with the mean free path, so that intermolecular collisions may be disregarded. Assume that B and A_1 have the same temperature as the gas and that A_2 is heated to a slightly higher temperature $T' = T + \delta T$.

a) Calculate the force acting on the movable plate B assuming that all molecules reach thermal equilibrium with the wall from which they are reflected (infinitely rough wall) and that the plates have equal areas A.

b) Calculate the pressure of the gas from this force (ultra-vacuum pressure gage).

Chapter IV

IV.1. During a certain experimental measurement the result depends in a random way (i. e. either $+ \varepsilon$ or $-\varepsilon$) on a large number n of equal, mutually independent elementary errors. Show that the probability of obtaining an experimental result which deviates from the true one by x is given by $W = a \cdot \exp(-x^2/2\,n\,\varepsilon^2)$ for large values of n. In order to make the derivation more lucid the student may visualize Galton's board: the probability that a sphere impinging on a nail will move to the left of it is the same as that for it to roll over to the right.

IV.2. Assume that the experimental error for a single measurement is not $\pm \varepsilon$, $\varepsilon = $ const, as in the preceding problem, but that it can vary in a certain interval. Let $f_1(x)\,dx$ denote the probability of the error to fall inside the interval $(x, x + dx)$.

a) Derive an expression for the probability $f_n(x)\,dx$ when n independent errors of the same kind are superimposed on each other;

b) Prove that all functions $f_n(x)$ are Gaussian if f_1 is Gaussian. Derive an expression for the spread of the Gaussian curve.

c) Derive an expression for f_n for large values of n when $f_1(x) = 1$ for $|x| < \varepsilon$ and $= 0$ for $|x| > \varepsilon$. Plot the curves f_1, f_2, and f_3 and deduce from it the geometrical form of these functions for a polydimensional cube.

IV.3. Compute the number W of permutations of N molecules,

a) when all molecules have the same velocity $+ \xi$ as regards magnitude and direction;

b) when one half each of them have the velocities $+ \xi$ and $- \xi$ respectively;

c) when one sixth each have the velocities $\pm \xi$, $\pm \eta$, and $\pm \zeta$, respectively. Prove that for $N \to \infty$ each successive distribution is infinitely more probable than the preceding one.

IV.4. A very small mirror is suspended from a quartz strand whose elastic constant is D, and reflects a beam of light in such a way that the angular amplitudes caused by the impacts due to surrounding molecules (Brownian motion) can be read on a suitable scale. The position of equilibrium is at $\phi = 0$ ($\phi = $ angular amplitude). The probability of finding the mirror at an amplitude between ϕ and $\phi + d\phi$ is given by

$$W\,d\phi = a\,\mathrm{e}^{-E_{pot}/kT}\,d\phi, \qquad E_{pot} = \tfrac{1}{2}D\,\phi^2,$$

in accordance with the law of equipartition. From the observed value of $\overline{\phi^2}$ it is possible to determine the Boltzmann constant k. Calculate the numerical value of the Loschmidt-Avogadro number from the following data obtained at $T = 287\ ^\circ\mathrm{K}$: $D = 9.43 \times 10^{-9}$ dyne cm; $\overline{\phi^2} = 4.18 \times 10^{-6}$ using the known value for the universal gas constant, $R = 8.32 \times 10^7$ erg/deg mol (Kappler).

IV.5. Consider a cubic crystal containing $N = 10^{21}$ atoms. The cohesion energy per surface atoms is 9 eV. Calculate the ratio of cohesion energy to thermal energy. Calculate the size of the crystal for which the two are equal.

IV.6. a) Show that the partition function (33.3) of a rotator can be deduced from the definition of the theta function

$$\vartheta_2\left(z\left|\frac{i\,q}{\pi}\right.\right) = 2\sum_{n=0}^{\infty}\mathrm{e}^{-(n+\frac{1}{2})^2 q}\cos(2n+1)\pi z$$

and that we have

$$Z(q) = -\frac{1}{\pi^2} e^{q/4} \int_0^\infty \frac{\partial \vartheta_2\left(z\left|\frac{iq}{\pi}\right.\right)}{\partial z} \frac{dz}{z}.$$

b) Making use of the transformation formula

$$\vartheta_2\left(z\left|\frac{iq}{\pi}\right.\right) = \left(\frac{\pi}{q}\right)^{1/2} e^{-\pi^2 z^2/q} \vartheta_0\left(\frac{-i\pi z}{q}\left|\frac{i\pi}{q}\right.\right)$$

(which is related to that in eq. (15.8) of Vol. VI for $\vartheta = \vartheta_3$) where

$$\vartheta_0(z|\tau) = \sum_{n=-\infty}^{+\infty} (-1)^n \exp\left[i\pi(n^2 \tau + 2 n z)\right]$$

derive the equation

$$Z(q) = \frac{1}{q} e^{q/4} \cdot (\pi q)^{-1/2} \oint \frac{t}{\sin t} e^{-t^2/q} dt$$

in which $\oint$ denotes Cauchy's principal value of the integral. In the process of derivation it is necessary to make use of the expansion of $t/\sin t$ into partial fractions known from the theory of meromorphic functions.

c) Calculate the principal value of the integral in (b): α) by expanding $1/\sin t$ in powers of e^{it}, β) by expanding $t/\sin t$ in powers of t. How far can this be regarded as a proof of the transformation formula $\vartheta_2 \to \vartheta_0$ given in (b)?

d) The expansion (c, β) is a semi-convergent series which gives good approximate values of q. Calculate the molar energy and the molar specific heat of rotation from the first term of this expansion and compare with the result in Sec. 33. Determine the value of q above which c_v is accurate to within 1%.

e) Examine the influence of the higher-order terms on the molar specific heat and compare the two series at $q = 0.458$, $e^{2q} = 2.5$.

IV.7. *Bose Einstein distribution.*

Let f_i, f_i', f_i'', ..., $f_i^{(n)}$, ... denote the probabilities of finding 0, 1, 2, ..., n, ... particles in the i-th phase cell, with the additional conditions:

$$\sum_n f_i^{(n)} = 1, \quad \sum_{n,i} n f_i^{(n)} = N, \quad \sum_{n,i} n f_i^{(n)} \varepsilon_i = U.$$

Calculate the distribution function $n_i = \sum_n n f_i^{(n)}$ at equilibrium if the logarithm of the thermodynamic probability is given by

$$\log W = - \sum_i (f_i^0 \log f_i^0 + f_i' \log f_i' + f_i'' \log f_i'' + \ldots)$$

i. e. as a sum of Boltzmann terms.

IV.8. *Density fluctuations.*

Calculate the density fluctuations in space in a volume element ΔV of a gas which occupies a vessel of volume V, stipulating that the mean density is constant throughout the volume.

IV.9. *Density fluctuations at the critical point.*

Consider a real gas composed of N molecules occupying a volume V.

a) Write down van der Waals' equation for this case and calculate the critical parameters.

b) Calculate the logarithm of the partition function by a thermodynamical argument, using the thermal and caloric ($c_v = 3/2\, R\, T$) equations of state, and taking into account that a real gas behaves like a perfect gas at high temperatures T and low densities N/V.

c) Determine the relative mean fluctuation of the number of particles in volume ΔV making use of the equation

$$\left(\frac{\Delta n}{\overline{n}}\right)^2 = - \frac{(V/N)\, \Delta V}{N\, \partial^2 \log Z / \partial N^2}.$$

Perform the calculation at the critical point assuming small deviations from perfect-gas behavior.

d) Justify this formula with the aid of eq. (40.15) in the text.

Chapter V

V.1. Making use of the energy and momentum equations, deduce the transformation equation for the velocities ($\mathbf{v}_1, \mathbf{v}_2$) before collision and ($\mathbf{v}_1', \mathbf{v}_2'$) after collision when the masses are m_1 and m_2 respectively. Show that Liouville's theorem about the equality of space cells is applicable to this transformation. What is the mean value of the ratio of the energy differences after and before collision?

When evaluating the mean, it is necessary to note that all directions of the central axis are equally probable and that the velocities $\mathbf{v}_1, \mathbf{v}_2$ before the impact are statistically independent.

Hints for the solution of problems

I.1. From $z = z(x, y)$ it follows that

(1) $$dz = \left(\frac{\partial z}{\partial x}\right)_y dx + \left(\frac{\partial z}{\partial y}\right)_x dy$$

for arbitrary changes dx, dy. Assume now that dx and dy are connected in a way to render $dz = 0$. Their ratio must then be equal to $(\partial y/\partial x)_z$ and it follows from (1) that

$$\left(\frac{\partial y}{\partial x}\right)_z = -\left(\frac{\partial z}{\partial x}\right)_y \bigg/ \left(\frac{\partial z}{\partial y}\right)_x,$$

from which the required proposition can easily be deduced.

Applying this result to the functional relation implied by the equation of state $p = p(T, V)$ we have

(2) $$\left(\frac{\partial p}{\partial T}\right)_V \left(\frac{\partial T}{\partial V}\right)_p \left(\frac{\partial V}{\partial p}\right)_T = -1$$

or, when the definitions in (1.5) and (1.6) are taken into account

(3) $$-p\beta \cdot \frac{1}{V\alpha} \cdot \frac{1}{V\varkappa} = -1.$$

In other words $\beta p = \alpha \varkappa V^2$.

I.2. The heating of the room from 0 C to 20 C requires the addition of $c_p \times 20$ deg units of heat. Now $(c_p - c_v)_{mol} = R$ and for diatomic gases we have $c_p/c_v = 1.4$. Hence

(1) $c_{p\,mol} = 3.5\,R,$ (2) $c_p = 3.5\dfrac{R}{\mu} = 3.5 \times \dfrac{2}{29}\,\dfrac{\text{cal}}{\text{g deg}}.$

The value for R was taken from (4.8) and the mean of the molar weights of N_2 and O_2 was used for μ. In order to reduce (2) to a unit of volume it is necessary to multiply by the mean density of the air

$$\rho = 1.25 \text{ kg/m}^3.$$

Multiplying by the temperature increment of 20 C we obtain from (2)

(3) $$\rho \times c_p \times 20 \text{ deg} = \frac{1.25 \times 3.5 \times 2 \times 20}{29} \sim 60\,\frac{\text{kcal}}{\text{m}^3}.$$

It will be noted that too many units of volume of air have been heated owing to its expansion, and that some air escapes because $p = \text{const}$. The intensity of heating required to maintain a temperature of 20 C depends on the tightness of walls and windows and on their heat conductivity and cannot be included in our calculation.

According to Emden (*cf.* Sec. 6 F) the energy imparted to the room on heating is partly lost with the internal energy of the escaping air. Emden's eq. (6.19) can be criticized on the ground that it leaves out of account that the definition of energy leaves a constant undefined and must be replaced by

(4) $$u - u_0 = c_v(T - T_0).$$

Here T_0 denotes some temperature above the point of liquefaction at which the perfect gas law is still valid. Since $u_1 = \rho u$ we have

(5) $$u_1 = c_v \rho T + \rho(u_0 - c_v T_0).$$

The first term on the right-hand side is independent of temperature in view of the relation $\rho T = \mu p/R$, but not the second. The latter is positive because of the large value of the latent heat and considerably exceeds the first term in magnitude. It is seen from this argument that Emden was not justified in restricting himself to the first term in his eqs. (19), (20). It is possible to show that the second term *decreases* with T (because of the relation $\rho = \mu p/RT$). The energy density does not remain constant, as Emden suggests, but even decreases on heating. Thus the remarkable conclusion concerning the preponderance of entropy over energy applies *a fortiori*.

I.3. Assuming that the working fluid in a Carnot cycle is a perfect gas we can write per mol for the isothermal processes $1 \to 2$ and $3 \to 4$ (Q denotes heat added):

$$du = 0, \qquad dq = p\,dv, \qquad Q_1 = \int_1^2 p\,dv = R T_1 \log \frac{v_2}{v_1},$$

$$du = 0, \qquad dq = p\,dv, \qquad Q_2 = \int_4^3 p\,dv = R T_2 \log \frac{v_3}{v_4}$$

so that

(1) $$\frac{Q_1}{Q_2} = \frac{T_1}{T_2} \frac{\log v_2/v_1}{\log v_3/v_4}.$$

On the other hand applying eq. (5.3 a) to the isentropic processes $2 \to 3$ and $4 \to 1$ we have

(2) $$T_1 v_2^{\gamma-1} = T_2 v_3^{\gamma-1} \quad \text{and} \quad T_2 v_4^{\gamma-1} = T_1 v_1^{\gamma-1}.$$

Eliminating T_1 and T_2 we can prove that the ratios v_2/v_1 and v_3/v_4 are equal. Equation (1) transforms into (6.8). In view of (6.7) we see that in fact $\phi(\theta) = T_1$ as postulated in eq. (6.7 a).

I.4. The heat capacity C is defined as the quantity of heat required to increase the temperature of a body by 1 degree. Since the specific heat is referred to a unit of mass, we have $C = M c_v$ if the volume is constant, M denoting the mass of the body under consideration.

In all problems of heat conduction $dW = 0$ because the volume is constant so that $dU = dQ$. In such cases the material theory of heat may be used. This leads to the "mixing rule"

(1) $$T = \alpha_1 T_1 + \alpha_2 T_2, \qquad \alpha_1 = \frac{C_1}{C_1 + C_2}, \qquad \alpha_2 = \frac{C_2}{C_1 + C_2}.$$

According to eq. (5.10) the change in the entropy of the perfect gases 1 and 2 at constant volume is given by

$$\Delta S_1 = C_1 \log \frac{T}{T_1}, \qquad \Delta S_2 = C_2 \log \frac{T}{T_2}.$$

Since entropies are additive we may write:

$$\Delta S = \Delta S_1 + \Delta S_2 = (C_1 + C_2) \log T - C_1 \log T_1 - C_2 \log T_2.$$

The Second Law states that in an isolated system $\Delta S > 0$. Dividing by $C_1 + C_2$ and making use of (1) we find:

(2) $\qquad \alpha_1 T_1 + \alpha_2 T_2 > T_1^{\alpha_1} \times T_2^{\alpha_2} \qquad$ with $\qquad \alpha_1 + \alpha_2 = 1$.

When $\alpha_1 = \alpha_2 = \tfrac{1}{2}$ we obtain the well known rule that "arithmetical mean > geometrical mean".

Equation (2) states: If, on forming an arithmetical mean, the two quantities T_1, T_2 are weighted with the factors α_1, α_2 then on forming the geometrical mean it is necessary to take the weighting factors into account exponentially. The sign $>$ must be replaced by $=$ only in the trivial case when $T_1 = T_2$.

For n gases which are free to exchange heat among each other eq. (2) is replaced by

(3) $\qquad \alpha_1 T_1 + \ldots + \alpha_n T_n > T_1^{\alpha_1} \ldots T_n^{\alpha_n} \qquad$ with $\qquad \alpha_1 + \ldots + \alpha_n = 1$ [1].

The student may try to apply the same method to other irreversible processes!

I.5. The initial state will be assumed to be given by T_0, v_0, p_0. Evaluating the integral of $p\,dV$ between the limits V_0 and $2 V_0$ for

a) $p = p_0$, b) $p = p_0 \dfrac{v_0}{v}$, c) $p = p_0 \dfrac{v_0^\gamma}{v^\gamma}$

---

[1] Algebraic proofs of this proposition can be found in the following — Pólya and Szegö: Aufgaben und Lehrsätze, Springer 1925, or Hardy, Littleword and Pólya: Inequalities, Cambridge Univ. Press, 1934.

we obtain, consecutively, the following expressions for work

a) RT_0,   b) $RT_0 \log 2$,   c) $RT_0 \dfrac{1-2^{1-\gamma}}{\gamma-1}$.

The quantity of heat required can be found from the integral of $dq = c_v\,dT + p\,dv = c_p\,dT - v\,dp$. Hence, in the same order as before:

a) $RT_0 \dfrac{c_p}{R}$,   b) $RT_0 \log 2$,   c) 0.

The change in entropy in each case is found from $s = c_v \log T + R \log v + \text{const.}$, and is:

a) $c_p \log 2$,   b) $R \log 2$,   c) 0.

I.6. From (7.8) and (7.8 a) we deduce

$$ds = \frac{1}{T}(du + p\,dv) = \frac{c_v}{T}dT + \left(\frac{\partial p}{\partial T}\right)_v dv.$$

Expressing $(dp/dT)_v$ with the aid of eq. (2) in the solution of Prob. I.1. and introducing the coefficient of thermal expansion α and that of compressibility $\varkappa$, we have

$$ds = \frac{c_v}{T}dT + \frac{\alpha}{\varkappa}dv.$$

The slope of the isentrope in the T, v-diagram is found by putting $ds = 0$. Hence

$$\left(\frac{\partial v}{\partial T}\right)_s = -\frac{c_v \varkappa}{\alpha T}.$$

It is seen that the slope of the isentrope is positive to the left of the minimum in the v vs. T curve because $\alpha < 0$ there. To the right $\alpha > 0$ and the slope is negative, whereas at the minimum itself the isentrope is parallel to the v-axis. Making a qualitative sketch of the isobars in a T, v-diagram it is easy to convince oneself that there are no isentropes which intersect both the 2 and 6 degree isotherms, provided that the process is carried out in a range of pressures where the isobars possess a minimum of volume between 2 C and 6 C.

I.7. At constant v we have $dq = c_v\,dT = c_v(\partial T/\partial p)_v\,dp$. At constant pressure p we have $dq = c_p\,dT = c_p(\partial T/\partial v)_p\,dv$. Hence in general

$$T\,ds = dq = c_p\left(\frac{\partial T}{\partial v}\right)_p dv + c_v\left(\frac{\partial T}{\partial p}\right)_v dp.$$

The isentropic compressibility is obtained by putting $ds = 0$ in (1). Hence from (1) we have

$$\left(\frac{\partial v}{\partial p}\right)_s = -\frac{c_v}{c_p}\left(\frac{\partial T}{\partial p}\right)_v \left(\frac{\partial v}{\partial T}\right)_p = +\frac{c_v}{c_p}\left(\frac{\partial v}{\partial p}\right)_T,$$

where eq. (2) from Problem I.1. has been utilized.

I.8. From (7.8) and (7.8 a) we have

$$dq = du + p\,dv = c_v\,dT + T\left(\frac{\partial p}{\partial T}\right)_v dv\,;$$

During isothermal compression $dT = 0$; further we may write $dv = (\partial v/\partial T)_T\,dp$. Then eq. (2) from Problem I.1. yields

$$dq = T\left(\frac{\partial p}{\partial T}\right)_v \left(\frac{\partial v}{\partial p}\right)_T dp = -T\left(\frac{\partial v}{\partial T}\right)_p dp.$$

Assuming a mean constant coefficient of expansion in the interval of pressure under consideration we find that the heat *added* is:

$$\int dq = -T v\,\alpha\,\varDelta\,p = -293 \text{ deg} \times 1\,\text{dm}^3 \times 2 \times (10^{-4}/\text{deg}) \times (19\,\text{kg/cm}^2) =$$
$$= -1.113 \text{ liter} \times \text{atm} = -26.1 \text{ cal}$$

(since 1 liter $\times$ atm = 1 dm^3 $\times$ kp cm^{-2} = 23.43 cal).
The work done on the system is

$$-\int p\,dV = -\int p\left(\frac{\partial V}{\partial p}\right)_T = +V \varkappa \varDelta \left(\frac{p^2}{2}\right)$$

$$= 1\,\text{dm}^3 \times 0.5 \times (10^{-4}/\text{at}) \times \tfrac{1}{2}(400-1)\,\text{at}^2$$

$$= 10^{-2} \text{ liter} \times \text{atm} = 0.23 \text{ cal}.$$

I.9. See Vol. II, Sec. 7.

I.10. Euler's equation, Vol. II, eq. (11.5) can be written

(1)
$$\operatorname{grad}\frac{\mathbf{v}^2}{2} = -\frac{1}{\rho}\operatorname{grad} p$$

if we neglect external forces, assume $\partial \mathbf{v}/\partial t = 0$ and curl $\mathbf{v} = 0$, i. e. steady and irrotational flow, and if we take account of eq. (6) *loc. cit.* The condition of incompressibility, *loc. cit.* (4 a), must now be replaced by the relation between p and ρ for isentropic processes

(2)
$$p = p_0(\rho/\rho_0)^\gamma.$$

Eliminating ρ from (1) with the aid of (2) we can integrate (1) to give:

$$(3) \qquad \frac{v^2}{2} - \frac{v_0^2}{2} = \frac{p_0}{\rho_0} \frac{\gamma}{\gamma-1} \left[1 - \left(\frac{p}{p_0}\right)^{\frac{\gamma-1}{\gamma}}\right] = c_p T_0 \left[1 - \left(\frac{p}{p_0}\right)^{\frac{\gamma-1}{\gamma}}\right].$$

According to (5.3 a) we have $(p/p_0)^{(\gamma-1)/\gamma} = T/T_0$. Hence

$$\frac{v^2}{2} - \frac{v_0^2}{2} = c_p T_0 - c_p T.$$

This shows that, in fact, the difference between the kinetic energies per gram is equal to the difference in the specific enthalpies per gram, $c_p T_0$ and $c_p T$, of the gas.

In our example $c_p = 0.49$ cal g^{-1} deg^{-1}, $\gamma = 1.33$. Hence (3) yields $v = 880$ m/sec if we assume $v_0 \approx 0$. This value may be compared with the mean molecular velocity of 770 m/sec for steam at 300 C (*cf.* Sec. 22). — The temperature after expansion is 110 C.

I.11. At first we shall assume an arbitrary equation of state. The state of the gas in cell i of volume V_i will be determined by indicating the temperature T_i and the specific volume v_i. The mass enclosed in a cell is then equal to V_i/v_i. The specific internal energy in cell i will be denoted by $u_i(T_i, v_i)$ so that the internal energy of the gas contained by cell i is $\dfrac{V_i}{v_i} u_i(T_i, v_i)$, its potential energy being $\dfrac{V_i}{v_i} g z_i$, where g denotes the gravitational acceleration. Let U denote the total energy, N the total mass in the system. Then

$$N = \sum_i \frac{V_i}{v_i}, \qquad U = \sum_i \frac{V_i}{v_i} [u_i(T_i, v_i) + g z_i].$$

In a similar way we can calculate the total entropy

$$S = \sum_i \frac{V_i}{v_i} s_i(T_i, v_i),$$

where $s_i(T_i, v_i)$ is the specific entropy for the state prevailing in cell i.

Now, at equilibrium S is a maximum with respect to all variations of the T_i, v_i which are compatible with the constant values of N and U. Introducing two Lagrangian multipliers λ and μ we must also have

$$\delta S \equiv \delta \left[\sum_i \frac{V_i}{v_i} s_i(T_i, v_i) + \lambda \left(N - \sum_i \frac{V_i}{v_i}\right) + \mu \left(U - \sum_i \frac{V_i}{v_i} [u_i(T_i, v_i) + g z_i]\right) \right] = 0$$

for arbitrary variations of the quantities T_i, v_i, λ, μ. This yields

$$(1) \qquad \frac{V_i}{v_i} \left(\frac{\partial s_i}{\partial T_i} - \mu \frac{\partial v_i}{\partial T_i}\right) = 0,$$

(2) $$\frac{V_i}{v_i^2}\left(-s_i + v_i \frac{\partial s_i}{\partial v_i}\right) + \lambda \frac{V_i}{v_i^2} + \mu \frac{V_i}{v_i^2}\left(u_i + g z_i - v_i \frac{\partial u_i}{\partial v_i}\right) = 0.$$

It follows from (1) that $\mu = 1/T_i$ if we take into account that $\dfrac{\partial s_i}{\partial T_i} = \dfrac{1}{T_i}\left(\dfrac{\partial u_i}{\partial T_i}\right)_{v_i}$, see (7.1). In other words all T_i's are equal, and we may put $T_i = T$. From the two thermodynamic relations (7.7) and (7.8) and from (2) we find that

$$\left(\frac{\partial s_i}{\partial v_i}\right)_{T_i} = \left(\frac{\partial p_i}{\partial T_i}\right)_{v_i}, \qquad \left(\frac{\partial u_i}{\partial v_i}\right)_{T_i} = T_i\left(\frac{\partial p_i}{\partial T_i}\right)_{v_i} - p_i,$$

and after a short calculation

$$g(T, v_i) \equiv u_i + g z_i - T s_i + v_i p_i = -\lambda T$$

The left-hand side contains the specific Gibbs potential for the state prevailing in cell i, the potential energy having been added to the internal energy. The right-hand side is independent of i, which must also be true of the left-hand side. In the case of a perfect gas (for the sake of simplicity we shall assume that the specific heat is independent of temperature)

$$u_i = c_v T + u_0, \qquad s_i = c_v \log T + \frac{R}{\mu} \log v_i + s_0, \qquad p_i = \frac{R T}{\mu v_i},$$

where now μ denotes molecular weight.

Since $g(T, v_i)$ is independent of i and so of z, we have

$$v(z) = v_0 e^{\mu g z/RT}, \qquad p(z) = p_0 e^{-\mu g z/RT}$$

For a height Δz defined by

$$\mu g \Delta z = R T$$

we have finally

$$\Delta z = \frac{R T}{\mu g} = \frac{p_0 v_0}{g} = \frac{p_0}{\rho_0 g}.$$

It is seen to be equal to the mean height

$$\Delta z = \int_0^\infty z\, p(z)\, dz \bigg/ \int_0^\infty p(z)\, dz.$$

With $p_0 = 10^6$ dyne cm^{-2}, $\rho_0 = 0.0012939$ g cm^{-3}, $g = 981$ cm sec^{-2} Δz has the numerical value of

$$\Delta z = 8 \times 10^5 \text{ cm} = 8 \text{ km}.$$

When a gas expands in a vertical direction at a height which is considerably below 8 km the decrease in pressure with altitude is small; for a difference in altitude of 80 m it is only 1% but it must be taken into account when barometric readings published by meterological stations are reduced to sea level.
I.12.

1. Substituting p from the equation of state into the thermodynamic relation $(\partial u/\partial v)_T = -p + T(\partial p/\partial T)_v$, we obtain the following partial differential equation for u:

$$\frac{\partial u}{\partial v} = -\frac{\alpha}{v} u + \frac{\alpha T}{v} \frac{\partial u}{\partial T}$$

whose general solution is:

$$u = v^{-\alpha} \phi(T v^\alpha);$$

there $\phi(T v^\alpha)$ is an arbitrary function, as is easy to verify by substitution. The statement about entropy is obtained without difficulty from

$$T\, ds = du + p\, dv.$$

2. If u/v depends only on T the expression $v^{-\alpha-1} \phi(T v^\alpha)$ must be independent of v which means that $\phi(T v^\alpha) = \sigma \cdot (T v^\alpha)^{(\alpha+1)/\alpha}$ where σ is a constant; this proves the assertion. For black body radiation we have $p = \frac{1}{3} u/v$, i. e. $\alpha = \frac{1}{3}$, whence we deduce the Stefan-Boltzmann law $u/v = \sigma T^4$. For $\alpha = 2/3$ we have $u/v = \sigma T^{5/2}$ which is a known result from Einstein's condensation theory, cf. (38.27).

3. a) We have $p = \alpha\, C\, T^m v^{\alpha m - \alpha - 1}$; $(\partial p/\partial v)_T \leq 0$ means

$$\alpha(m-1) - 1 \leq 0.$$

b) We have

$$v = \left(\frac{p}{\alpha C}\right)^{\frac{1}{\alpha(m-1)-1}} \times T^{\frac{m}{-\alpha(m-1)+1}}.$$

If we are to have $v > 0$ for $T \to 0$ and any p, we must also have

$$\alpha(m-1) - 1 \geq 0.$$

c) The two inequalities in a) and b) are then, and only then, satisfied simultaneously if $\alpha(m-1) - 1 = 0$, i. e. if $m = 1 + 1/\alpha$ so that

$$u = C\, T^{\frac{\alpha+1}{\alpha}} \times v; \qquad p = \alpha\, C \times T^{\frac{\alpha+1}{\alpha}}.$$

From $u = C\, T^m v^{\alpha(m-1)} > 0$ and $< \infty$ for $T = 0$ we conclude that $m = 0$ so that $u = v^{-\alpha} \times$ const and $p\, v^{\alpha+1} =$ const. With $\alpha = 2/3$ we are led to the relation between the pressure and the volume of electron gas in (39.10)

PROBLEMS

II.1. It follows from (14.11 b) that

(1) $$d(p\,V) = p\,dV + V\,dp = S\,dT + p\,dV + \sum_i n_i\,d\mu_i.$$

From (1) we deduce the relations

$$\left(\frac{\partial S}{\partial V}\right)_{T,\mu_i} = \left(\frac{\partial p}{\partial T}\right)_{V,\mu_i}; \quad \left(\frac{\partial S}{\partial \mu_i}\right)_{T,V,\mu_j} = \left(\frac{\partial n_i}{\partial T}\right)_{V,\mu_j}; \quad \left(\frac{\partial n_i}{\partial \mu_j}\right)_{T,V,\mu_k} = \left(\frac{\partial n_j}{\partial \mu_i}\right)_{T,V,\mu_k}.$$

II.2. From the definition $r = \Delta h$ we find

$$dr = \Delta\left\{\left(\frac{\partial h}{\partial T}\right)_p dT + \left(\frac{\partial h}{\partial p}\right)_T dp\right\},$$

by taking the differential along the vapor-pressure curve $\phi(p, T) = 0$. Since dp and dT both derive from $\phi = 0$, we obtain

(1) $$\left(\frac{\partial r}{\partial T}\right)_\phi = \Delta c_p + \frac{r}{T\,\Delta v}\,\Delta\left(\frac{\partial h}{\partial p}\right)_T,$$

by making use of (4.11) together with the Clapeyron equation. According to the Table in Sec. 7 we have

$$dh = T\,ds + v\,dp \quad \text{and hence} \quad \left(\frac{\partial h}{\partial p}\right)_T = T\left(\frac{\partial s}{\partial p}\right)_T + v.$$

From the same Table we take the relation

$$\left(\frac{\partial s}{\partial p}\right)_T = -\left(\frac{\partial v}{\partial T}\right)_p; \quad \text{hence} \quad \left(\frac{\partial h}{\partial p}\right)_T = v - T\left(\frac{\partial v}{\partial T}\right)_p = v - T\,v\,\alpha,$$

where α denotes the coefficient of thermal expansion. Consequently,

(2) $$\Delta\left(\frac{\partial h}{\partial p}\right)_T = \Delta v - T\,\Delta(v\,\alpha).$$

Substitution into (1) yields

(3) $$\left(\frac{dr}{dT}\right)_\phi = \Delta c_p + \frac{r}{T} - r\,\frac{\Delta(v\alpha)}{\Delta v}.$$

It will be noted that the preceding derivation introduces no approximations as regards the behavior of the liquid or vapor.

II.3. We shall denote the mass of the steam by x, and that of the water by $1\text{ kg} - x$. The volume of the steam becomes $x \times v_2$ and that of the water is $(1\text{ kg} - x)\,v_1$. Thus x is determined by the equation

$$x\,v_2 + (1\text{ kg} - x)\,v_1 = 20\text{ dm}^3$$

PROBLEMS 365

and hence

$$x = (20 \text{ dm}^3 - v_1 \times 1 \text{ kg})/(v_2 - v_1)$$

so that $x = 0.169$ g at 10 C, $x = 149.7$ g at 200 C.

The quantities of heat added during the three subsidiary processes are:

a) Condensing from 0.169 g at 10 C: $\quad -0.169 \text{ g} \times 591.6 \dfrac{\text{cal}}{\text{g}} = -100 \text{ cal.}$

b) Heating from 10 C to 200 C: $\quad 1 \text{ kg} \times 190 \text{ deg} \times 1 \text{ cal g}^{-1} \text{ deg}^{-1} = 190{,}000 \text{ cal.}$

c) Evaporation of 149.7 g at 200 C: $\quad 149.7 \text{ g} \times 463.5 \dfrac{\text{cal}}{\text{g}} = 69{,}400 \text{ cal.}$

$$\overline{259{,}300 \text{ cal.}}$$

The work of compression or expansion is: a) $+ 19 \text{ dm}^3 \times 0.0125 \text{ kp cm}^{-2} = 0.238 \text{ dm}^3 \text{ kp cm}^{-2}$, c) $-18.84 \text{ dm}^3 \times 15.86 \text{ kp cm}^{-2} = -298.8 \text{ dm}^3 \text{ kp cm}^{-2}$; in all $-298.6 \text{ dm}^3 \text{ kp cm}^{-2} = -7000 \text{ cal}$. The total energy to be added is thus equal to 252.3 kcal.

II.4. We have

$$\frac{dT}{T} = \frac{v_{sat} - v_{liq}}{r} dp.$$

Integrate first between 0 and 100 C. Let T_0 denote the absolute temperature of the ice point, the absolute temperature of the boiling point of water at 760 torr being defined as $T_0 = 100$ C. Let, further, p_0 and p_{100} denote the vapor pressures of the fluid under consideration at 0 C and 100 C respectively. We then find:

(1) $$\log \frac{T_0 + 100 \text{ deg}}{T_0} = \int_{p_0}^{p_{100}} \frac{v_{vap} - v_{liq}}{r} dp.$$

The integral on the right-hand side can be evaluated numerically on the assumptions given which leads to an equation for T_0. The absolute temperature T which corresponds to any other vapor pressure is given by

(2) $$\log \frac{T}{T_0} = \int_{p_0}^{p} \frac{v_{vap} - v_{liq}}{r} dp.$$

According to the general propositions of thermodynamics, T_0 is independent of the liquid selected. In the same way T from (2) is the same for different liquids provided that they are in thermal equilibrium.

II.5. The variation in the vapor pressure in each of the intervals of 10 C is too large to permit the substitution of $\Delta p/\Delta T$ for dp/dT. Furthermore, according to the terms of the problem it is necessary to integrate the Clausius-Clapeyron equation

(1) $$\frac{dp}{dT} = \frac{r}{T(v_{vap}-v_{liq})}$$

exactly, assuming that r is constant in the interval of 10 C, and that the vapor behaves like a perfect gas. The quantities r, v_{vap}, and v_{liq} will be referred to one mol. Hence $v_{vap} = RT/p$, and v_{liq} may be neglected with respect to v_{vap}. Thus we obtain from (1) that $d \log p = -r/R \cdot d(1/T)$ and

$$\log \frac{p_2}{p_1} = \frac{r}{R}\left(\frac{1}{T_1} - \frac{1}{T_2}\right)$$

or

$$\frac{r}{R} = \frac{T_1 T_2}{T_2 - T_1}\left(\log \frac{p_2}{p_1}\right)^{-1}.$$

Substituting the numerical values for the two intervals, we have

$$r = 7413\,R \quad \text{and} \quad r = 7045\,R.$$

The assumption that the latent heat is constant does not apply over large temperature intervals; the above numerical values must be regarded as average values over each of the intervals; they will be assumed to refer to 55 C and 305 C or to 328 K and 578 K respectively. Linear interpolation now yields:

(2) $$r = (7896\,\text{C} - 1.472\,T)\,R.$$

In order to determine the boiling point at 760 torr we can integrate (1) more accurately by substituting (2). Thus

$$d \log p = + (7896\,\text{C} - 1.472\,T)\frac{dT}{T^2}$$

and

(3) $$\log \frac{p}{p_2} = 7896\,\text{C}\left(\frac{1}{T_2} - \frac{1}{T}\right) - 1.472 \log \frac{T}{T_2}.$$

Here p_2 denotes the vapor pressure of 305 torr at 310 C = 583 K. Substituting the value of 760 torr for p we can determine the boiling point T at that pressure from (3). We obtain

$$\frac{1000\,\text{C}}{T} = 1.600 - 0.1864 \log \frac{T}{583\,\text{C}}.$$

This transcendental equation is best solved by the method of successive approximations by assuming an approximate value for the unknown temperature on the right-hand side and by calculating T on the left-hand side. The new value may again be inserted on the right-hand side and the process may be repeated.

The first approximation may be chosen at, say, 630 K leading to $T = 630.7$ K. Repeated substitution yields the same value of T again, including the first decimal. We conclude, therefore, that the boiling point is at 630.7 K or 357.7 C.

II.6. We refer to eq. (13.13) and proceed by forming the variation

$$(1) \qquad \delta G = \sum_i \delta n_i \left\{ g_i(T, p) - R T \log \frac{n}{n_i} \right\}.$$

Since transitions $0 \longleftrightarrow 1$ and $0 \longleftrightarrow 2$ are possible we can vary n_1 and n_2 independently, and we must have $\delta n_0 = -\delta n_1 - \delta n_2$. The problem considered in Sec. 13 involved only one chemical reaction and so only one of the n_i's could be varied arbitrarily; the remaining δn_j's were determined.

It follows from (1) that

$$(2) \qquad g_1 - g_0 = R T \log \frac{n_0}{n_1}; \qquad g_2 - g_0 = R T \log \frac{n_0}{n_2}.$$

Since $g_i(T, p) = u_i(T, p) - T s_i(T, p) + p v_i$, where $v_0 = v_1 = v_2$, and the entropy constants are equal, $s_0 = s_1 = s_2$, we find $g_1 - g_0 = u_1 - u_0 = L(\varepsilon_1 - \varepsilon_0)$; $g_2 - g_0 = u_2 - u_0 = L(\varepsilon_2 - \varepsilon_0)$, where L denotes the number of molecules in the mass to which g_i has been referred, i. e. per mol in our case. Hence

$$(3) \qquad \frac{n_1}{n_0} = \exp\left\{-\frac{L(\varepsilon_1 - \varepsilon_0)}{R T}\right\}; \qquad \frac{n_2}{n_0} = \exp\left\{-\frac{L(\varepsilon_2 - \varepsilon_0)}{R T}\right\}.$$

This is the Maxwell-Boltzmann law, and we have deduced it here from thermodynamical considerations. We shall deduce it once more with the aid of the methods of statistical mechanics in Sec. 29. The inclusion of transitions $1 \longleftrightarrow 2$ makes no change in the argument. The generalization to any number of excited states is immediate.

II.7. a) The n_i's will now be assumed to refer to our arbitrary composition. The equilibrium values from the preceding problem will be denoted by $\bar{n}_i$. Thus

$$\mu_1 - \mu_0 = g_1 - g_0 - R T \log \frac{n_0}{n_1} = R T \log \frac{\bar{n}_0}{\bar{n}_1} - R T \log \frac{n_0}{n_1} = R T \left(\log \frac{n_1}{\bar{n}_1} - \log \frac{n_0}{\bar{n}_0} \right)$$

or, for $|n_1 - \bar{n}_1| \ll \bar{n}_1$, i. e. in the neighborhood of equilibrium

$$(1) \qquad \mu_1 - \mu_0 = R T \left(\frac{n_1 - \bar{n}_1}{\bar{n}_1} - \frac{n_0 - \bar{n}_0}{\bar{n}_0} \right), \qquad \mu_2 - \mu_0 = R T \left(\frac{n_2 - \bar{n}_2}{\bar{n}_2} - \frac{n_0 - \bar{n}_0}{\bar{n}_0} \right).$$

b) The variation in the n_i's with time is given by the differential equations

(2)
$$\begin{cases} \dfrac{dn_0}{dt} = -k_{00} n_0 + k_{01} n_1 + k_{02} n_2 \\[4pt] \dfrac{dn_1}{dt} = k_{10} n_0 - k_{11} n_1 + k_{12} n_2 \\[4pt] \dfrac{dn_2}{dt} = k_{20} n_0 + k_{21} n_1 - k_{22} n_2 \end{cases}$$

The sum of all n_i's is constant irrespective of their individual values. Hence

(3) $\quad k_{00} = k_{10} + k_{20}, \qquad k_{11} = k_{01} + k_{21}, \qquad k_{22} = k_{02} + k_{12}.$

The first of these two relations states that the mass lost by state 0 per second, namely $k_{00} n_0$ will be found as $k_{10} n_0$ and $k_{20} n_0$ at states 1 and 2. The same can be said of the remaining two relations.

c) Equilibrium means that $dn_i/dt = 0$, or that

(4)
$$\begin{cases} k_{00} \overline{n}_0 = k_{01} \overline{n}_1 + k_{02} \overline{n}_2 \\ k_{11} \overline{n}_1 = k_{10} \overline{n}_0 + k_{12} \overline{n}_2 \\ k_{22} \overline{n}_2 = k_{20} \overline{n}_0 + k_{21} \overline{n}_1. \end{cases}$$

The state of equilibrium is thus seen not to be one in which no more transitions occur. Transitions which cause the mass of one component to increase turn out to be as frequent as those causing an opposite effect. Of the three equations in (4) at most two are mutually independent because of the constraint in (3). They determine the ratios $\overline{n}_1/\overline{n}_0$ and $\overline{n}_2/\overline{n}_0$ in terms of the k_{ik}'s which are, in turn, independent of composition as assumed at the outset. Recalling eq. (3) of the preceding solution it is seen that the coefficients k_{ik} satisfy two more relations in addition to the relations (3) above. Having proved that the ratios of the $\overline{n}_i$ are independent of composition we have shown the plausibility of our assumption (2), even if we have not actually proved it. The same conclusion would remain true even if the right-hand sides of (2) were replaced by arbitrary functions, each being a linear combination of the n_i, provided that the functions vanish for the argument 0.

d) It follows from (1) and from $(n - \overline{n}_0) + (n_1 - \overline{n}_1) + (n_2 - \overline{n}_2) = 0$ (preservation of total mass) that

$$RT \frac{n_0 - \overline{n}_0}{\overline{n}_0} = -\frac{\overline{n}_1}{n} (\mu_1 - \mu_0) - \frac{\overline{n}_2}{n} (\mu_2 - \mu_0),$$

$$RT \frac{n_1 - \overline{n}_1}{\overline{n}_1} = \left(1 - \frac{\overline{n}_1}{n}\right)(\mu_1 - \mu_0) - \frac{\overline{n}_2}{n}(\mu_2 - \mu_0),$$

$$RT \frac{n_2 - \overline{n}_2}{\overline{n}_2} = -\frac{\overline{n}_1}{n}(\mu_1 - \mu_0) - \left(1 - \frac{\overline{n}_2}{n}\right)(\mu_2 - \mu_0);$$

because $n = n_0 + n_1 + n_2 = \overline{n}_0 + \overline{n}_1 + \overline{n}_2$. We substitute this in the right-hand side of (2) having replaced all n_i's by $n_i - \overline{n}_i$, this being permissible in view of (4). We then obtain:

$$RT\frac{dn_1}{dt} = -k_{11}\overline{n}_1(\mu_1 - \mu_0) + k_{12}\overline{n}_2(\mu_2 - \mu_0)$$

$$RT\frac{dn_2}{dt} = k_{21}\overline{n}_1(\mu_1 - \mu_0) - k_{22}\overline{n}_2(\mu_2 - \mu_0),$$

taking (4) into account once more. We now compare this expression with (21.26). Onsager's reciprocal relation $a_{12} = a_{21}$ is equivalent to:

(5) $$k_{12}\overline{n}_2 = k_{21}\overline{n}_1.$$

Making use of (3) and (4) we deduce the further relations

(6) $$k_{01}\overline{n}_1 = k_{10}\overline{n}_0, \quad k_{02}\overline{n}_2 = k_{20}\overline{n}_0.$$

Now, $k_{12} n_2$ denotes the mass transferred from component 2 to component 1 per second, $k_{21} n_1$ denotes the mass transferred per second from component 1 to component 2. Onsager's reciprocal relation leads to the conclusion that these two expressions are equal at equilibrium. In other words the reaction $1 \rightleftarrows 2$ is in equilibrium on its own, independently of the fact that the reactions $0 \rightleftarrows 1$ and

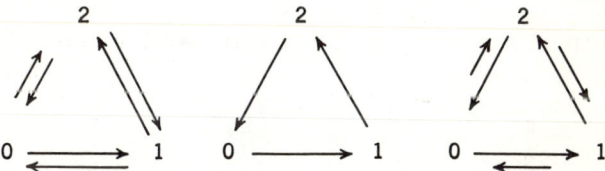

Fig. 40. Illustrating detailed, cyclic, and mixed equilibrium.

$0 \rightleftarrows 2$ take place simultaneously. In the same way (6) shows that the remaining reactions are in equilibrium on their own. The preceding proposition enjoys a much wider area of validity and is known in chemistry as the principle of detailed equilibrium.

In addition to detailed equilibrium we could also imagine cyclic or mixed equilibrium. The three cases have been shown schematically in Fig. 38. The length of the arrows is assumed to be proportional to the frequency of the individual transitions. The case of detailed equilibrium is the only one of the three which is unaffected when one transition (say 1, 2) is impeded in *both* directions, for example by the introduction of a decatalyzer. In the case of cyclic equilibrium the suppresion of, say, the reaction $1 \rightarrow 2$ would, at first, leave the reactions $2 \rightarrow 0$ and $0 \rightarrow 1$ unaffected so that the masses of components 0 and 1 would increase compared with equilibrium. The same is true for the case of mixed equilibrium. According to our laws of thermodynamics a state of equilibrium cannot be affected by the introduction of a catalyzer or decatalyzer, once it has set in, and provided that no constraints are removed. Hence the principle of detailed equilibrium can also be stated as follows: There can be no decatalyzer which would in different ways affect a reaction and its opposite, such as would, for example, suppress one and not the other.

The ideas expressed here in the form of the principle of detailed equilibrium will be seen to recur in the kinetic theory of gases, or in the electron theory of metals. In the last analysis it is always a consequence of the equality in the contributions of both quantum mechanical matrix elements for the transition between two steady states.

III.1. a) Assume that the velocity of the piston before impact is v. On impact its velocity will change sign and

(1) $$M v = m c.$$

The time of rise t_s for the piston is equal to $t_s = v/g$, the return time is $t_r = 2 v/g$ and the latter must be equal to that for the sphere, $t_r' = 2 l/c$. Hence

(2) $$v c = l g.$$

The mean force on the piston is

(3) $$K = \frac{2 m c}{t_r} = m g \frac{c}{v} = M g,$$

as seen from (1). Equations (1) and (2) (A = cross-sectional area) also show that

$$m c^2 = M v c = M g l = \frac{M g}{A} \cdot A l,$$

or, introducing the pressure $p = M g/A$ and the volume $V = A l$:

(4) $$p \cdot V = m c^2 = 2 U.$$

instead of the value of $2/3\ U$ for a three-dimensional gas.

b) Equation (1), elevation and return time of the piston, and eq. (3) remain unaltered. The length l in (2) should be replaced by $l - 2 r$. It follows from the modified eq. (2) that

$$m c^2 = M g(l - 2 r)$$

or

(5) $$p(V - 2 r A) = 2 U.$$

It is seen that the volume V is decreased by a constant term in the same way as in the van der Waals equation. In the present case the term to be subtracted is equal to a layer of the thickness of the sphere covering the whole cross-section.

c) Imagining that the piston is very heavy (so that the change of its velocity may be ignored), we may apply the law of reflection. If c is the velocity before impact and c' that after it, we have

$$c - V_p = c' + V_p.$$

The change in kinetic energy is

$$\frac{1}{2} m(c'^2 - c^2) \approx -2 m c V_p.$$

During a time interval, Δt, this quantity of energy is transferred $\frac{\Delta t}{t_r} = \frac{c \Delta l}{2 l}$ times; substituting the distance $\Delta x = V_p \Delta t$ traversed by the piston, we have:

(6) $\quad -\Delta W = -2 m c V_p \cdot \dfrac{c \Delta t}{2 l} = -\dfrac{m c^2}{l} \cdot \Delta x = -M g \Delta x = -p \Delta V.$

III.2. a) Taking the logarithmic derivative of

$$\phi(c) = 4 \pi c^2 \left(\frac{m}{2 \pi k T} \right)^{3/2} e^{-\frac{m c^2}{2 k T}}$$

with respect to c, we have

$$\frac{\phi'}{\phi} = \frac{2}{c} - \frac{2 m c}{2 k T};$$

putting $\phi' = 0$ we find that

(1) $\quad c_w = \left(\dfrac{2 k T}{m} \right)^{1/2}$

which agrees with (23.10).

b) Recalling that $\bar{c} = \int_0^\infty c \phi \, dc$ and putting $m c^2 / 2 k t = \xi^2$, we have

$$\bar{c} = 4 \left(\frac{2 k T}{\pi m} \right)^{1/2} \int_0^\infty e^{-\xi^2} \xi^3 \, d\xi.$$

The integral can be reduced to one of Laplace's type and its numerical value is $\frac{1}{2} \times 1!$. Hence

(2) $\quad \bar{c} = \left(\dfrac{8 k T}{\pi m} \right)^{1/2}.$

c) Recalling that $\overline{c^2} = \int_0^\infty c^2 \phi(c)\, dc$ we have:

$$\overline{c^2} = 4 \cdot \frac{2kT}{m} \cdot \frac{1}{\sqrt{\pi}} \int_0^\infty e^{-\xi^2} \xi^4\, d\xi.$$

We compute

$$\int_0^\infty e^{-\gamma \xi^2} \xi^4\, d\xi = \frac{\sqrt{\pi}}{2} \frac{d^2}{d\gamma^2} \frac{1}{\sqrt{\gamma}} = \frac{3}{8} \sqrt{\pi}\, \gamma^{-5/2}.$$

Consequently

(3) $$\overline{c^2} = \frac{3kT}{m}$$

The velocity squares form the proportion

(4) $$\overline{c^2} : \overline{c}^2 : c_w^2 = 3 : \frac{8}{\pi} : 2$$

which leads to eq. (23.11).

III.3. The number of molecules per second coming from the cylinder in Fig. 23 on p. 170 is equal to

$$dZ = \sigma\, c_z \cdot n \left(\frac{m}{2\pi kT}\right)^{3/2} e^{-\frac{m}{2kT}(c_x^2 + c_y^2 + c_z^2)} dc_x\, dc_y\, dc_z.$$

The total number is obtained by integrating over c_x, c_y from $-\infty$ to $+\infty$ and over c_z from $c_0 = 12000$ m/sec to ∞. Introducing cylindrical polar co-ordinates and putting

$$\left(\frac{m}{2kT}\right)^{1/2} \cdot (c_x, c_y, c_z) = (\rho \cos \phi, \rho \sin \phi, \zeta),$$

we have

$$Z = n\sigma \left(\frac{2kT}{m}\right)^{1/2} \frac{1}{\pi^{3/2}} \int_0^\infty e^{-\rho^2} \rho\, d\rho\, d\phi \int_{\left(\frac{m}{2kT}\right)^{1/2} c_0}^\infty e^{-\zeta^2} \zeta\, d\zeta = n\sigma \left(\frac{kT}{2\pi m}\right)^{1/2} \times \exp\left(-\frac{mc_0^2}{2kT}\right).$$

Introducing the mean velocity

$$\overline{c} = 4\left(\frac{kT}{2\pi m}\right)^{1/2}$$

we find

$$Z = \frac{n\,\sigma\,\overline{c}}{4} \exp\left(-\frac{m\,c_0^2}{2\,k\,T}\right).$$

The numerical result is: $\overline{c} = 1690$ m/sec, $1/4\,n\,\sigma\,\overline{c} = 8.45 \times 10^{23}$; $m\,c_0^2/2\,k\,T = 64$. Hence $Z = 5 \times 10^{-3}$ sec^{-1}. A particle possessing such a very high velocity arrives about once every 3 minutes.

III.4. See Sec. 27 D, eq. (14). The probability that the 6 will show up for the first time after the

        1.   2.   3.   ... k'th... throw

is, respectively

$$W_k = \frac{1}{6},\ \frac{5}{6} \times \frac{1}{6},\ \left(\frac{5}{6}\right)^2 \times \frac{1}{6}, \ldots, \left(\frac{5}{6}\right)^{k-1} \times \frac{1}{6}, \ldots$$

The calculations of the mean values of $k, k^2, \ldots$ is best performed with the aid of the generating function

$$f(t) = \sum_{k=1}^{\infty} W_k\, t^k = \frac{(1-p)\,t}{1-p\,t}.$$

Here $p = 5/6$ and it follows at once that

$$f(1) = \sum_{k=1}^{\infty} W_k = 1$$

and the required mean values are given by the derivatives

$$f'(1) = \sum k\, W_k = \overline{k},$$

$$f''(1) = \sum k\cdot(k-1)\, W_k = \overline{k^2} - \overline{k}.$$

It is convenient to represent them with the aid of the following logarithmic derivatives

$$\overline{k} = \frac{f'(1)}{f(1)},$$

$$\overline{k^2} - \overline{k} = \left(\frac{f'(t)}{f(t)}\right)'_{t=1} + \frac{f'(1)}{f(1)}.$$

Hence

$$\overline{k} = \left(\frac{1}{t} + \frac{p}{1-p\,t}\right)_{t=1} = \frac{1}{1-p} = 6$$

and

$$(\Delta k)^2 = \overline{(k-\bar{k})^2} = \left(-\frac{1}{t^2} + \frac{p^2}{(1-pt)^2}\right)_{t=1} + \frac{1}{1-p} = \frac{p}{(1-p)^2} = 30.$$

Thus the result is:

$$\bar{k} \pm \Delta k = 6 \pm \sqrt{30} = 6 \pm 5.5.$$

III.5. The energy equation can be written

(1) $$\frac{1}{2}m(\dot{x}^2 + \dot{y}^2 + \dot{z}^2) = E + A e^{-\alpha x} - B e^{-2\alpha x}.$$

The velocity components $\dot{y}$ and $\dot{z}$ are constant. Putting $\frac{1}{2}m(\dot{y}^2 + \dot{z}^2) = E_{tg}$ (tangential energy) and $E - E_{tg} = E_n$, we find that E_{tg} is constant, so that

(2) $$\frac{1}{2}m\dot{x}^2 = E_n + A e^{-\alpha x} - B e^{-2\alpha x}.$$

This equation can be integrated to give

$$t - t_0 = \left(\frac{m}{2}\right)^{1/2} \int (E_n + A e^{-\alpha x} - B e^{-2\alpha x})^{-1/2} dx.$$

Substituting $e^{\alpha x} = \xi$ we have

$$t - t_0 = \frac{1}{\alpha}\left(\frac{m}{2}\right)^{1/2} \int (E_n \xi^2 + A \xi - B)^{-1/2} d\xi.$$

The substitutions $\xi = \zeta - A/2 E_n$, $\zeta_0^2 = \frac{B}{E_n} + \frac{A^2}{4 E_n^2}$ lead to:

$$t - t_0 = \frac{1}{\alpha}\left(\frac{m}{2 E_n}\right)^{1/2} \int (\zeta^2 - \zeta_0^2)^{-1/2} d\zeta.$$

Substituting $\zeta = \zeta_0 \cosh \lambda$, we have

$$\alpha \left(\frac{2 E_n}{m}\right)^{1/2} (t - t_0) = \lambda = \cosh^{-1} \frac{\zeta}{\zeta_0}$$

or

$$\zeta = \zeta_0 \cosh \alpha \left(\frac{2 E_n}{m}\right)^{1/2} (t - t_0),$$

i. e:

$$\xi = e^{\alpha x} = \zeta_0 \cosh \alpha \left(\frac{2 E_n}{m}\right)^{1/2} (t - t_0) - \frac{A}{2 E_n}$$

or

$$x = \frac{1}{\alpha} \log \left\{ \zeta_0 \cosh \alpha \left(\frac{2 E_n}{m}\right)^{1/2} (t-t_0) - \frac{A}{2 E_n} \right\}$$

and

(3) $$\dot{x} = \left(\frac{2 E_n}{m}\right)^{1/2} \cdot \frac{\sinh \alpha \left(\frac{2 E_n}{m}\right)^{1/2} (t-t_0)}{\cosh \alpha \left(\frac{2 E_n}{m}\right)^{1/2} (t-t_0) - \frac{A}{2 E_n \zeta_0}}.$$

The change in momentum for case a) is

*(4) $$\Delta p_0 = m[\dot{x}(+\infty) - \dot{x}(-\infty)] = 2(2 m E_n)^{1/2}.$$

This is the same expression as that for impact on a rigid wall, so that there is no difference as far as pressure is concerned. Deviations may, however, be expected in case b) when the influence of the wall extends over a distance which is comparable with the mean free path. Let τ denote the average time between two collisions. Then, according to (2), we have

(5) $$\Delta p = m[\dot{x}(t_0 + \tau) - \dot{x}(t_0 - \tau)] = \Delta p_0 \frac{\sinh \alpha \left(\frac{2 E_n}{m}\right)^{1/2} \tau}{\cosh \alpha \left(\frac{2 E_n}{m}\right)^{1/2} \cdot \tau - \frac{A}{2 E_n \zeta_0}},$$

or, as a first approximation (for large τ's):

(5 a) $$\Delta p = \Delta p_0 \left(1 + \frac{A}{E_n \zeta_0} e^{-\alpha \tau (2 E_n/m)^{1/2}}\right).$$

According to eq. (1), $1/\alpha$ is of the order of the range of influence of wall forces. The latter have a marked influence on Δp when the second term in the brackets of (5 a) ceases to be negligible compared with 1. This occurs when

$$\frac{1}{\alpha} \geq \left(\frac{2 E_n}{m}\right)^{1/2} \cdot \tau = \left(\frac{2 E_n}{m c^2}\right)^{1/2} \cdot l$$

i. e. only when the range of influence of the forces is at least equal to the mean free path l.

III.6. a) With reference to Fig. 23 on p. 170 we can write that the number of particles from the cylinder shown in the sketch is

$$dv = \sigma c \cos \theta \cdot m \left(\frac{m}{2 \pi k T}\right)^{3/2} \cdot e^{-mc^2/2kT} c^2 \, dc \sin \vartheta \, d\vartheta \, d\phi.$$

Integration over the half-space gives

$$v = \sigma n \left(\frac{kT}{2\pi m}\right)^{1/2} \cdot 2 \int_0^\infty e^{-\rho^2} \rho^3 \, d\rho.$$

The integral, inclusive of the factor 2, is equal to $\int_0^\infty e^{-t} \cdot t \, dt = 1!$. Expressing n with the aid of pressure, we obtain

$$v = \sigma p (2\pi m k T)^{-1/2}.$$

If $p_2 > p_1$, more particles will flow from left to right than in the opposite direction. The difference is

(1) $$\Delta v = \sigma (2\pi m k T)^{-1/2} \cdot \Delta p.$$

b) ΔW can be calculated from $\tfrac{1}{2} m c^2 \, dv$. Calculating W in a way analogous to v, we have

$$W = \sigma v \cdot \frac{m}{2} \cdot \frac{2kT}{m} \cdot 2! = \sigma p \left(\frac{2kT}{\pi m}\right)^{1/2}$$

Hence

(2) $$\Delta W = \sigma \left(\frac{2kT}{\pi m}\right)^{1/2} \cdot \Delta p.$$

c) The average energy transferred per particle is:

(3) $$\frac{\Delta W}{\Delta v} = 2kT > \frac{3}{2}kT.$$

d) It is larger than $3/2\, kT$ because the particles possessing high energy per second arrive from a larger volume.

e) The flow of matter causes the pressure to change. According to the inequality in (3) the right-hand chamber becomes cooled whereas the left-hand compartment is heated. It is, therefore, necessary to control the pressures and to provide for an exchange of heat (temperature bath).

III.7. Let n_1, n_2 and n_0 denote the densities of molecules reflecting from A_1, A_2 and B respectively. Mass equilibrium will prevail when $n_1 \overline{c_z} = n_0 \overline{c_z} = n_2 \overline{c_z'}$. Owing to isotropy this is equivalent to

(1) $$n_1 \overline{c} = n_0 \overline{c} = n_2 \overline{c'},$$

where $\overline{c}$ and $\overline{c'}$ denote the mean velocities at temperatures T and T'.

The equilibrium pressure p is given by $p = \tfrac{1}{3} m n c^2$, where $n = 2n_1 = 2n_0$ so that $p = \tfrac{2}{3} m n_1 \overline{c^2}$. The recoils due to the molecules travelling away from

B are equal on both sides and cancel. The forces which act on B and which are not compensated are solely due to the arriving molecules. The resultant force is

(2) $$F = \frac{2}{3} m A (n_2 \overline{c'^2} - n_1 \overline{c^2}) = p A \left(\frac{n_2}{n_1} \frac{\overline{c'^2}}{\overline{c^2}} - 1 \right).$$

According to (1) it follows that

$$F = p A \left(\frac{\overline{c}}{\overline{c'}} \cdot \frac{\overline{c'^2}}{\overline{c^2}} - 1 \right).$$

Since $\overline{c} \sim \sqrt{T}$, $\overline{c^2} \sim T$, we have

(3) $$F = p A \left[\left(1 + \frac{\delta T}{T} \right)^{1/2} - 1 \right] \approx \frac{p A \, \delta T}{2 T}$$

or

(4) $$p = \frac{2 F T}{A \, \delta T}.$$

IV.1. The following table indicates the frequencies of the respective errors:

Error =	-4ε	-3ε	-2ε	$-\varepsilon$	0	$+\varepsilon$	$+2\varepsilon$	$+3\varepsilon$	$+4\varepsilon$
$n = 0$					1				
1				$\frac{1}{2}$		$\frac{1}{2}$			
2			$\frac{1}{4}$		$\frac{2}{4}$		$\frac{1}{4}$		
3		$\frac{1}{8}$		$\frac{3}{8}$		$\frac{3}{8}$		$\frac{1}{8}$	
4	$\frac{1}{16}$		$\frac{4}{16}$		$\frac{6}{16}$		$\frac{4}{16}$		$\frac{1}{16}$

In general the probability of a given error is given by the coefficients of the binomial expansion. If n is the number of individual errors and if k is the number of positive errors, then the probability of this case is given by

$$w_{n,k} = \frac{1}{2^n} \frac{n!}{k!(n-k)!}$$

and the *magnitude* of the error is:

$$f_{n,k} = k \varepsilon + (n-k)(-\varepsilon) = (2k - n) \varepsilon = x.$$

Instead of k we introduce the magnitude of the error, x, so that

$$k = \frac{x}{2\varepsilon} + \frac{n}{2}$$

and (with $dk = 1$):

$$dw = \frac{1}{2^n} \frac{n!\, dk}{k!(n-k)!} = \frac{1}{2^n} \frac{n!}{\left(\frac{n}{2} + \frac{x}{2\varepsilon}\right)! \left(\frac{n}{2} - \frac{x}{2\varepsilon}\right)!} \frac{dx}{2\varepsilon}.$$

The right-hand side is evaluated with the aid of Stirling's formula

$$n! = (2\pi n)^{1/2} \left(\frac{n}{e}\right)^n.$$

Consequently:

$$dw = \frac{1}{\sqrt{2\pi}} \left(\frac{4n}{n^2 - x^2/\varepsilon^2}\right)^{1/2} \frac{dx/2\varepsilon}{\left(1 + \frac{x}{n\varepsilon}\right)^{\frac{n}{2} + \frac{x}{2\varepsilon}} \left(1 - \frac{x}{n\varepsilon}\right)^{\frac{n}{2} - \frac{x}{2\varepsilon}}}.$$

The logarithms of the factors in the denominator are

$$\frac{n}{2}\left(1 \pm \frac{x}{n\varepsilon}\right) \log\left(1 \pm \frac{x}{n\varepsilon}\right) = \frac{n}{2}\left(1 \pm \frac{x}{n\varepsilon}\right)\left(\pm \frac{x}{n\varepsilon} - \frac{x^2}{2n^2\varepsilon^2}\right) = \pm \frac{x}{2\varepsilon} + \frac{x^2}{4n\varepsilon^2}.$$

and their sum is:

$$\frac{x^2}{2n\varepsilon^2}.$$

Thus the product in the denominator gives $\exp(x^2/2n\varepsilon^2)$, so that

$$dw = (2\pi n \varepsilon^2)^{-1/2} \exp(-x^2/2n\varepsilon^2) \cdot dx,$$

q. e. d. Putting $x = \xi(2n\varepsilon^2)^{1/2}$, we see that the integral with respect to dw is

$$\int dw = \frac{1}{\sqrt{\pi}} \int \exp(-\xi^2)\, d\xi = 1.$$

IV.2. a) When two statistically independent errors of the same kind are superimposed then the probability for the errors to lie within the intervals $(x', x' + dx')$ and $(x'', x'' + dx'')$ is

$$f_1(x')\, f_1(x'')\, dx'\, dx''.$$

The total error is $x' + x'' = x$. Since we are interested in the probability of a definite total error we put $x'' = x - x'$ and integrate with respect to x'. Thus

(1) $$f_2(x) = \int f_1(x') f_1(x - x') \, dx'.$$

In general, when an n'th error is superimposed on $(n-1)$ errors, we have

(2) $$f_n(x) = \int f_1(x') f_{n-1}(x - x') \, dx'.$$

b) When $f_n = (\pi a_n)^{-\frac{1}{2}} e^{-a_n x^2}$ (with $\int f_n \, dx = 1$) it follows from (2) that

(3) $$\frac{1}{a_n} = \frac{1}{a_{n-1}} + \frac{1}{a_1} = \frac{n}{a_1}.$$

Gaussian functions are seen to satisfy eq. (2) and the half-width increases like $n^{-\frac{1}{2}}$.

c) In order to answer the question we notice that in this case eq. (3) assumes the form

$$f_n(x) = \frac{1}{2\varepsilon} \int_{-\varepsilon}^{+\varepsilon} f_{n-1}(x - x') \, dx'.$$

Putting

$$f_n(x) = A_n e^{i\lambda x},$$

we find a complete system. It follows that for an arbitrary value of λ we have

$$A_n(\lambda) = A_{n-1}(\lambda) \cdot \frac{1}{2\varepsilon} \int_{-\varepsilon}^{+\varepsilon} e^{-i\lambda x'} \, dx' = A_{n-1}(\lambda) \left(\frac{\sin \lambda \varepsilon}{\lambda \varepsilon} \right)$$

and

(4) $$A_n(\lambda) = C(\lambda) \left(\frac{\sin \lambda \varepsilon}{\lambda \varepsilon} \right)^{n-1}.$$

The required solution is found by expanding $f_1(x)$ into a Fourier series:

$$f_1(x) = \int C(\lambda) e^{-i\lambda x} \, d\lambda.$$

Applying Fourier's integral theorem (*cf.* Vol. VI, Sec. 4, eq. (13)), we have

$$C(\lambda) = \frac{1}{2\pi} \int f_1(x) e^{-i\lambda x} \, dx = \frac{1}{4\pi\varepsilon} \int_{-\varepsilon}^{+\varepsilon} e^{-i\lambda x} \, dx = \frac{1}{2\pi} \frac{\sin \lambda \varepsilon}{\lambda \varepsilon}.$$

Hence

(5)
$$A_n(\lambda) = \frac{1}{2\pi}\left(\frac{\sin \lambda \varepsilon}{\lambda \varepsilon}\right)^n$$

represent the Fourier components of f_n. The function itself is:

(6)
$$f_n(x) = \frac{1}{2\pi}\int \left(\frac{\sin \lambda \varepsilon}{\lambda \varepsilon}\right)^n e^{i\lambda x}\, d\lambda.$$

For large values of n the expression $\left(\dfrac{\sin \lambda \varepsilon}{\lambda \varepsilon}\right)^n$ differs markedly from 0 only in the immediate neighborhood of the zero-point. Hence we may replace this factor by the osculating Gaussian bell-curve

$$\left(\frac{\sin \lambda \varepsilon}{\lambda \varepsilon}\right)^n \approx \exp\left(-\frac{n\varepsilon^2}{6}\lambda^2\right)$$

Now the integral (6) can be evaluated. We have

$$f_n(x) = \frac{1}{2\pi}\int \exp\left[-\frac{n\varepsilon^2}{6}\left(\lambda - \frac{3\,i\,x}{n\varepsilon^2}\right)^2\right] d\lambda \cdot \exp(-3x^2/2n\varepsilon^2)$$

or

(7)
$$f_n(x) = \left(\frac{3}{2\pi n \varepsilon^2}\right)^{1/2} \exp(-3x^2/2n\varepsilon^2).$$

The result is a Gaussian distribution, in analogy with the result in IV.1.

We now compute the integrals (6) for $n = 1, 2, 3$. Putting $\lambda \varepsilon = t$, $\xi = x/\varepsilon$, we obtain generally:

$$f_n = \frac{1}{2\pi\varepsilon}\int_{-\infty}^{+\infty} \frac{\sin^n t}{t^n} \cos \xi t\, dt.$$

On partial integration we find that

$$f_n = \frac{1}{2\pi\varepsilon}\cdot\frac{1}{(n-1)!}\int_{-\infty}^{+\infty} \frac{dt}{t}\frac{d^{n-1}}{dt^{n-1}}(\sin^n t \cos \xi t).$$

The functions operated on by the differential operator are:

$$\sin t \cos \xi t \qquad\qquad \text{for} \quad n = 1,$$

$$\sin^2 t \cos \xi t = \frac{1}{2}(1 - \cos 2t)\cos \xi t \qquad \text{for} \quad n = 2,$$

$$\sin^3 t \cos \xi t = \frac{1}{4}(3\sin t - \sin 3t)\cos \xi t \qquad \text{for} \quad n = 3.$$

The 0-th, 1st and 2nd derivatives of the respective functions are:

$n = 1 \quad \sin t \cos \xi t,$

$\quad 2 \quad \sin 2t \cos \xi t - \dfrac{1}{2} \xi (1 - \cos 2t) \sin \xi t,$

$\quad 3 \quad -\dfrac{3}{4} (\sin t - 3 \sin 3t) \cos \xi t - \dfrac{3}{2} \xi (\cos t - \cos 3t) \sin \xi t$

$\qquad -\dfrac{1}{4} \xi^2 (3 \sin t - \sin 3t) \cos \xi t.$

The remaining integrals are all of the same type (Dirichlet's discontinuous factor). We now write down the results of the integration for the individual terms and for the intervals in which they do *not* vanish:

Case $n = 1$: $\qquad = 1 \qquad$ for $\qquad |\xi| < 1,$

which is our starting point.

Case $n = 2$: $\qquad = 1 \qquad$ for $\qquad |\xi| < 2,$

$\qquad\qquad\qquad -\dfrac{1}{2} |\xi| \qquad$ for $\qquad$ all ξ's,

$\qquad\qquad\qquad +\dfrac{1}{2} |\xi| \qquad$ for $\qquad |\xi| > 2.$

It follows that

$\qquad\qquad 2 \varepsilon f_2 = 1 - \dfrac{1}{2} |\xi| \qquad$ for $\qquad |\xi| < 2,$

Case $n = 3$: $\qquad = -\dfrac{3}{4} \qquad$ for $\qquad |\xi| < 1,$

$\qquad\qquad\qquad \dfrac{9}{4} \qquad$ for $\qquad |\xi| < 3,$

$\qquad\qquad\qquad -\dfrac{3}{2} |\xi| \qquad$ for $\qquad |\xi| > 1,$

$\qquad\qquad\qquad +\dfrac{3}{2} |\xi| \qquad$ for $\qquad |\xi| > 3,$

$\qquad\qquad\qquad -\dfrac{3}{4} |\xi|^2 \qquad$ for $\qquad |\xi| < 1,$

$\qquad\qquad\qquad +\dfrac{1}{4} |\xi|^2 \qquad$ for $\qquad |\xi| < 3.$

Thus in the given intervals we have:

$$0 \leqslant |\xi| \leqslant 1 \qquad 4\varepsilon f_3 = \frac{3}{2} - \frac{1}{2}|\xi|^2,$$

$$1 \leqslant |\xi| \leqslant 3 \qquad \frac{9}{4} - \frac{3}{2}|\xi| + \frac{1}{4}|\xi|^2,$$

$$3 \leqslant |\xi| \leqslant \infty \qquad 0.$$

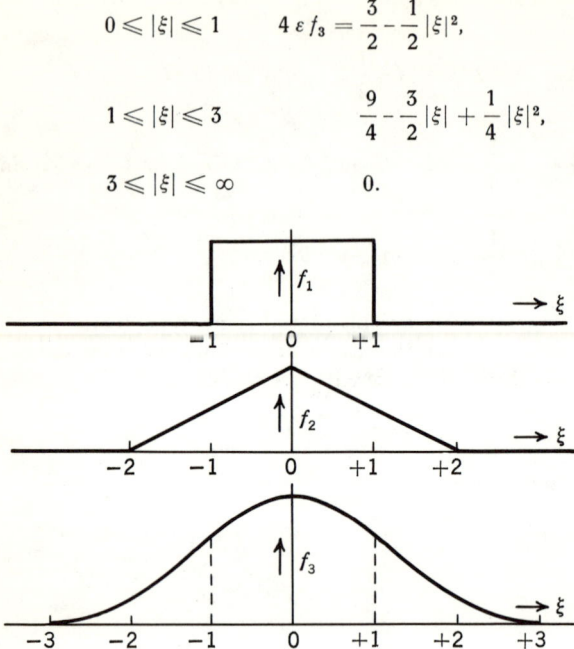

Fig. 41. Principal cross-section functions for polydimensional cubes, drawn for $n = 1, 2, 3$.

The functions f_1, f_2, f_3 are seen plotted in Fig. 41. The function f_1 consists of a horizontal straight segment, f_2 consists of two inclined straight segments, f_3 consists of three parabolic segments, etc. The geometrical interpretation of these functions can be given at once. Since f_2 arises from the superposition of two constant functions f_1, we see that inside a square whose sides are 2ε we have uniform coverage. We wish to determine the strips of constant total error. They are given by the segments cut out of the square by the straight lines at right angles to the diagonal DD in Fig. 41 a. Their lengths are given by f_2. Correspondingly f_3 gives sectional areas in a cube at right angles to a principal diagonal, etc.

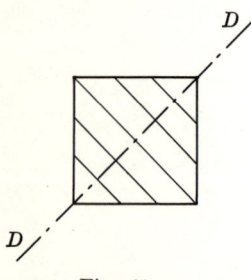

Fig. 41 a.
Principal cross-sections of a square.

IV.3. If N particles are distributed over n compartments containing $N_1, N_2, \ldots, N_n$ particles each, we have

$$W = \frac{N!}{N_1! \ldots N_n!}.$$

Thus, with the aid of Stirling's formula (Sec. 294 a) we have for the three cases respectively:

$$\text{a)} \quad W_a = \frac{N!}{N!} = 1,$$

$$\text{b)} \quad W_b = \frac{N!}{(N/2)!^2} = (2\pi N)^{1/2} \frac{2^N}{\pi N},$$

$$\text{c)} \quad W_c = \frac{N!}{(N/6)!^6} = (2\pi N)^{1/2} \left(\frac{3}{\pi N}\right)^3 \cdot 6^N,$$

and hence the ratios:

$$\frac{W_b}{W_a} = \left(\frac{2}{\pi N}\right)^{1/2} \cdot 2^N \to \infty \quad \text{for} \quad N \to \infty.$$

$$\frac{W_c}{W_b} = \frac{27}{\pi^2 N^2} \cdot 3^N \to \infty \quad \text{for} \quad N \to \infty.$$

IV.4. We have

$$\overline{\phi^2} = -\frac{d}{d\gamma} \log \int_{-\infty}^{+\infty} e^{-\gamma \phi^2} d\phi = \frac{1}{2\gamma}, \quad \gamma = \frac{D}{2kT},$$

and it follows that

$$\overline{\phi^2} = \frac{kT}{D} = \frac{RT}{LD}.$$

The Loschmidt-Avogadro number is

$$L = \frac{RT}{D\overline{\phi^2}} = \frac{8.32 \times 287}{9.43 \times 4.18} \times 10^{22} \text{ mol}^{-1} = 6.05 \times 10^{23} \text{ mol}^{-1}.$$

IV.5. We put $N = Z^3$, $Z = 10^7$. The number of particles on the surface is then $N_S = 6Z^2$. The thermal and surface energies are, respectively

$$U = 3Z^3 kT \quad \text{and} \quad U_S = 6Z^2 \text{ eV}.$$

For $T = 290 \, K$ we have:

$$\frac{U}{U_0} = \frac{ZkT}{2eV} = \frac{10^7 \times 400 \times 10^{-23}}{2 \times 1.6 \times 10^{-19} \times 9} = \frac{10^6}{72} \approx 14000.$$

Consequently, $U \approx U_0$ if $Z = 10^7/14000 = 700$; this means that the number of particles is $N = 3.4 \times 10^8$, the linear dimension $l = 2 \times 10^{-8} Z$ cm $= 1.4 \times 10^{-5}$ cm.

IV.6. a) We have

$$\int_0^\infty \frac{\partial \vartheta_2}{\partial z} \frac{dz}{z} = -2\pi e^{-\frac{1}{4}q} \sum_{n=0}^\infty (2n+1) e^{-n(n+1)q} \int_0^\infty \frac{\sin(2n+1)\pi z}{z} dz.$$

It can be seen that all integrals under the sign of summation are equal to $\frac{1}{2}\pi$.

b) Substituting the series ϑ_0 into the transformation equation we have

$$\vartheta_2\left(z \left| \frac{iq}{\pi} \right.\right) = \left(\frac{\pi}{q}\right)^{1/2} \sum_{n=-\infty}^{+\infty} (-1)^n \exp\left[-\frac{\pi^2}{q}(z-n)^2\right].$$

Consequently

$$Z(q) = -\frac{1}{\pi^2} e^{\frac{1}{4}q} \left(\frac{\pi}{q}\right)^{1/2} \cdot \left(\frac{-2\pi^2}{q}\right) \cdot \lim_{\varepsilon \to 0} \sum_{n=-\infty}^{+\infty} (-1)^n \int_\varepsilon^\infty \frac{z-n}{z} \exp\left[-\frac{\pi^2}{q}(z-n)^2\right] dz.$$

A simultaneous change in the signs of z and n only changes the limits of integration, so that

$$\lim_{\varepsilon \to 0} \int_\varepsilon^\infty = \lim_{\varepsilon \to 0} \int_{-\infty}^{-\varepsilon} = \frac{1}{2} \oint_{-\infty}^{+\infty}.$$

Hence

$$Z(q) = \frac{1}{q} e^{\frac{1}{4}q} \cdot \left(\frac{\pi}{q}\right)^{1/2} \cdot \sum_{n=-\infty}^{+\infty} (-1)^n \oint_{-\infty}^{+\infty} \frac{z-n}{z} \exp\left[-\frac{\pi^2}{q}(z-n)^2\right] dz.$$

The transformation $z = t/\pi + n$ gives:

$$Z(q) = \frac{1}{q} e^{\frac{1}{4}q} \cdot (\pi q)^{-1/2} \cdot \oint_{-\infty}^{+\infty} \left(\sum_{n=-\infty}^{+\infty} (-1)^n \frac{t}{t+n\pi}\right) e^{-t^2/q} dt.$$

This proves the proposition because of the relation:

$$\sum_{n=-\infty}^{+\infty} (-1)^n \frac{t}{t+n\pi} = \frac{t}{\sin t}.$$

c) We have, formally

$$\frac{1}{\sin t} = \frac{2 i e^{-it}}{1 - e^{-2it}} = 2i \sum_{k=0}^{\infty} e^{-(2k+1)it}$$

$$= \frac{-2 i e^{+it}}{1 - e^{+2it}} = -2i \sum_{k=0}^{\infty} e^{+(2k+1)it}$$

$$= -i \sum_{k=0}^{\infty} (e^{+(2k+1)it} - e^{-(2k+1)it}) = 2 \sum_{k=0}^{\infty} \sin(2k+1)t.$$

The integrals in $Z(q)$ [see (b)] can be transformed as follows:

$$2 \int_{-\infty}^{+\infty} t \sin(2k+1) t \cdot e^{-t^2/q} \, dt = 2 \int_{0}^{\infty} \sin(2k+1) t \cdot e^{-t^2/q} \, dt^2,$$

Putting $\tau = t^2$ we obtain an integral of the type of a Laplace transformation, namely:

$$2 \int_{0}^{\infty} \sin(2k+1) \sqrt{\tau} \cdot e^{-\tau/q} \, d\tau = q(\pi q)^{1/2} (2k+1) \cdot \exp\left(-(k+\tfrac{1}{2})^2 q\right).$$

The integral can be easily evaluated or taken from tables of Laplace transforms (see e. g. W. Magnus, F. Oberhettinger "Formeln und Sätze für die speziellen Funktionen der mathematischen Physik, Springer, Berlin 1943", or Bateman manuscript project, ed. by A. Erdélyi, W. Magnus, F. Oberhettinger and F. G. Tricomi, McGraw-Hill, New York, 1954/5).

It follows that

$$(\alpha) \qquad Z(q) = \sum_{k=0}^{\infty} (2k+1) e^{-k(k+1)q}.$$

In this manner we obtain eq. (34.3) which was derived from the transformed representation. Consequently, it becomes superfluous to prove the transformation equation. It is worth noting that it can also be deduced from the transformation equation in Vol. VI., eq. (15.8).

When q is very small, the only significant contribution to the integral comes from a small interval at $t = 0$ (its length being $\Delta t = q$). In this case we may use the expansion

$$\frac{t}{\sin t} = 1 \bigg/ \left(1 - \frac{t^2}{3!} + \frac{t^4}{5!} - \frac{t^6}{7!} + \ldots\right) = 1 + \frac{1}{6} t^2 + \frac{7}{360} t^4 + \frac{31}{15120} t^6 + \ldots.$$

Inserting this into $Z(q)$ and putting for the integrals

$$J_n = \int_{-\infty}^{+\infty} e^{-\gamma t^2} t^{2n}\, dt = \pi^{1/2} \left(-\frac{d}{d\gamma}\right)^n \gamma^{-1/2},$$

or, in particular,

$$J_0 = \left(\frac{\pi}{\gamma}\right)^{1/2}, \quad J_1 = \frac{1}{2\gamma}\left(\frac{\pi}{\gamma}\right)^{1/2}, \quad J_2 = \frac{3}{4\gamma^2}\left(\frac{\pi}{\gamma}\right)^{1/2}, \quad J_3 = \frac{15}{8\gamma^3}\left(\frac{\pi}{\gamma}\right)^{1/2}$$

we have:

(β) $$Z(q) = \frac{1}{q} e^{\frac{1}{4} q} \left[1 + \frac{1}{12} q + \frac{7}{480} q^2 + \frac{31}{8064} q^3 + \cdots \right].$$

Expanding $e^{\frac{1}{4}q}$ and multiplying the series term by term we obtain:

(β') $$Z(q) = \frac{1}{q}\left[1 + \frac{1}{3} q + \frac{1}{15} q^2 + \frac{4}{315} q^3 + \cdots \right]$$

and

(β'') $$\log Z(q) = -\log q + \frac{1}{3} q + \frac{1}{90} q^2 + \frac{8}{2835} q^3 + \cdots$$

d) Differentiating Z we obtain:

$$u = R T^2 \frac{d \log Z}{dT} = -R \Theta \frac{d \log Z}{dq},$$

$$c_v = \frac{\partial u}{\partial T} = -\frac{q^2}{\Theta}\frac{\partial u}{\partial q} = R q^2 \frac{d^2 \log Z}{dq^2}.$$

Hence

(γ) $$u = R \Theta \left[\frac{1}{q} - \frac{1}{3} - \frac{1}{45} q - \frac{8}{945} q^2 - \cdots \right],$$

$$c_v = R \left[1 + \frac{1}{45} q^2 + \frac{16}{945} q^3 + \cdots \right].$$

To a first approximation (except for a constant energy term in the same way as in Sec. 33), we find that:

$$u = R T - \frac{1}{3} R \Theta, \quad c_v = R.$$

The accuracy is better than 1% if $q^2/45 < 1/100$, or if $q < 0.67$.

e) The following table gives an estimate of the accuracy which can be achieved with a small number of terms:

Magnitude of the	First	Second	Third term in c_v
for $q = 0.3$	1	0.002	0.000467
for $q = 0.5$	1	0.008	0.00372
for $q = 0.9$	1	0.018	0.01263
for $q = 1.2$	1	0.032	0.0299

From $q = 0.6$ onwards the third term becomes of the same order as the second and cannot be neglected any more. Thus the asymptotic expansion can be used up to $q = 0.5$ approximately (i. e. for $T > 2\Theta$). From this value onwards it is preferable to use eq. (α). Equation (β) shows, however, that for large temperatures c_v tends to the value R very fast, and that it approaches it from above.

In conclusion we propose to calculate c_v from the two expansions for $e^{2q} = 2.5$, i. e. for $q = 0.458$, $q^2 = 0.20977$, $q^3 = 0.09607$. The asymptotic series gives

$$c_v = 1.006\, R$$

when terms up to and including the third are taken into account.

The value from eq. (33.3) can be calculated from (γ):

$$c_v = R\, q^2 \frac{d^2 \log Z}{dq^2} = R\, q^2 \left(\frac{Z''}{Z} - \frac{Z'^2}{Z^2} \right).$$

Here

$$Z = 1 + 3\,e^{-2q} + 5\,e^{-6q} + 7\,e^{-12q} + 9\,e^{-20q} + 11\,e^{-30q},$$

$$-Z' = 6\,e^{-2q} + 30\,e^{-6q} + 84\,e^{-12q} + 180\,e^{-20q} + 330\,e^{-30q},$$

$$Z'' = 12\,e^{-2q} + 180\,e^{-6q} + 1008\,e^{-12q} + 3600\,e^{-20q} + 9900\,e^{-30q}.$$

Substituting the preceding value for q we have

$$Z = 2.549; \quad Z' = 4.683; \quad Z'' = 20.836;$$

and

$$c_v = 1.0064\, R.$$

Since the series (α) takes into account all terms which make a marked contribution, it is seen that the fourth term of the asymptotic series (β) would have to be taken into account. However, the two series join to within 0.05% and the molar specific heat is seen to be well represented by taking only a small number of terms in each series.

IV.7. It is necessary to make log W a maximum under the conditions indicated. Taking these into account with the aid of Lagrange's method of parameters we find that it is necessary to determine the maximum of

$$-\sum_{n,i}\{f_i^{(n)}\log f_i^{(n)} + \lambda_i f_i^{(n)} + \alpha n f_i^{(n)} + \beta n \varepsilon_i f_i^{(n)}\}$$

The derivatives with respect to all $f_i^{(n)}$'s must vanish. Thus

$$\log f_i^{(n)} + 1 + \lambda_i + \alpha n + \beta n \varepsilon_i = 0$$

or

(1) $$f_i^{(n)} = e^{-1-\lambda_i} \cdot e^{-(\alpha+\beta\varepsilon_i)n}.$$

The condition $\sum f_i^{(n)} = 1$ gives:

$$e^{-1-\lambda_i}/(1-e^{-\alpha-\beta\varepsilon_i}) = 1$$

and eq. (1) assumes the following form:

(2) $$f_i^{(n)} = (1-e^{-\alpha-\beta\varepsilon_i}) e^{-(\alpha+\beta\varepsilon_i)n}.$$

The distribution function follows from

$$n_i = \sum_n n f_i^{(n)} = -(1-e^{-\alpha-\beta\varepsilon_i})\frac{d}{d\alpha}\sum_n e^{-(\alpha+\beta\varepsilon_i)n}.$$

Thus we have

$$n_i = -(1-e^{-\alpha-\beta\varepsilon_i})\frac{d}{d\alpha}\left(\frac{1}{1-e^{-\alpha-\beta\varepsilon_i}}\right).$$

This is the Bose-Einstein distribution:

(3) $$n_i = \frac{1}{e^{\alpha+\beta\varepsilon_i} - 1}$$

and proves the assertion.

The constants α and β can be determined from the remaining conditions which now appear in the simpler form:

(4) $$\sum n_i = N, \quad \sum n_i \varepsilon_i = U.$$

IV.8. The volume V of the vessel consists of the parts $V_1 = \Delta V$ and $V_2 = V - \Delta V$. The probability of finding a molecule in V_1 or V_2 is V_1/V or V_2/V, respectively, since equal volume elements are associated with equal probabilities. The probability of finding N_1 particles in V_1 and N_2 in V_2 becomes

$$W(N_1, N_2) = \frac{N!}{N_1! N_2!}\left(\frac{V_1}{V}\right)^{N_1}\left(\frac{V_2}{V}\right)^{N_2}.$$

We calculate, further, in an obvious way that

$$\overline{N_1} = N \frac{V_1}{V}, \qquad \overline{N_1(N_1-1)} = N(N-1)\frac{V_1^2}{V^2}.$$

It follows that

$$(\Delta N_1)^2 = \overline{N_1^2} - \overline{N_1}^2 = N \frac{V_1 V_2}{V^2}$$

or

$$\frac{\Delta N_1}{\overline{N_1}} = \sqrt{\frac{V_2}{N V_1}} = \sqrt{\frac{V - \Delta V}{N \Delta V}}.$$

IV.9. a) The van der Waals equation for N particles in volume V can be written:

(1) $$\left[p + A\left(\frac{N}{V}\right)^2\right]\left(\frac{V}{N} - B\right) = kT.$$

The constants A and B are related to the constants a and b in the original van der Waals equation; we have

(1 a) $$A = \frac{a}{L^2}; \qquad B = \frac{b}{L}.$$

(L = the Loschmidt-Avogadro number). The critical parameters are obtained in the same way as in Sec. 9 from:

$$\left(\frac{\partial p}{\partial V} - \frac{2AN^2}{V^3}\right)\left(\frac{V}{N} - B\right) + \frac{1}{N}\left(p + \frac{AN^2}{V^2}\right) = 0,$$

$$\left(\frac{\partial^2 p}{\partial V^2} + \frac{6AN^2}{V^4}\right)\left(\frac{V}{N} - B\right) + \frac{2}{N}\left(\frac{\partial p}{\partial V} - \frac{2AN^2}{V^3}\right) = 0,$$

putting $\partial p/\partial V = 0$ and $\partial^2 p/\partial V^2 = 0$.

Starting with the last equation we obtain

(2) $$V_{crit} = 3NB, \qquad p_{crit} = \frac{A}{27B^2}; \qquad kT_{crit} = \frac{8A}{27B}.$$

b) The logarithm of the partition function is a thermodynamic parameter of the form

(3) $$d\log Z = -U d\beta + \beta p\, dV.$$

Substituting the thermal equation of state for p, we obtain:

$$\left(\frac{\partial \log Z}{\partial V}\right)_\beta = \beta p = \frac{N}{V - NB} - \beta A \frac{N^2}{V^2}.$$

On integrating we have:

(4) $$\log Z = N\left[\log(V - NB) + \beta A \frac{N}{V}\right] + N f(\beta).$$

Here $f(\beta)$ is an arbitrary function of β.

Evaluating the energy from this expression with the aid of eq. (3), we obtain, further,

$$U = -\left(\frac{\partial \log Z}{\partial \beta}\right)_v = A \frac{N^2}{V} - N f'(\beta).$$

The heat capacity at constant volume is, therefore,

$$\left(\frac{\partial U}{\partial T}\right)_v = -k \beta^2 \left(\frac{\partial U}{\partial \beta}\right)_v = k \beta^2 N f''(\beta) = \frac{N}{L} c_v \to \frac{3}{2} N k.$$

The last expression is valid asymptotically for the limiting case of $B \to 0$. Hence:

$$f''(\beta) \to \frac{3}{2 \beta^2}$$

and integration yields the asymptotic expressions

$$f'(\beta) \to -\frac{3}{2\beta} + C_1,$$

$$f(\beta) \to -\frac{3}{2} \log \beta + C_1 \beta + C_2.$$

Thus, for $\log Z$, we obtain

(5) $$\log Z \to N\left[\log(V - NB) + \beta A \frac{N}{V} - \frac{3}{2} \log \beta\right] + N C_1 \beta + N C_2.$$

In the limiting cases of $\beta \to 0$, $N/V \to 0$, we have the simplified forms:

(5 a) $$\log Z = N \log N + N\left(\log \frac{V}{N} - \frac{3}{2} \log \beta\right) + N C_1 \beta + N C_2.$$

This expression must be identical with that for a perfect gas:

$$(\log Z)_{perf. gas} = N\left[1 + \log \frac{V}{N} - \frac{3}{2} \log \beta + \log\left(\frac{2\pi m}{h^2}\right)^{3/2}\right],$$

and it follows that $C_1 = 0$ and that

$$N \log N + N C_2 = N\left[1 + \log\left(\frac{2\pi m}{h^2}\right)^{3/2}\right].$$

Finally, we obtain

(6) $$\log Z = N\left[1 + A\beta\frac{N}{V} + \log\left(\frac{V}{N} - B\right)\left(\frac{2\pi m}{\beta h^2}\right)^{3/2} + g(\beta)\right].$$

Here $g(\beta)$ denotes that part of $f(\beta)$ which vanishes for $\beta \to 0$.

c) Equation (6) gives at once:

$$\frac{\partial \log Z}{\partial N} = \frac{\log Z}{N} + \left(A\beta\frac{N}{V} - \frac{V/N}{V/N - B}\right).$$

Repeated differentiation with respect to N and multiplication by $-N$ gives:

(7) $$-N\frac{\partial^2 \log Z}{\partial N^2} = \frac{\log Z}{N} - \left(\frac{\log Z}{N} + A\beta\frac{N}{V} - \frac{V}{V - NB}\right) - A\beta\frac{N}{V} + \frac{BN/V}{\left(1 - \frac{BN}{V}\right)^2}$$

$$= \frac{1}{\left(1 - \frac{BN}{V}\right)^2} - \frac{2BN}{V}\cdot\frac{A}{BkT}.$$

In the limiting case of a perfect gas $(A, B \to 0)$, we have:

(7 a) $$-N\frac{\partial^2 \log Z}{\partial N^2} = 1.$$

The relative mean fluctuation for this case agrees with that in eq. (40.18).

When A and B are to be taken into account only in the first approximation, we have

(7 b) $$-N\frac{\partial^2 \log Z}{\partial N^2} = 1 + \frac{2BN}{V}\left(1 - \frac{A}{BkT}\right).$$

Substituting the critical parameters from eq. (2), namely $BN/V = 1/3$, $A/BkT = 27/8$, we have

(7 c) $$-N\frac{\partial^2 \log Z}{\partial N^2} = 0.$$

The relative mean fluctuation becomes:

(8) $$\left(\overline{\frac{\Delta n}{n}}\right)^2 = -\frac{V/N \Delta V}{N\, \partial^2 \log Z/\partial N^2} \to \infty,$$

i. e. infinitely large. Large fluctuations cause strong scattering of light (cf. vol. IV, Sec. 33) which is evidenced by the strong opalescence of real gases near their critical point.

d) According to eq. (40.15), we may write

$$(\Delta n)^2 = \frac{1}{\beta^2} \sum_{i,k}{}' \left(\frac{\partial^2 \log Z}{\partial \varepsilon_i \partial \varepsilon_k}\right)_{\alpha, \beta, \varepsilon_j} = \frac{1}{\beta^2} \left(\sum_{i,k}{}' \frac{\partial^2 \Phi}{\partial \varepsilon_i \partial \varepsilon_k}\right)_{\alpha, \beta, \varepsilon_j}$$

because α must be kept constant in addition to β (*cf.* Sec. 40, in particular, eq. (12)). The symbol Σ' denotes that the sum must be taken over all phase cells in ΔV. According to eq. (38.4) the last equation may be written:

(9) $$(\Delta n)^2 = -\frac{1}{\beta} \sum{}' \left(\frac{\partial \overline{n}}{\partial \varepsilon_i}\right)_{\alpha, \beta, \varepsilon_j}.$$

Here $\overline{n}$ denotes the mean number of molecules in ΔV $\left(\overline{n} = \frac{\Delta V}{V} \cdot N\right)$. It follows from eq. (38.7) that

$$d\overline{\phi} = -\overline{n}\, d\alpha - \overline{u}\, d\beta - \beta \sum{}' \overline{n}_i\, d\varepsilon_i$$

($\overline{\phi}$ is equivalent to $\Phi = \log Y$ from Sec. 38, eq. (1), referred to volume ΔV) and

(10) $$\left(\frac{\partial \overline{n}}{\partial \varepsilon_i}\right)_{\alpha, \beta, \varepsilon_j} = \beta \left(\frac{\partial \overline{n}_i}{\partial \alpha}\right)_{\beta, \varepsilon_i}.$$

It follows from (9) that

(9 a) $$(\Delta n)^2 = -\sum_i{}' \frac{\partial \overline{n}_i}{\partial \alpha} = -\frac{\partial \overline{n}}{\partial \alpha}.$$

Reverting to the partition function by means of a Legendre transformation

$$d(\overline{\phi} + \alpha \overline{n}) = \alpha\, d\overline{n} - \overline{u}\, d\beta - \beta \sum_i{}' \overline{n}_i\, d\varepsilon_i = \frac{\Delta V}{V} d \log Z,$$

we have:

$$\alpha = \frac{\Delta V}{V} \cdot \left(\frac{\partial \log Z}{\partial \overline{n}}\right)_{\beta, \varepsilon_i} = \left(\frac{\partial \log Z}{\partial N}\right)_{\beta, \varepsilon_i}.$$

This shows that eq. (9 a) can be written:

$$(\Delta n)^2 = -\frac{1}{\partial \alpha / \partial \overline{n}} = -\frac{V}{\Delta V} \frac{1}{\partial^2 \log Z / \partial N^2}.$$

Dividing by the square of $\overline{n} = N \Delta V / V$, we obtain the fluctuation equation indicated in the statement of the problem.

PROBLEMS

V.1. It can be verified directly that the momentum and energy equations are satisfied by:

(5)
$$\mathbf{v}_1' = \mathbf{v}_1 + \frac{2m_2}{m_1 + m_2} (\mathbf{V}\,\mathbf{e})\,\mathbf{e},$$

$$\mathbf{v}_2' = \mathbf{v}_2 - \frac{2m_1}{m_1 + m_2} (\mathbf{V}\,\mathbf{e})\,\mathbf{e}.$$

$\mathbf{V} = \mathbf{v}_2 - \mathbf{v}_1$ denotes the relative velocity. $\mathbf{V} \sim \mathbf{e}$ gives central impact. In order to prove the validity of Liouville's theorem on the equality of the velocity space cells it is necessary to show that the transformation matrix of the six-dimensional space of the combined velocities $\mathbf{v}_1$ and $\mathbf{v}_2$ has a determinant -1 (**e** should be placed along one axis!).

According to (5) the difference in the energies after impact is

(6)
$$\tfrac{1}{2} m_1 \mathbf{v}_1'^2 - \tfrac{1}{2} m_2 \mathbf{v}_2'^2 = \tfrac{1}{2} m_1 \mathbf{v}_1^2 - \tfrac{1}{2} m_2 \mathbf{v}_2^2 - \frac{4 m_1 m_2}{(m_1 + m_2)^2} (\mathbf{v}_1 - \mathbf{v}_2 \cdot \mathbf{e})(m_1 \mathbf{v}_1 + m_2 \mathbf{v}_2 \cdot \mathbf{e}).$$

Making use of the condition that all directions are equally probable we find that the *mean value* of a product of the form $(\mathbf{A}\,\mathbf{e})(\mathbf{B}\,\mathbf{e})$ taken over $\mathbf{e}$ is:

$$\overline{(\mathbf{A}\,\mathbf{e})(\mathbf{B}\,\mathbf{e})}^{(\mathbf{e})} = \tfrac{1}{3} \mathbf{A}\,\mathbf{B}.$$

Inserting this value into eq. (6), we have:

(7)
$$E_1' - E_2' = E_1 - E_2 - \frac{1}{3} \frac{4 m_1 m_2}{(m_1 + m_2)^2} [(\mathbf{v}_1 - \mathbf{v}_2) \cdot (m_1 \mathbf{v}_1 + m_2 \mathbf{v}_2)].$$

Since the direction of $\mathbf{v}_2$ is independent of that of $\mathbf{v}_1$ the mean value with respect to $\mathbf{v}_2$ becomes:

$$\overline{(\mathbf{v}_1\,\mathbf{v}_2)}^{(\mathbf{v}_2/v_2)} = 0.$$

Thus in the last bracket in eq. (7) only the terms $m_1 \mathbf{v}_1^2 - m_2 \mathbf{v}_2^2 = 2(E_1 - E_2)$ remain. Hence

$$\frac{E_1' - E_2'}{E_1 - E_2} = 1 - \frac{2}{3} \cdot \frac{4 m_1 m_2}{(m_1 + m_2)^2}.$$

For $m_1 = m_2$ the ratio becomes equal to 1/3, for $m_1 \ll m_2$ it becomes equal to $1 - 8 m_1/3 m_2$. This result means that on the average the difference in the kinetic energy of the two particles taking part in the collision decreases continuously. Consequently the mean kinetic energy of the translational motion becomes equally distributed.

The law of equipartition is valid for any molecules and not only for molecules of one kind.

Index

A

Absolute temperature scale 30
  zero 70
  impossibility of reaching 74
Absorptive power 138
Adiabatic process 19 f.
Affinity, chemical 47, 72
Ammonia, synthesis of 86
Aristoteles 13
Atmosphere 8
  equilibrium of 348
Atomic beams, method of 179
Avogadro 10, 11
  's law 10
  number see Loschmidt-Avogadro number

B

Bar 8
Barometric formula 180
Becker, R. 133
Bernoulli, D. 169
Berthelot, D. 72
Bethe, H. A. 277
Black body radiation 135
de Boer, J. 57
Boiling point depression 111
  rise in 111
Boltzmann, L. 55, 140, 162, 169, 205, 207, 212, 213
  's combinatorial method 211
  's constant 141, 213, 252
  distribution 251
  factor 179, 223
  's principle 213 f.
  's statistics 124
Born, M. 39, 212, 293, 302
Bosch, C. 86
Bose, S. N. 261
Bose-Einstein distribution 266, 354
  gas 272 f.
  statistics 73, 261 f., 266 f.

Boyle's law 8, 21
Boyle and Mariotte, law of see Boyle's law
Braun-Le Chatelier principle 46
Brown, R. 181
Brownian motion 181

C

Caloric condition (for perfect gas) 15, 46
Calorie 5, 6
Canonical distribution 256
Carathéodory, C. 39, 40
  's proof 39
Carnot, L. 22
Carnot, S. 22, 26, 28, 43
  cycle 27 f.
  's ratio 30, 68
Casimir, H. B. G. 166
Catalyst see catalyzer
Catalyzer 13, 370
Change of state see process
  virtual 48
Chapman, S. 298
Characteristic equation see equation of state
Charles' law (the law of Gay-Lussac) 9
Chemical constants 87 f.
  potentials 87 f.
Chlorine – hydrogen mixture 86
Clapeyron's equation 44, 68, 97 f.
Claret, from cold cellar 41
Clausius, R. 25, 26, 27, 38, 39, 44, 154, 169, 182, 197
  postulate 27
Clausius, K. 73, 80, 163, 298
Cloud chamber 103
Coefficient of tension 3, 43, 74
Cohesion energy 353
  forces (van der Waals) 197
Collision equation for electrons 333 f.
  equation for molecules 293 f., 323 f.
  integral 299 f.
  invariants 306

Collision moments 314, 327 f.
Collisions, law of 297 f.
Compressibility 4
Conduction of heat 204, 340, 342 f.
   in anisotropic body 155
Conductivity, thermal 333 f.
Contact potential 159
Continuity, equation of 153
Couette flow 317, 332
Critical point 67
Curie constant 123, 127, 192
   's law for paramagnetic substances 123
   point 121, 127, 192
   temperature see Curie point
Curie-Weiss law 127
Cycle 4, 22

D

Dalton, J. 10
   's law 10, 77
Daniell cell 115
Darwin, C. G. 212, 259
Darwin-Fowler method 212, 259
Debye, P. 70, 74, 125, 135, 247
   's theory of specific heat 247 f.
Degeneration, characteristic temperature of 280
   degree of 270
   of gases 266 f., 274, 277
Degrees of freedom, additional in retarded equilibrium 50
   dying out of (freezing) 234, 240
   thermal 3, 18, 239
   thermodynamic 108
Delbrück, M. 253
Demagnetization 74 f., 134
Density fluctuations 291, 355
Dewar 20
Diesel engine 37
Diesselhorst, H. 277
Dieterici, C. 196
Differential, perfect 2, 14, 31
Diffusion 206
   increase in entropy due to 80
   thermal 163
Dilute solutions 92 f.
Dipole 145

Dirac, P. A. M. 261
Dirichlet's discontinuous factor 381
Dissociation of steam 81
Doppler's effect 179
Döring, W. 133
Drude, P. 276, 344
   's equation 343
   's method 276
Dufour, L. 163
Duhem-Margule conditions 90
Dulong-Petit rule 236, 248

E

Efficiency 27 f.
Effusion, thermal 163
Ehrenfest, P. 47, 208, 210
Ehrenfest, T. 47, 208, 210
Einbinder, H. 349
Einstein, A. 181, 240, 261
   condensation 273, 349
   's equation 183
   's reversal of Boltzmann's principle 213
Electric field strength, impressed 160
Electro-motive force 117
Electron gas in metals 276
Elementary cells in phase space 212, 219
Emden, R. 40, 357
Emmisivity 138
Energetics 13
Energy 13, 40 f.
   conservation of 13, 153
   dissipation of 165 f.
   distribution 179
   flux of 17, 24, 339
   limiting 279, 342
Ensemble, canonical 267
   grand canonical 267, 290
   microcanonical 267
Enskog, D. 168, 295, 322
Enthalpy 13, 16, 24, 42 f.
Entropy 25 f.
   additivity of 34
   and probability 213
   at absolute zero 71
   constant 230
   decrease of 41
   flux of 154, 165
   increase 36

Entropy increase in diffusion 80
  local generation of 152 f., 166, 328, 336
  source density 328, 336
  theorem 34, 320 f.
Equation of state 3, 10
  Dieterici 188
  Gay-Lussac 9
  magnetic 121 f.
  perfect gas 9 f.
  van der Waals 55 f.
Equilibrium conditions 48
  cyclic 369
  detailed 351
  liquid and gas 65
  retarded 36
  state of 19
  thermal 1, 34
  thermodynamic 1, 47 f.
  unconstrained 47
  unstable 66
Equipartition, law of 174, 238
Equivalence of heat and work 6, 13
Eucken, A. 234, 332
Euler's equation 317
Evaporation, latent heat of see latent heat
Exner, F. 187
Extensive quantity 3, 42

F

Faraday's equivalent charge 114
Fermi, E. 261
Fermi-Dirac distribution 266
  gas, complete degeneracy 277 f.
  statistics 73, 261 f., 266 f.
Ferro-magnetism 121 f.
  Weiss' theory of 125
Fluctuations 39, 181
  higher powers 291
  mean square of 286
Fluid dynamics, fundamental equation of 310 f.
Flux of current 339 f.
Fourier, T. B. 153
Fowler, R. H. 1, 212, 260
Free energy 42 f., 53
  enthalpy 42 f.
  enthalpy per electron 158, 278
Freezing of degrees of freedom 234, 240

G

Galvanic cell, emf 113
Γ-space 207 f.
Gas degeneration 73, 266 f.
  diatomic 230 f.
  mixture 77 f., 226
  monatomic 227
  polyatomic 233
  perfect see perfect gas
  thermometer 68
  universal constant 9, 12
Gaussian distribution 178
Gay-Lussac, J. 10, 20, 71
  's law see Charles' law
  's experiment 20, 25
Giauque, W. J. 70, 135
Gibbs, W. J. 44, 47, 48, 87, 212, 250
  condition 250 f.
  -Helmholtz fundamental equation 72, 118
  paradox 80
  phase rule 106 f.
Giorgi's M.K.S. system 7
Glaser, W. 141
Grad, H. 307, 323
Gram-mol 11
Guldberg, C. M. 81, 83
Guldberg and Waage, law of 77 f.

H

H-theorem 302 f.
de Haas, G. L. 70
de Haas-Lorentz, Mrs. 47
Haber, F. 86
von Haller, A. 6
Hardy 358
Heat, conduction of 204, 340, 342 f.
  "content" 7
  equivalent, mechanical 6
  flux 154, 319
  substance 26
  transfer 27
  quantity of 6
Heating 6
von Helmholtz 13, 44, 54, 213
Helmholtz-Gibbs law 118
Henry's law of absorption 112

INDEX

Hertz's dipole 145
Hertzfeld, K. F. 323
Hilbert, D. 205
van 't Hoff, J. H. 93, 111
    's equation of state 93

I

Ideal gas *see* perfect gas
Impressed electric field strength 160
Indicator diagram 5, 103
Integrating denominator 2, 31
Intensive quantity 3, 42
Internal transformations 163
Inversion curve 61
Irreversible processes 19 f., 152 f.
Isentrope 26
Isentropic process 20 f.
Ising, E. 187
Isolated system 34
Isotherm 21, 23

J

Jäger, G. 277
Jaumann, G. 165
Jordan, P. 259
Joule, J. P. 22, 23
    heat 6
Joule-Kelvin process 23 f., 47, 60

K

Kamerlingh-Onnes, H. 63, 125, 135
Kapitza, P. 63
Kappler, E. 185, 186, 353
Kelvin, Lord *see* Thomson, W.
    's clouds 233
    's postulate 27
    's temperature scale 10, 68 f.
Kilogram-mol 11
Kinetic theory of gases 169 f., 293 f.
Kirchhoff, G. R. 136
    's law 136
Knudsen, M. 199
Kronecker's symbol 312
Krönig, A. 169
Kundt, A. 203
Kurlbaum, F. 149

L

Langevin, P. 122, 181
    's function 124, 187 f.
Latent heat 99
    of fusion 100
von Laue, M. 141
Law of mass action 77 f.
Law of thermodynamics, Zeroth 1
    First 13 f.
    Second 26 f.
    Second, differential form 155
    Third 71 f.
    very large numbers 286
Legendre transformation 42 f.
Light, pressure due to 139
Linde's regenerative process 61
Liouville's theorem 207 f.
Liquefaction of gases 60 f.
Liquid line 65, 97
Littlewood, J. E. 358
Local mean values 294
London, F. 272
Lorentz, H. A. 96, 141, 344
    number 344
Loschmidt-Avogadro number 114
Lummer, O. 149

M

Mach, E. 71
McLewis, W. 16, 44
Magnetization, work of 122
Magneto-caloric effect 70, 134
Magnetostriction 122
Mass action, law of 77 f., 81 f.
Mathematical expectation 206
Maximum of entropy 47
Maxwell, J. C. 43, 46, 65, 139, 169, 175, 202
    's line 64
    relations *see* reciprocity relations
Maxwellian distribution 178
Maxwell-Boltzmann velocity distribution
    174 f., 224, 306 f.
    law 207 f.
Mayer, R. 6, 13, 22
Mean free path 197 f.
Mechanical equivalent of heat 6
    theory of heat 13
Meissner, W. 62, 63

INDEX

Meixner, T. 322
Melting curve 101, 104
    point depression 101, 111
Michelson, A. A. 179
Microbalance 185, 353
Microcanonical distribution 256
microscopic reversibility 163, 167
Microstate 209
Mittasch, A. 86
Mixing term (entropy) 81
Mixture, perfect 90
MKS System 7
Mol 10
Molar heat 15
    of quantum-mechanical oscillator 239
                    rotator 245
    volume 11
Molecule, diatomic 234
    polyatomic 236
    volume of 192, 193
Molecular chaos 301
    velocity 173
    weight 12
        determination of 11
Molière, G. 253
Momentum, transfer of 200
Multiple proportions 10
μ-space 207 f.

N

Natural vibrations 246
    phenomena and Second law 36
Navier-Stokes equations 317
Near-equilibrium processes 152 f.
Nernst, W. 11, 71
    's law for electron gas 285
    's Third Law 71 f., 275, 289
Numbers, very large, law of 286

O

Ohms' law 160, 341
Onsager, L. 156
    's reciprocal relations 155 f.
Oscillator, energy of 150
    linear 237
osmotic pressure 92 f.
Ostwald, W. 11, 27, 47

P

Paramagnetic properties 121 f.
    substances 187
Partial pressures, Dalton's law 77
Partition function 218, 237, 248 f.
Paschen, F. 149
Pauli's principle 279
Peltier effect 161 f., 340 f.
Perfect differential 2, 31
Perfect gas 8 f.
    equation 169
Permanent magnet 127
Permutability 214
Permutations 353
Perpetual motion machine of the first
        kind 27
            second kind 27
Perrin, J. 180
Pfaff's differential 2, 42
Pfeffer, W. 78, 92
Phase 63 f., 96
    elements in Γ-space 208, 210
        equality of probability
    equilibrium 100, 103 f.
    rule 106 f.
    space of molecules 210
Phases, coexistence of 63
Physical space 174
Planck, M. 27, 38, 39, 47, 111, 145, 148,
    213, 247
    's law of radiation 145 f., 247
    's quantum of action 143, 150, 211, 237
Pohl, R. W. 7
Poisson's equation 20
Polya, G. 358
Polytrope, equation of 21
porous plug experiment 23
Potentials, chemical 87 f., 158
    electrochemical 113
    thermodynamic 42 f.
Pressure, kinetic definition of 171
Prime mover 28
Pringsheim, E. 149
Probability maximum 217
    (of a state) 214
Process, adiabatic 19 f.
    irreversible 22 f., 152 f.
    isobaric 4

INDEX

Process, isochoric 4, 221
  isothermal 4
  reversible 19 f., 51
Property, thermodynamic 1

Q

Quantization of rotational energy 242 f.
  vibrational energy 237 f.
Quantum of action 143, 150, 211, 237
  states, equal probability 255
  statistics 253 f.
  statistics of identical particles 257

R

Radiation, black body 135
  laws of 245 f.
Ramsay, W. 77
Randall, M. 16, 44
Rankine, H. O. 13
Raoult, F. M. 111
  's law for dilute solutions 108
Rayleigh, Lord 77, 142, 243
  -Jeans law 247
  radiation law 142, 149
Reaction, heat of 119
  at absolute zero 92
Reciprocal relations *see* Onsager
Reciprocity relations 42
Refrigerator 28
Regelation of ice 101
Relaxation phenomena 163
Release, processes of 13
Reversibility 19
  microscopic 163, 167
Reversible process 19 f.
Richardson's effect 283
Rotational energy (specific heat) 242 f., 330
Rubens, H. 149
Rumford, Count (B. Thompson) 5

S

Sackur, O. 229
  -Tetrode formula 229, 258
Saddle point method 262
Saha, M. N. 1
Sauter, F. 192

von Schiller, F. 234
Schrödinger, E. 260
Second Law and direction of natural phenomena 36
Semi-permeable membrane 78, 92
Separation of gases 78
Shear flow *see* Couette flow
von Smoluchowski, M. 185
Solid, molar specific heat of 240
  theory of 245
Sommerfeld, A. 163, 277, 293
Soret, Ch. 156
Sound, velocity of 174
Specific heat 6, 16, 17, 227 f., 234 f.
  of metal electrons 283
  of rigid molecules 227
  of saturated steam 101
  of solid bodies 234, 247
  of vibrating molecules 234
Specific heats at absolute zero 74 f.
  c_H and c_M 129 f.
  difference $c_p - c_v$ 45
  ratio c_p/c_v 18
Spin 281
Spontaneous magnetization 127
Srivartava, B. N. 1
Statistical mechanics 207 f.
  Gibbs' method 212
Steam engine 37, 96 f.
  line *see* vapor line
  vapor-pressure curve 96 f., 100
Stefan-Boltzmann law 139, 152, 248
Stern, O. 179
Stokes' theorem 2
Sublimation 104 f.
  curve 104
Superposition of errors 352 f.
Supersaturation 66
Synthesis of ammonia 86
System, isolated 34
Szegö, G. 358

T

Tamman, G. 104
Temperature 1 f.
  absolute 30
  characteristic 237, 243, 247
  dimension of 3

Temperature, kinetic definition of 172
  Kelvin scale 10
  scale, absolute 30
  thermodynamic scale 350
Tension, coefficient of 3, 43, 74
Tetrode, H. 229
Thermal death of universe 38
  diffusion 163
  effusion 163
  equilibrium 1
  expansion, coefficient of 3, 45, 73
    value at absolute zero 73
  force, absolute 159, 340, 342
Thermodynamic potentials 42 f.
  extremum properties 50
  table 44
Thermodynamic property 1
  system 3
Thermoelectric phenomena 157
Thompson, B. see Rumford
Thomson, W. (Lord Kelvin) 13, 23, 26, 68, 160, 161, 162, 163, 227
  's effect 160
Total heat 16
Transfer of electrons 333 f.
Transport phenomena in gases 312 f.
Triple point 104 f.

U

Ultra-vacuum gage 352
Undercooling 66
Units, system of 7

V

Vapor line 65, 97
Vapor-pressure curve 96 f., 104
  decrease in 110

Velocity space 170, 174
Vibrations, natural 246
Viscosity 168, 200, 328 f.
"*vis viva*" 13
Voigt, W. 156
Volume of a molecule 193

W

Waage, P. 81, 83
van der Waals, J. D. 55, 57, 58
  constant, statistical interpretation of 192
  equation 55 f.
  gas 33, 55 f.
Wald, F. 41
Waldmann, L. 80, 163, 295
Warburg, E. 203
Water, anomalous behavior of 46
  phases of 63, 96
Watt, J. 5
Weighting factor, quantum statistics 242
Weiss, P. 125
  domains 126
  theory of ferromagnetic phenomena 125
Wiedemann-Franz law 205, 276, 333, 343
Wien, W. 140, 149
  's law 140, 141, 143
Wilson, C.T.R. 103
  's cloud chamber 103
Woltjer, H. R. 125
Work 4
Work, maximum 52 f.

Z

Zermelo, E. 237